# Turning the Corner:

*Energy*

*Solutions*

*For the*

*21ˢᵗ Century*

Here's to a brighter energy future and a healthier world!

Mark McLaughlin
Oct 2001

Alternative Energy Institute, Inc.
P.O. Box 7074 • Tahoe City, CA 96145

LIBRARY
NSCC, PICTOU CAMPUS
39 ACADIA AVE.
STELLARTON, NS B0K 1S0 CANADA

TURNING THE CORNER: Energy Solutions for the Twenty-first Century

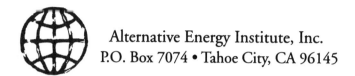

Alternative Energy Institute, Inc.
P.O. Box 7074 • Tahoe City, CA 96145

All Rights Reserved

© 2001 by Alternative Energy Institute, Inc.
First Edition

Library of Congress Catalog Number: 99-095243
ISBN 0-9673118-2-9

Printed in the United States of America on recycled paper

# TABLE OF CONTENTS

Foreword .................................................................................... vii
1 Road Map to the Future ........................................................ 1

## Section I — Future Shock: The Coming Energy Crisis ............ 9
2 The Wolf Is at the Door ...................................................... 15
3 Coal ..................................................................................... 47
4 Natural Gas ......................................................................... 61
5 Nuclear Energy ................................................................... 79

## Section II — Renewable Energy: A Traditional Bridge to an Alternative Future ........ 99
6 Solar Energy ..................................................................... 105
7 Wind Energy .................................................................... 115
8 Hydrogen & Fuel Cells ..................................................... 123
9 Hydropower ..................................................................... 137
10 Geothermal Power ........................................................... 147
11 Biomass Energy ................................................................ 155
12 Tidal & Ocean Thermal ................................................... 165

## Section III — The Energy Revolution: Technologies on the Horizon ........ 175
13 Cold Fusion: Fact or Fiction? .......................................... 183
14 Zero-Point Energy: Sailing The Sea of Energy ............... 207
15 Universal Forces: Blackholes, Electrogravitics, & Deep Space Propulsion ...... 239
16 The Race for New Energy ............................................... 275

## Section IV — Arcs, Sparks, & Electrons: Accelerating Into The Future ........ 291
17 Transition: Life Beyond the Oil Patch ............................ 305
18 Twenty-first Century Solutions: Turning the Corner .... 325

Bibliography ........................................................................... 367
Index ....................................................................................... 373

# ACKNOWLEDGMENTS

This book would not have come to life without the help and assistance of many people. It would be impossible to name them all, but Alternative Energy Institute, Inc. would especially like to thank the following industry experts: Walter Youngquist, Richard Duncan, Colin Campbell, Jean Laherrere, L. F. "Buzz" Ivanhoe, Ed Storms, Ken Shoulders, Eugene Mallove, Tom Valone, James Woodward, David Hamilton, Thomas Bearden, David Wallman, Charles Platt, Hal Fox, and many more.

In addition we would like to express our appreciation to Edee Olson for her insightful research, Scott Ackley for his assistance in prepress layout, and Nina Riley for her expert editing skills. Any errors still remaining are our own.

This book was created with Adobe Pagemaker 6.5 software using Adobe Garamond 11 point on 14 point leading for the body text and URW Grotesk 10 point on 13 point leading for the sidebars.

# THE AUTHORS

**Dohn Riley** graduated from Massachusetts Institute of Technology in 1970 with a Bachelor of Science in Earth Sciences. In 1976 he received a Masters from Stanford University in Geophysics. Dohn has worked as a geophysical consultant in a variety of fields and taught college classes in Physical Sciences for 22 years for Sierra College in California. He has been Executive Director of Alternative Energy Institute, Inc. since its inception in 1997. Dohn also currently owns and manages a graphic design business, Riley Works, located in Tahoe City, California.

**Mark McLaughlin**, a professional researcher and writer with more than 200 published articles, trained as a historian and cultural geographer at the University of Nevada, Reno. McLaughlin's work appears regularly in California and Nevada newspapers; he was awarded the Nevada State Press writing award five times. Author of three books, McLaughlin frequently writes historical articles for such magazines as *Sierra Heritage*, *Nevada*, *Weatherwise*, and his work will appear in the Grolier Educational *2002 Science Annual*. McLaughlin is a professional lecturer, a frequent guest on regional television and radio programs, and a consultant for *The History Channel*. Mark McLaughlin owns Mic Mac Publishing, located at Lake Tahoe, California.

# FOREWORD

This book was created with you in mind to provide information on which to base intelligent decisions that society will be facing soon. In order to ensure that this book is as accurate as possible, all of the topical coverage was reviewed by a leading expert in the relevant field. Frontier science is often controversial and subject to arrogant dismissal, but we believe that it is imperative to develop an encouraging atmosphere for research that may help us as a society in the inevitable transition away from nonrenewable fuels. We encourage you to examine the selected bibliography and to follow our numerous footnotes to learn as much as you can about these vital issues.

The energy field changes quickly, and, in order to keep pace with it, we have coupled this book with a portion of our website in order to provide you with the latest developments. In addition, there will be a forum for you to express your theories and ideas, and share your thoughts with other people. We will also post any errata of which we become aware on the Alternative Energy Institute, Inc., (AEI) website, and we ask you to let us know if you find any information that you believe may be incorrect or outdated. As you can see in the footnotes, we have documented this book with many links to other websites and Internet resources. It is the nature of the Web to be fluid and constantly changing. As we went to press all links were valid, but we assume that over time some of the links will become nonfunctional. A portion of the AEI website will be devoted to providing alternative links (where possible) to replace those that fail.

All of the information and updates to *Turning the Corner: Solutions for the 21$^{st}$ Century* will be accessible at this location:

<www.altenergy.org/solutions>

# Chapter 1
# Road Map to the Future

Can you imagine a world where energy is produced without negative environmental impacts? A life on a healthy planet with a clean atmosphere, safe from polluting hydrocarbons such as oil, coal, and natural gas, or the dangers of nuclear radiation? Can you envision a time when on-site renewable and alternative energy systems provide a more equitable distribution of electricity and a better standard of living for all of the world's people? Do you foresee a future where a full-fledged energy revolution has minimized the human impact on the biosphere and changed everyone's life for the better? Ready or not, this exciting frontier is right around the corner and heading your way. Fuel cells, electric-hybrid vehicles, and high-tech, energy-efficient homes and appliances are just the tip of the iceberg and represent the first surge of an energy revolution that will transform the human condition in the 21st century.

A new age is dawning, and clean, renewable energy systems are the vanguard of the next energy revolution. Behind the wind turbines and solar panels will come a new understanding of physics and exotic energy technologies that derive power from the invisible sea of energy that permeates the Universe. We are literally immersed in this zero-point energy, which, if tapped, will solve the energy crisis forever. Existing reserves of fossil

*We don't inherit the Earth from our ancestors, we borrow it from our children.*

fuels are a one-time gift from the planet and are now being consumed as if there were no tomorrow, with little consideration for future generations. If scientists are successful in their efforts to gather energy from the vacuum of space, smokestacks and air pollution will be a distant memory, and a clean and viable global ecosystem will be passed on to each succeeding generation. After all, we do not inherit the Earth from our ancestors; we borrow it from our children.

Optimists claim that there is enough oil worldwide to last another 40 years. Many experts consider it to be much less. In some ways, the shrill, hyperbole warning of petroleum's imminent demise is a bit misleading. We are not running out of fossil fuels per se but running out of liquid fuels that are cheap to produce. There are still significant oil and natural gas reserves, but they are diminishing and will be more costly to extract. There are also coal reserves if we can find a way to use them without damaging the environment. Unfortunately, fossil fuels are killing us and destroying our planet's health. We live in troubled times, and, if scientists are right, the next stretch of road looks rough indeed. Climatologists warn of rapid climate change and global warming caused by increasing concentrations of greenhouse gases in the atmosphere. Geologists caution that in the next couple of decades, the planet's petroleum and natural gas supplies will reach their high-point of production and decline forever after. Coal is a gross polluter, and nuclear fission is uneconomical when capital investment, radioactive waste management, and long-term storage of hazardous materials are factored in. A burgeoning human population is overloading the planet's ability to support it, and disturbing trends indicate that we are stressing the biosphere to the breaking point. Unless these destructive activities are reversed, future generations will struggle to survive on a planet suffering from depleted resources, polluted air and water, and unpredictable weather patterns.

*The critical issue of high-level waste management relates not only to the US but also to the global community.*

Our increasing demand for electricity and transportation fuel is gutting the planet's limited natural resources, and the byproducts of our fossil-fuel-based economy have degraded the global environment. Drilling for oil in environmentally sensitive areas does not make for an intelligent comprehensive energy policy, nor does gutting federal funding for energy-efficiency R&D and renewable energy research. Attempts to maintain the illusion of perpetual low-cost energy have distorted US energy markets for decades, leaving American consumers with homes, appliances, personal vehicles, and equipment that would be expensive to operate if energy costs suddenly increased. It's time to rein in America's profligate rate of energy and resource consumption and waste; we must take responsibility for our environmental impact.

*Driving industrialized economies with fossil fuels has polluted the biosphere and contributed to global warming.*

We are speeding toward a watershed event when the industrialized world's century-old, fossil-fuel-based economy will begin the transformation toward a cleaner, more sustainable one driven by renewable and other alternative power sources. In the 21$^{st}$ century, it is imperative that the United States, as the greatest energy consumer, polluter, and contributor to greenhouse gas emissions, develops a sound energy strategy for the future. By setting an example for greater energy conservation and efficiency, the US has a unique opportunity to provide critically needed global environmental leadership. A strategic long-range energy plan should encourage supply diversity through research and development and offer a more equitable tax and subsidy structure that will open markets to clean, renewable, and carbonless alternatives. The federal government should offer tax credits for companies developing and using renewable energy and energy efficiency technologies, and remove subsidies from the fossil-fuel power industry. The economic arguments against these changes hold no water. Wind energy systems are already directly cost-competitive with many currently installed fossil-fuel technologies that are degrading the biosphere. US health costs associated with air pollution are estimated at $100 billion per year.

In order to develop new-energy systems that will complement wind and solar power to help replace fossil fuels and nuclear power, the United States must implement a national energy policy that provides an informed and balanced review of the full range of new- and emerging-energy technologies that are struggling due to lack of R&D funding and professional organization. There is currently a movement in the US Department of Energy to initiate a review program to critically analyze potential new-energy and breakthrough propulsion technologies, but government approval and adequate funding are far from guaranteed. Experts doubt that renewable resources like wind and solar power will support the energy-hungry industrialized nations, let alone a world population with more than six billion people and growing fast. Aggressive and coordinated research and development in revolutionary energy systems are necessary to sustain the quality of life associated with an energy-intensive lifestyle.

*Compared to conventional vehicles, EVs offer zero to minimal smog-forming emissions.*

*Turning the Corner: Energy Solutions for the 21$^{st}$ Century* entices the imagination with a thorough overview of the many potential breakthrough technologies being pursued by free-thinking inventors, engineers, and physicists. Professional research into new-energy and futuristic propulsion systems is being conducted by both private and government sectors around the world. Due to the innovative nature of this type of experimental research, delays, disappointments, and dead-ends are to be expected, but the destination remains the same: decentralized, limitless, and emission-free energy systems, as well as advanced propulsion engines that utilize gravity-modification devices or burn no fuel.

This book offers a concise review of the current state of research into these radically different energy systems in a clear, straightforward, and reader-friendly text for the nonexpert. This revolutionary work is extensively footnoted with references listed at the end of each chapter. Internet links offer the interested reader instant electronic access to the latest information regarding the subject matter under discussion. Simply type the reference Uniform Resource Locator (URL) into your computer's address window, and you will access the website or webpage referenced by the footnote. If the link is broken, you will find a replacement link at Alternative Energy Institute's website <www.altenergy.org>.

Certain topics of interest covered in the book are also linked to the Alternative Energy Institute website. The extensive AEI website offers a huge compendium of information related to diverse energy and energy-related topics. The wide-ranging website also contains energy conference coverage, inventor profiles and photographs, informational updates, and book errata. *Turning the Corner: Energy Solutions for the 21st Century* has a unique and innovative format that gives this Internet-linked book an enduring quality as a prime research resource for the latest developments in the futuristic world of emerging energy and propulsion technologies.

*An energy crisis in California in 2001 made front-page news.*

Other special features in the book are relevant sidebar quotes gleaned from traditional print media such as books, science journals, magazines, and mainstream newspapers, which offer the reader a contemporary look at how energy experts, authors, politicians, and industry representatives perceive the energy revolution. These quotes are referenced at the end of each chapter for readers interested in reviewing the relevant book, article, or journal.

*Turning the Corner: Energy Solutions for the 21st Century* is divided into four parts.

**Section One** reviews the benefits and liabilities associated with burning fossil fuels and using nuclear power. Fossil fuels in one form or another have driven the world economy for nearly 200 years, but proven reserves of these nonrenewable energy sources are ultimately limited. Continuing to rely on fossil fuels for energy is a throwback to the 19th and 20th centuries.

**Section Two** profiles the latest developments in renewable energy technologies like wind and solar power, as well as other lesser known renewable technologies such as biomass gasification and tidal and ocean thermal power systems. Wind and solar power have been expanding their market share by more than 30% a year but, at this time, provide only a tiny fraction of the world's energy budget. These clean, renewable energy systems will contribute to the ever-increasing demand for electricity, but it will be a serious technological challenge to replace the transportation power derived from energy-rich petroleum and the myriad of vital products like medicine, pesticides, and fertilizer gained from oil.

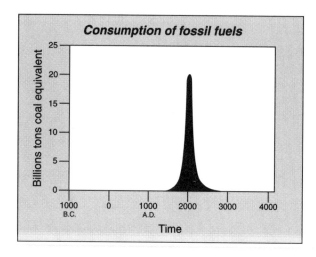

*The era of Carbon Man is ending.*

**Section Three** focuses on potential breakthrough energy and propulsion systems being pursued by professional researchers in the United States and in laboratories around the world. If realized, the commercialization of these intriguing emerging technologies could remove the terms *energy crisis* and *poverty* from our daily vocabulary. Researchers at the frontier of physics and science are studying Advanced Electromagnetic Theory; Mach's Principle and Impulse Engines; Anomalous Gravitational/Inertial Mass Effects; and Zero-Point-Field Energy Extraction, among other ideas. The exotic nomenclature may sound like it was lifted from a *Star Trek* episode, but the theories and science behind these technologies represent the seeds of future mainstream science.

**Section Four** describes the transition from fossil fuels to a world economy based on hydrogen and electricity as primary energy carriers. The final chapter in this section expands on AEI's primary solutions for making this transition feasible and sustainable. The road to success relies on an informed and proactive voting public, visionary politicians willing to advance their constituent's environmental protection demands, and an intelligent comprehensive federal energy policy that places the planet's long-term health in front of short-term gains for the dirty fossil-fuel industry.

Native American spiritual elder Oren Lyons drives home the point: "Man sometimes thinks he's been elevated to be the controller, the ruler. But he is not. He's only a part of the whole. Man's job is not to exploit but to oversee, to be a steward. Man has responsibility, not power."[1] Our role is compelling and clear; our children and grandchildren are counting on us to make the right decisions today.

## FOOTNOTE

[1] Steve Wall and Harvey Arden, *Wisdomkeepers: Meetings with Native American Spiritual Elders* (Hillsboro, Oregon: Beyond Words Publishing, Inc., 1990), p. 67. Quote by Oren Lyons — Onondaga Clan.

# Section I

# Future Shock: The Coming Energy Crisis

TWO HUNDRED YEARS AGO, the world experienced an energy revolution that launched the Industrial Age. The catalyst to this epochal shift was ordinary black coal, an energy-rich hydrocarbon that supplanted wood as the primary fuel. The energy stored in coal gave enterprising inventors and industrialists the power they needed to process steel, propel steamships, and energize machines. A century later, the industrialized world's thirst for energy had increased tremendously. Petroleum and natural gas (methane) were exploited as versatile and high quality energy products, and they soon joined coal as principal fuels. More recently, scientists have tapped uranium to fuel nuclear reactors and provide atomic energy. We've reaped the benefits, but at what cost, and when will they be depleted?

Today, cheap energy is the lifeblood of American society. It is essential to our quality of life and has been the crucial underpinning to the nation's long economic success. But there is a dangerous dark side to relying on non-renewable resources like coal, oil, natural gas, or uranium ore to heat and cool our homes, fuel our cars and planes, and generate electricity. The supply of these fuels is physically limited, and their use threatens our health and environment. Fears of global warming aside, burning fossil fuel releases chemicals and particulates that can cause cancer, brain and nerve damage, birth defects, lung injury, and breathing problems. The toxic stew released by combusting hydrocarbons also pollutes the air and water and causes

> And future wars, Chinese analysts agree, will be fought over oil, not ideology.[A]
> — *Gannett News Services*

> Decades-long scientific observations have convinced most nations that human activities are producing environmental changes on a global scale — including alterations in climate, land productivity, water resources, ecological systems, and the atmosphere.[B]
> — *National Research Council*

acid rain and smog. Nuclear energy, once touted as "too cheap to meter," has never been economically successful, and fear of disasters like the Chernobyl reactor meltdown has virtually shut the industry down in the US and Europe. Inexpensive and seemingly abundant non-renewable energy fueled the 20th century economy, but geologists, climatologists, politicians, and environmentalists are warning that the honeymoon may soon be over.

Coal is the most abundant of the carbon-based fossil fuels, but it is also a leading threat to human health and the environment. Coal currently provides 24% of the world's primary energy requirements and, in 1999, generated 57% of all the electricity used in the US (Most Americans are unaware that coal and nuclear fission together provide about 77% of the country's electricity needs today. Oil is principally used as a transportation fuel.) Existing coal reserves may be large, but they won't last forever, and the health and environmental costs limit its potential as an acceptable primary fuel in the future. Burning coal currently accounts for 43% of all annual global carbon emissions, about 2.7 billion tons.

The top ten most air-polluted cities in the world — nine in China, one in India —all use coal as a primary energy source. Atmospheric scientists have tracked large dust clouds of particulates and sulfur from Asia to the United States' west coast. In the US, coal reserves surpass those of oil and natural gas by about 200 years and can be mined domestically, but using coal simply because there is plenty of it would be a serious mistake. Countries like India and China are struggling with how to supply their growing populations with electricity but, at the same time, are scaling back using their reserves of coal to generate power. The air pollution, acid rain, greenhouse gas emissions, and other health dangers associated with processing vast amounts of coal into electricity take their toll on countless people around the world. Western governments rarely discuss "coal" and the "future" in the same sentence anymore, but burning coal has become a global problem that respects no international boundaries.

Cheap and abundant oil is an intoxicating elixir that the world's industrial nations have guzzled down as if there is no

> Hundreds of Ukrainian [coal] miners are killed each year in accidents — 274 died in 1999, down from 360 in 1998.[C]
> — *Los Angeles Times*

> Applying the technologies of horizontal drilling and multiple zone well completion clearly allows many reservoirs to be drained at a far faster rate, thus accelerating the depletion level once peak production has been reached.[D]
> — *Middle East Insight*

tomorrow. Energy-rich petroleum currently accounts for 40% of the world's energy, but many geologists anticipate an oil supply crisis sometime within the next two decades when global demand will exceed supply. While some argue that huge deposits of oil may lie undetected in far-off locations of the globe, experts point out that there is only so much crude in the world, and the industry has found about 90% of it. In fact, a large number of the giant, older fields — which anchor the world's hydrocarbon production base — have now started to decline. The world's burgeoning population is based on food grown with petroleum-based fertilizers and cultivated by machines running on cheap fuel.

In 1950, the US was producing half the world's oil. Fifty years later, we don't produce half our *own* oil. Domestic production peaked in 1970. Originally, America was blessed with about 260 billion barrels; only one country, Saudi Arabia, had more. Although the US is now the world's third largest producer, about 65% of our known oil has been burned. It's downhill from here. Baby boomers born after World War II will have consumed 70% of America's total petroleum heritage in just one generation. The US has 4% of the world's people but slurps down 25% of the world's oil. If the Chinese annually consumed oil at the same per capita rate as Americans, there would be none available for anyone else. What historians will someday call the Oil Era will last only about 250 years. In year 2000, we are closer to the end than to its beginning.

Natural gas (methane) is being touted by energy providers as an abundant clean fuel for the 21st century. It is forecasted to be the fastest growing primary energy source because it burns cleaner than coal or oil and is not as controversial as nuclear power. But this resource is also non-renewable, and the Department of Energy states that the United States has only 3.5% of the world's total natural gas reserves — enough to last about 65 years. More than 70% of the world's proved natural gas reserves are located in the politically risky Persian Gulf and former Soviet Union. After 2020, the bulk of the world's remaining supplies of both oil and natural gas will be centered there.

> When compared to people in other countries, Americans seem lucky to pay close to $1.50 per gallon of gasoline; gas costs drivers in Hong Kong $5.40 a gallon, England $5.00, France $4.50, and Japan $3.46.[E]
> — *Associated Press*

> Abundant and cheap oil helped to propel the United States rapidly to its high "oil standard" of living enjoyed today but maintained now only by increasing oil imports.[F]
> — *Duncan and Youngquist*

> [Natural] gas is not the economic or social equivalent of crude due to the inherent convenience, safety and flexibility of oil.[G]
> — *World Oil*

> Without continued advances in supply technology, the projected growth in [natural] gas consumption...cannot occur.<sup>H</sup>
>
> — *Power*

> In a big setback for nuclear energy in Japan, a utility last month dropped plans for a nuclear power plant at the urging of the governor of the prefecture set to host the site.<sup>I</sup>
>
> — *Engineering News-Record*

> The Irish and Danish governments have expressed concern about nuclear contamination of the Irish Sea and said they would take joint action today to press for closing the [British] plant.<sup>J</sup>
>
> — *New York Times*

According to the Energy Information Administration (EIA), natural gas provides 27% of the energy used today. Similar to their consumption of oil, Americans consume more than their share of natural gas; in 1997, the United States used 28% of the world's total production. Consumption in the US and Canada is expected to grow 50% by 2020, enforced in part by the Clean Air Act Amendments of 1990, which were designed to curb acid rain, toxic emissions, and urban air pollution. Due to these efforts, natural gas-powered vehicles, super-efficient boilers, and fuel cells are now reaching the market. Compared to the combustion of oil and coal, natural gas combustion is relatively benign as a contributor to air pollution. American troops have already shed blood in the Middle East protecting our oil interests, but today the costs of "cheap" Persian Gulf petroleum exceed $100 billion per year. Relying on natural gas for a primary energy source has similar costs and risks. The upcoming peak and subsequent plateau for America's domestic natural gas reserves will arrive two decades from now, and, once again, we will be forced to rely on Persian Gulf supplies.

Nuclear energy presents similar problems to those associated with non-renewable fossil fuels. The planet's supply of uranium is limited, and mining the ore is hazardous to human health. Worse, the radioactive waste byproducts are a lethal long-term danger to the environment. In 1999, nuclear energy provided about 17% of the world's electricity, but splitting the atom to boil water is like using a chainsaw to cut butter. The health and environmental costs of using atomic energy have become serious obstacles to the industry. Disposal of radioactive waste has proven to be a much greater problem than originally estimated by industry experts. Nuclear power does not contribute to air pollution and greenhouse gas emissions, but a good solution to safely storing tons and tons of radioactive waste, a nuclear byproduct that remains dangerous to all life-forms for thousands of years, remains elusive.

In the 1950s and 60s, atomic energy was hailed as an unlimited panacea to the pollution problems generated by fossil fuels and was destined to be so cheap that electric companies wouldn't even put meters on houses. But technical engineering difficulties and unexpected equipment and system failures have dispelled the technological hubris that accompanied the dawn of the Nuclear Age. Today, there is little support among Americans and Europeans for nuclear energy. Nobody wants a nuclear reactor in the backyard, and frightening publicity regarding reactor meltdowns at Chernobyl and Three Mile Island has only enforced these fears. Health officials estimate that at least 4,365 people who took part in the Chernobyl cleanup have died in the Ukraine. The risk from nuclear power plants and waste disposal sites demands a vigilance and longevity of our social institutions that is unprecedented. It is a serious responsibility that will ultimately affect the health and safety of future generations.

Coal, natural gas, and uranium are alternative non-renewable energy resources to cheap oil, but each has advantages and limitations, and none is as versatile as petroleum. The replacement of oil will require a mix of energy sources, including polluting non-renewables like coal, natural gas, uranium, as well as clean renewable energy such as solar and wind power. This adjustment will involve substantial reorganization of the world's economic structure and significant lifestyle changes in the industrialized countries. It will take decades before alternative renewable energy sources replace even half the energy generated by fossil fuels.

Economists like to point out that the world contains enormous caches of unconventional oil that can substitute for crude oil as soon as the price becomes competitive. It is true that resources of heavy oil, tar sands, and shale deposits exist in large quantities. But the industry will be hard-pressed for the time and money needed to ramp up production of unconventional oil quickly enough to forestall an economic crisis. Yesterday's energy supply cannot feed, clothe, and house tomorrow's population. And yesterday's energy supply surely can't fuel tomorrow's technological advances. Experts that point out the approaching end of Hydrocarbon Man are not pessimists or alarmists; they are simply saying NOW is the time to plan, lest the end of cheap fossil fuels be an unprecedented disaster in human history. Think of your children's children's children. What legacy do you want to leave — abundance and health or a depleted, sickened world?

Nuclear and hydropower, long considered to be low-impact environmentally friendly power sources, are proving to be not so friendly and downright dangerous in the case of nuclear power.[K]
— *Offshore*

A federal judge has thrown out a lawsuit filed by a company that has spent 16 years trying to open what would be the nation's fourth active nuclear waste dump.[L]
— *Associated Press*

Barring a global recession, it seems most likely that world production of conventional oil will peak during the first decade of the 21st century.[M]
— *Scientific American*

SIDEBAR FOOTNOTES:

A  John Omicinski, "Chinese elite anticipate US decline," Pentagon says. *Gannett News Service* article published in *The Sacramento Bee*, February 20, 2000, p. A-16.

B  "Global Change" National Research Council *NewsReport* (Spring-Summer 1998), p. 9.

C  Maura Reynolds, "Ukraine coal mine blasts kill 80," *Los Angeles Times* article published in *The Sacramento Bee*, March 12, 2000, p. A-14.

D  Matthew R. Simmons, "Has Technology Created $10 Oil?" *Middle East Insight* (May-June 1999), p. 37.

E  Dirk Beveridge, "2 bucks a gallon looks OK to some," *Associated Press* article published in *The Sacramento Bee*, March 10, 2000, p. F-3.

F  Richard C. Duncan and Walter Youngquist, "Encircling the Peak of World Oil Production," *Natural Resources Research*, Vol. 8, No. 3, 1999, p. 223.

G  L. F. Ivanhoe, "Future world oil supplies: There is a finite limit," *World Oil* (October 1995), p. 82.

H  Cate Jones, "Don't underestimate challenges of bringing gas to powerplants," *Power* (January/February 2000), p. 28.

I  "Local Opposition Kills Japan Nuke," *Engineering News-Record* (March 20, 2000), p. 21.

J  *New York Times*, "Troubled British nuke facility admits sabotage," article published in *The Sacramento Bee*, March 27, 2000, p. A-6.

K  Leonard Le Blanc and Victor Schmidt, "Future of hydrocarbons secure, but supply, demand remain erratic." *Offshore* (December 1999), p. 42.

L  *Associated Press*, "Suit seeking nuclear dump in Mojave Desert is tossed," article published in *The Sacramento Bee*, March 27, 2000, p. A-3.

M  Colin J. Campbell and Jean H. Laherrere, "The End of Cheap Oil," *Scientific American* (March 1998), p. 81.

# Fossil Fuels: The Wolf Is at the Door

## 2

Fossil fuels have propelled civilization to the greatest economic prosperity in history, but after 150 years, the free ride is almost over, and the relentless combustion of oil is now threatening our very existence. Cheap petroleum is the lifeblood of industry and transportation, but experts warn that the supply will fail to satisfy global demand within 10 to 20 years. US troops have already spilled blood on the oil-rich sands of the Iraqi desert. In geo-political terms, the desperate fight for oil in our future has already begun.

FOSSIL FUELS ARE A ONE-TIME ENERGY GIFT to the human race. Unfortunately, once they are gone, they are gone forever (or at least for millions of years — far longer than human history). The average age of the gasoline in your car's fuel tank is about 70 million years, yet we are using this precious resource as though it were unlimited. The powerful industrialized economies of Western Europe and the United States are based on these cheap, abundant fossil fuels. But, as the global population swells, humans worldwide will seek to improve their own standard of living. Even if allowance is made for improvements in energy efficiency, future generations will require enormous amounts of energy. Current trends indicate

*US oil production peaked in 1970.*

that most of this demand will be satisfied by burning fossil fuels such as coal, oil, and natural gas.

For decades, people have been aware that coal, oil, and natural gas are limited resources and that the damage caused by extracting and transporting them often causes negative environmental impact. But oil consumption has only increased, and there are more cars and trucks on the road than ever before. Now we know that burning these fossil fuels adversely affects the world's ecology and our global climate. From oil spills to contaminated ground water, using fossil fuels has caused great harm to the environment and humans. Burning this type of fuel releases stored carbon dioxide into the Earth's atmosphere, which then acts as a blanket, trapping heat and causing air and ocean temperatures to rise. Carbon dioxide is necessary to keep the planet warm enough to sustain life, but human activities have pumped so much $CO_2$ into the atmosphere that it is generating a worldwide effect. It is called global warming, and has significant consequences for humans.

Policymakers and scientists have acknowledged the links between fossil fuels, acid rain, air pollution, and human health. And even though recent medical research indicates that energy-related pollution threatens our health even more than previously suspected, industry and politicians are still dragging their feet to change policies that are slowly destroying the planet. Currently, there is a serious lack of political will for switching over to renewable energy sources in a manner that will minimize the unprecedented economic and social disruption expected to occur as supplies of cheap petroleum diminish. Many governments and industries refuse to acknowledge this imminent disaster for fear of triggering a political or economic crisis.

The world's most powerful governments and its wealthiest citizens may not want to address the looming oil crisis, but it's coming sooner than they think. Petroleum is the lifeblood of our civilization, but industry experts predict that the world's oil supply will reach its maximum production and midpoint of depletion sometime around the year 2010.[1] By then, "Hydrocarbon Man" will have consumed half of all the easily recoverable oil that ever existed on our planet. That

> Deepwater oil is valuable, useful, and much needed, especially by the United States, which now imports more than half of its requirements; but it's not going to have much global impact, at best, meeting a few years of world demand.[A]
>
> — Colin J. Campbell

peak foreshadows the final deadline for humanity's ultimate switch to the alternative sources of energy that must replace petroleum in the production decline that's sure to follow.[2] Even more disturbing for oil-dependent Western economies is the fact that in the near future, more than half of the planet's petroleum reserves will be owned and controlled by a handful of countries in the politically unstable Middle East.[3]

Currently, oil provides about 40% of the energy Americans consume (25% of the world's total) and 97% of our transportation fuels. In 1950, the US was producing half the world's oil. Fifty years later, we don't produce half our *own* oil. Domestic production peaked in 1970. Originally, America was blessed with about 260 billion barrels; only one country, Saudi Arabia, had more. Although the US remains the world's third largest producer, about 65% of our known oil has been burned. It's downhill from here. As global population and oil demand rise, more and more people will be competing for less and less oil. By 2015, when several Persian Gulf nations will be exporting significant quantities, a revitalized Organization of Petroleum Exporting Countries (OPEC) could control oil prices as it did in the 1970s and 1980s.[4] (The 11 countries that make up OPEC currently produce about 40% of the world's oil and hold more than 77% of the world's proven oil reserves. OPEC also contains nearly all of the world's excess oil production capacity.)

Despite predictions that world petroleum output will peak during the first decade of this century, oil consumption continues to grow. Globally, we were consuming half again as much oil in the late '90s as we were in the early '70s.[5] Computer models predict that OPEC oil exports will exceed non-OPEC production sometime between 2010 and 2020. The last time non-OPEC oil producing countries fell behind OPEC in production was in 1973. Although that imbalance was only temporary, it triggered a world oil crisis, produced mass panic, and generated long lines at gas stations worldwide. The 1970s crisis was short-lived because there were still plenty of oil and gas deposits available for commercial extraction. This time, however, there are few, if any, undiscovered oil reserves or basins large enough to have global impact. The OPEC/non-OPEC production crossover ratio

> The US is now the most thoroughly drilled country in the world, and is fast running out of places in which to make further major oil discoveries. The balance of power in terms of oil production has definitely left the United States, never to return.[8]
>
> — Walter Youngquist

> A mile under the ocean floor (Gulf of Mexico) may lie 15 billion barrels. It's a lot of oil, but only as much as the nation uses every 2.5 years.[C]
>
> — Randy Udall

during the mid-'70s represents only a small blip on the chart. After the next crossover, whenever it is, the imbalance will be permanent.

In 1998, the U.S. *imported* more oil than any other nation *used*. In fact, America imports more than Denmark, Finland, France, Germany, Greece, Italy, Norway, Spain, and Sweden collectively use. And there are no signs that the upward spiral will abate. Incredibly, Baby Boomers born after World War II will have burned up 70% of America's total oil supply in just one generation. Oil supply problems rarely concern US residents; but when there's a burp in the pipeline and prices spike, as in the gasoline crisis from 1973-1979, riots and war are usually not far behind. U.S. troops have already spilled blood on the oil-rich sands of the Iraqi desert. In geo-political terms, the desperate fight for oil futures may have already begun.

World oil experts fall into two camps, bulls and bears. Both agree we've used 800 billion barrels.[6] The bears think there are 1 trillion left and that production may peak by 2010. Bulls think there are 1.8 trillion left and that the peak won't come until 2020. If ten years is the only difference between them, it is time to prepare. Americans see no sinister shadow on the horizon; but unless governments and industry support alternative sources for their crucial energy needs, dark days are coming. The only question is how soon.[7]

> The world is not running out of oil. Or rather, not for a long time. What it is running out of is cheap oil, and soon.[D]
>
> — Colin J. Campbell

Energy dependence on fossil fuels is fraught with peril, and time is running out in the race to fund and implement cleaner, renewable, and alternative energy sources. Experts warn that installing the infrastructure of new energy systems will take decades. Some alternatives — solar, wind, geothermal, and other traditional renewable energy sources — are already available. But renewables by themselves will not solve the problem of diminishing petroleum production that will eventually fail to satisfy rising global demand. Other alternative energy sources, such as cold fusion and hydrogen fuel cells, offer great promise, but the research is still in the early stage of development.

In January 1999, President Clinton strongly urged Congress to increase funding for alternative energy research. Clinton had the right idea, but more effort must be made toward building a sustainable world economy that is not based on environmentally degrading and polluting fossil fuels. Many scientists have now acknowledged that fossil fuel greenhouse gases are a major contributor to the much-publicized problem of global warming. In fact, every stage in the production of a fossil fuel (e.g., gasoline) exposes the environment to potential and real harm through its proven toxic and carcinogenic nature. At the drill site and wellhead, leaks and spills may infiltrate the groundwater. During transportation and filling, spills and accidents leak oil into the environment. At the refinery, more toxic leakage occurs. In California, gasoline processed with clean air additives is contaminating precious groundwater supplies. Nearly one-third of South Lake Tahoe's municipal water wells have been shut down because of gasoline infiltration. Neighborhood gas stations and boating docks contribute additional spills. Vaporization and underground tank leaks add even more chemicals to the environment. And, of course, automobile exhaust gases pollute the air and contribute to respiratory ailments. Car and truck exhaust has been linked to the increasing levels of asthma in children.

*The US has 4% of the world's people, but Americans consume 25% of the world's oil.*

The control of fossil fuel has involved armed conflict and strange alliances throughout the history of its use. In every major conflict from the Second World War to the Persian Gulf War, oil has played an important role in deciding who becomes an ally and who becomes the enemy. The upcoming domination of the world's oil supply by a handful of Persian Gulf countries should not take Americans by surprise. Domestic petroleum production in the US peaked thirty years ago, and the country has imported more and more oil ever since. In 1998, more oil was imported to the United States from the Persian Gulf than in any year since 1977-78. This dangerous dependence on imported oil seriously influences US foreign and military policy in oil-rich regions like the Middle East. The politics and economic power of oil are overwhelming and frightening. In a decade or two, a few governments in the Persian Gulf will control world energy prices and commercial drilling access to those reserves. They will then have the potential power to cripple any other nation's economy. Protection of oil fields and pipelines is a priority with the American military. When it comes to oil and national security, the US military will first protect America's interests. The marines will go where the oil flows.

The steady supply of cheap oil during the last 100 years has fed America's love affair with the automobile. Unfortunately, the toxic gases emitted by millions of cars, trucks, buses, and SUVs have also contaminated the atmosphere. The air we breathe keeps us alive, but it can also kill. Millions of children die each year from acute respiratory infections caused by air pollution. Air pollution can make you sick, burn your eyes and nose, give you an itchy, irritated throat, or cause trouble with your breathing. The chemicals in polluted air can cause cancer, brain and nerve damage, birth defects, lung injury, and breathing problems. Air pollution harms the environment and the people, animals, forests, streams, and lakes that exist within it.[8]

> There have been nods and smiles and occasional symbolic moves, but the real effort and the real money have gone into the building and widening of more and more auto-oriented, sprawl-producing projects.[E]
> — Attorney Joe Brecher

Polluted air respects no boundaries; it can easily travel from its source in one state to another state. Air pollution is international; airborne contaminants originating in Mexico drift into the United States, and polluted air from the US is causing problems in southeastern Canada. Smog has become commonplace in many regions around the US. Cars, factories, and even products used in the home release smog-forming pollutants. These gases undergo chemical reactions when exposed to heat and sunlight. The reactions form ground-level ozone, the principal component of smog. In 1970, the US Congress passed the Clean Air Act. At that time, the Environmental Protection Agency developed health-based criteria in order to develop ways to protect humans and the environment from air pollutants.

As a possible solution to the problem of stationary-source air pollution, notably from coal-burning utilities, the Clean Air Act established ambient air quality standards for pollutants such as sulfur dioxide, particulates, and ozone. This legislation required the EPA and the states to create plans to achieve the new reduced standards for pollution by a specified deadline. In order to encourage cooperation in the face of the technological and economic difficulties of reducing air pollution, the Clean Air Act employed marketable emission permits. This system allows firms wishing to establish new facilities in restricted areas or to improve their existing pollution record to establish credits in a "banking" program. Under this system, firms can store emission credits for future use or, when they meet the new standards, sell these credits to other polluters not yet meeting the cleaner emission standards.[9]

In 1990, the US Congress passed an 800-page updated version of the Clean Air Act. The act is a federal law covering the entire United States. The law was designed to ensure that all Americans have the same basic health and environmental protections. A geographic area that meets or does better than the EPA's primary standard is called an attainment area; areas that don't meet the primary standard are called non-attainment areas. Despite the government's 28-year effort to reduce atmospheric pollution in the United States, it is estimated that about 90 million Americans currently live in non-attainment areas. Globally, millions of children die each year of acute respiratory infections linked with indoor (smoky cooking fuels) and outdoor (industrialization) pollution.[10] Air pollution is also affected by greenhouse gases and their influence on global warming. EPA estimates make it apparent that warmer weather will lead to more smog and ozone. The EPA has concluded that it will cost billions of dollars to curb increasing ozone levels.[11]

*Despite the 1990 Clean Air Act, about 90 million Americans live in non-attainment areas.*

Unfortunately, more Americans are driving to work; and though each of today's cars produces 60% to 80% less pollution than vehicles in the 1960s did, there are also more cars and trucks on the road. Currently, motor vehicles are responsible for up to half of the smog-forming oxides generated. Motor vehicles also release more than 50% of all hazardous air pollutants and up to 90% of the carbon monoxide found in urban air. Automobiles and the infrastructure needed to support them dominate American society. In 1990, 87% of all trips were made in automobiles. In one California study, 90% of emissions in a 7-mile auto trip were generated in the first mile. Traffic is responsible for 15% to 20% of carbon dioxide emissions. In many major European cities, bicycles now account for 20% to 30% of all short trips. In some Chinese cities, urban bicyclists account for more than half of all trips taken.[12]

> Los Angeles alone has more cars today than all of China.[F]
> — Worldwatch Institute

Despite stringent state regulations on both mobile and stationary pollution emitters, California fails to meet EPA air quality standards more often than any other state in the nation. Several factors are involved in the Golden State's degraded air quality, including geography and topography, climate, and large population. California's Central Valley counties currently experience worse ozone levels than New York, Houston, Philadelphia, and Chicago. California's poor air quality in the Central Valley affects more than human beings. Ozone is the primary atmospheric pollutant causing injury to vegetation in California. Estimates of crop losses in the Central Valley due to ambient ozone levels range from 10% to more than 20%. Together with other photochemical oxidants, ozone already causes estimated economic losses in the range of several hundred million dollars a year in California. Recent studies indicate that photosynthesis and growth rates in valuable tree crops, such as almonds, apricots, plums, prunes, pears, and apples, are substantially reduced when the plants are exposed to increased ambient levels of ozone, such as are found at the University of California's field station located near Fresno. In addition to agricultural concerns, as air quality deteriorates, skin cancers and eye and respiratory problems are expected to increase in Central Valley livestock, especially in light-colored cattle and sheep. Nitric acid deposition, the result of a chain of reactions from the burning of fossil fuels, causes damage in fisheries that draw water from watersheds, such as those in the Central Sierra Nevada.[13]

> Although California has made strides in reducing hazardous air pollution, the [report] found cancer-causing toxics at levels 426 times higher than health standards established by the 1990 federal Clean Air Act.[G]
> — Los Angeles Times

In the last two decades, improvements in gasoline efficiency for cars and trucks have helped reduce fuel consumption and the pollution emitted by each vehicle. In 1975,

Congress passed the CAFE (Corporate Average Fuel Economy) legislation, which set the standard of 27.5 miles per gallon for cars and 20.7 mpg for light trucks—including pickups, minivans, and SUVs. The CAFE standards are applied on a fleet-wide basis for each manufacturer; i.e., the fuel economy ratings for a manufacturer's entire line of passenger cars must average at least 27.5 miles per gallon for the manufacturer to comply with the standard. The practice has saved the nation an estimated 3 million barrels of oil a day. In 1999, for the fifth straight year, the House of Representatives barred the US government from even studying changes in the 25-year-old CAFE law. In September 1999, President Clinton received the signatures of 31 senators seeking help in raising CAFE standards. A rival group backing a continued freeze submitted 37 signatures.

Efficiency standards are a good place to start, but Americans are driving their cars more miles on more trips. In 1970, US residents traveled one trillion miles in motor vehicles; that figure is expected to reach four trillion miles by year 2000. People travel farther to work, and most people still drive to work alone. Sport utility vehicles (SUVs), minivans, and light trucks currently account for nearly 50% of all new vehicle sales in the United States. Because of the incredible popularity of gas-guzzling SUVs, the average fuel economy of new passenger cars in the United States is at the lowest point since 1980. An SUV that averages 14 miles per gallon (mpg) emits 70 tons of carbon dioxide over its lifetime, compared with 38 tons for a car getting 27.5 mpg. It is sad irony that the same folks who vigorously support family values and the environment are buying stylized motor vehicles that consume 33% more gasoline, emit 30% more carbon monoxide and hydrocarbons, and are less safe than nearly all sedans and minivans. In fact, all but the very largest SUVs are more dangerous than sedans and minivans for their passengers, and they are three times more likely to kill the occupants of the other car. In another ironic twist, even though SUV marketing evokes an image of the

*Hidden beneath a labyrinth of government subsidies and protectionism, the real cost of gasoline exceeds $5.00 a gallon.*

Lulled by low gas prices, California consumers have snapped up gas-guzzling sport utility vehicles that average a meager 12 miles to the gallon. The number of single-car commuters has gone up in recent years, not down.[H]

— *The Sacramento Bee*

rugged wilderness experience, Ford Motor Company states that 87% of Ford Explorer owners have never taken their vehicle off-road.

SUVs aren't the only problem. There are more than 5 million large trucks and buses on the road in 1999. According to the environmental advocacy group Clean Air Trust, large trucks and buses account for only 2.5% of all vehicles but produce 26% of the smog-causing nitrogen oxide generated by vehicles and more than half of the soot in urban air. Buses, trucks, and SUVs have not been held to the same environmental restrictions as automobiles, but that may be changing. The Environmental Protection Agency (EPA) is crafting new rules for tougher emission standards for SUVs and cleaner burning diesel fuel for heavy trucks. If the standards are enacted, the EPA will require that by 2007, smog-causing nitrogen oxide emissions from trucks be cut by 75% beyond what already will be required in 2004. Particulates, or microscopic soot, must be reduced by 90% over what is required today, and the sulfur content of diesel fuel must be cut to less than 40 parts per million, compared with an average of 500 parts per million today.

*Globally, automobiles are reproducing five times faster than people.*

Although leaded gas is being phased out, generating dramatic declines in air levels of lead, automobile fuel has actually become more polluting. Gasoline refiners change the formulas to make up for the loss of octane, and the refined product is more likely to release smog-forming vapors into the air. In addition, auto pollution control devices, which have been installed in cars since the 1970s, were required to work for only 50,000 miles; however, a car in the United States is usually driven for 100,000 miles or more.

We all complain about those smog-belching trucks and SUVs, but have you ever checked out the emissions from your own vehicle? You can find out how "green" your car is at the Environmental Defense Fund's green driving website at <www.edf.org/greencar>. If you are thinking about buying a new vehicle, the American Council for an Energy-Efficient Economy (ACEEE) provides a "Green Guide," which rates the environmental impact of every new passenger car, minivan, SUV, and pickup truck sold in the US. For more information, check out the ACEEE website at <www.aceee.org/greenercars>.

There are new and improved hybrid electric cars being produced, but American consumers feel that available models do not meet their needs, particularly those of families with active lifestyles. Honda is selling a two-seat hybrid coupe powered by a combination of gasoline and electricity. The Honda Insight reportedly gets 61 miles per gallon (mpg) in the city and 70 mpg on the highway. Unfortunately, the top ten fuel-miser vehicles in the EPA's 1999 Best Mileage Derby constitute only 0.57% of the US auto market.

In 1995, a study called the Index of Environmental Trends was published, which examined several environmental quality indicators. It may be the only study of its kind, and most of the information it utilized came from reliable data compiled by national governments. The conclusion was that, "despite 20 years of substantial effort, each of the nine countries [studied] has failed to reverse the trends of environmental destruction." Of particular note is that the study did not include "the vast range of hazardous air pollutants, called 'air toxics' in the United States," because "regulatory bodies in the nine countries have failed to comprehensively monitor or regulate most hazardous air pollutants." The study says, "There are roughly 48,000 industrial chemicals in the air in the United States, only a quarter of which are documented with toxicity data." It concludes that the reductions in carbon dioxide and nitrogen oxides that are necessary for environmental improvement "may require far more change than seems politically possible," such as monitoring private automobiles.[14]

Obviously, despite the good intentions of the Clean Air Act, the U.S. has a long way to go before all Americans will be breathing safe, clean air where they live, work, and play. In fact, it is unlikely that Americans will ever breathe clean

> Any controls that limit fossil fuel combustion...also limit emissions of the traditional combustion pollutants, such as microscopic particles. When inhaled, these dust-like particulates can aggravate respiratory and heart diseases.[i]
> — *Science News*

> It is now well accepted that winds can transport toxic materials of many kinds throughout the global atmosphere.[j]
> — *Gene E. Likens*

air while the US remains dependent on fossil fuels for its transportation energy.

Polluted air is not just bad to breathe. Though you can't see it or taste it, something in the rain is killing our forests, rivers, and lakes. It can cause hazy skies, damage your property, and harm your health. The nasty culprit is acid rain. In the United States, acid rain is primarily caused by emissions from big coal-burning power plants in the Midwest. Sulfur in the coal becomes sulfur dioxide when burned. Sulfur dioxide and nitrogen oxides released from tall smokestacks form acid compounds when they react with oxygen, oxidants, and water in the atmosphere. The acids are carried by prevailing winds toward the east coast and Canada to permeate the environment in gaseous or liquid form.[15] There are clear signs of its existence as its alteration of the environment is traceable from the minute it falls to Earth. In Vermont, rain has been measured with a pH of 3.8, which is almost 100 times more acidic than the 5.6 pH of natural rainwater.[16]

*Acid rain knows no boundaries.*

Electric utility plants in the U.S. release about 30% of annual nitrogen oxide emissions and about 70% of sulfur dioxide emissions. Overall, more than 20 million tons of these chemicals are spewed into the air every year. The resulting acid rain is contributing to forest degradation, especially in high-elevation spruce trees that grow on the ridges of the Appalachian Mountains from Maine to Georgia and on the west coast of California. The acidic precipitation impairs tree growth and may increase susceptibility to injury during winter. As it percolates through the soil, the acid can strip away vital plant nutrients through chemical reactions, thus changing the chemistry of sensitive soils and posing a threat to future forest productivity. Increased amounts of acid have been measured in lakes and streams from Maine to Florida. Adverse effects have been identified in the hundreds of Adirondack lakes that are currently too acidic for the successful survival of many of the native fish species.

If the chemicals released into the air do not encounter a "wet" environment and combine with rain, they remain in a "dry" state, generating a visible haze in regions such as the Grand Canyon in Arizona and the Great Smoky Mountains. In the eastern US, acid chemicals that sink to Earth in the form of gases or dust have been linked to breathing and lung problems in children and asthma sufferers.[17]

Acid rain is not just a problem in the United States. In Canada, it accounts for $1 billion worth of damage each year.[18] In Germany, acid rain is destroying the forests. In Norway, airborne pollution from abroad has acidified the rivers and lakes, wiping out much of the fish stocks of trout, salmon, and perch. Acid rain has ruined the fisheries in nearly 5,000 Norwegian lakes — a tragic loss of a natural heritage.[19] A 1993 scientific survey in Norway indicated that airborne pollution has damaged an approximate area of 86,000 square kilometers (40,000 square miles). That estimate has now increased to 120,000 square kilometers (almost 60,000 square miles), signifying that the country has reached a plateau in which the environment can no longer withstand acidification. Scientists can only imagine how many plants and animals are directly or indirectly affected by the deterioration of the different species they depend on.

Fortunately for future generations in the US, in 1990, Congress passed the Clean Air Act Amendments, which include an acid rain curtailment program. Currently, every major US power plant is subject to extensive state and federal regulations. Due to this legislation, emissions of sulfur dioxide, nitrogen oxides, and particulate matter from electric utility plants have been decreasing. In both 1995 and 1996, utilities reduced sulfur dioxide by 35-40% (more than 3 million tons each year) beyond requirements. Because of these strict regulatory mandates, in 2010, a utility burning a ton of coal will, on average, emit 80% less sulfur dioxide than in 1970. The Environmental Protection Agency recently reported that all 445 affected facilities demonstrated 100% compliance with the Acid Rain Program to reduce emissions.

> The problem of acid rain demands a regional solution. For instance, Japan has reduced sulfur emissions through gains in efficiency, heavier reliance on oil and nuclear power, and stringent pollution control laws, but it cannot stop China's emissions from drifting across the Sea of Japan.[K]
> 
> — Worldwatch

> Since the beginning of the twentieth century, human activities have added 925 billion tons of carbon dioxide ($CO_2$) to the atmosphere, taking concentrations of this heat-trapping gas to the highest levels in 160,000 years.[L]
>
> — Christopher Flavin

Although it is apparent that the new environmental regulations have reduced certain emissions, they have also generated grumbling by the electric utility industry. As lobbyist for the utility industry, the Edison Electric Institute (EEI) "vigorously opposes the stringency of EPA's new regulatory decisions and proposals regarding ozone, particulate matter, and regional haze. These decisions have been criticized widely as being based on questionable science while causing tremendous impacts to business and virtually every citizen in this nation. EEI also opposes attempts to deal with these air quality issues in any federal legislation dealing with restructuring of the industry."[20]

The electric utility industry is being forced by law to clean up its act. But there are also ways for individual citizens to help solve the serious acid rain problem. Consider this program that is helping protect New York State's six-million-acre Adirondack Park. The Adirondack Council, a non-profit environmental group that has been working since 1975 to protect open space in the Adirondacks, is planning for the future of the region through public education, research, and policy analysis. The Council is selling $50 gift certificates to make it possible to retire permanently one ton of sulfur pollutants that cause acid rain; the certificates are issued in the designated individual's name. This purchase ensures that those pollutants will never be emitted through a smokestack to eventually drift over Adirondack Park.[21]

Air pollution and acid rain are just two of the various environmental problems associated with the burning of fossil fuels. Melting glaciers, steamy summers, unusual weather patterns, rising oceans, and the migration of flora, fauna and deadly microbes may sound like a thrilling blockbuster film from Hollywood. However, according to most of the world's scientists, this may be just the beginning of what we can expect in the next century. The culprit behind this hellish scenario is global warming. A vast network of temperature measurements and anecdotal evidence proves that the climate is definitely heating up. Scientists and climatologists warn that burning coal and oil to generate electricity and to power motorized vehicles is contributing to this warming of the atmosphere. The combustion of coal and oil produces carbon dioxide gas, which traps the Sun's heat like the glass of a greenhouse. The ability to quantify the human influence on global climate is still emerging from the background noise of natural variability, but more and more climatologists can demonstrate that global warming is a legitimate concern. There are still many many scientists, industry lobbyists, and politicians who insist that global warming is not real or a potential threat; but their numbers are dwindling.[22]

According to the climatological data, the warmest decade on record was the 1980s, with seven of the eight warmest years up to 1990. The 1990s have been warmer still with 1990, 1995, and 1997 being the hottest in 100 years. In fact, 1998 was the hottest year in the hottest decade of the hottest century of the millennium.[23,24] Ice core samples from a US Geological Survey show that Antarctica is warmer than it has been during the last 4,000 years. Closer to home in the last 150 years, there has been a 73% reduction in the region of Glacier National Park covered by glaciers. Computer models indicate that the present warming rate will eliminate all glaciers in Glacier National Park by 2030. In fact, even with no additional warming over that which has already occurred in the Glacier Park area, the glaciers there are likely to be gone by 2100.

In September 1995, the United Nation's Intergovernmental Panel on Climate Change warned that the earth had entered a period of climatic instability likely to cause "widespread economic, social, and environmental dislocation over the next century." Scientists believe that rising temperatures will drive a more vigorous hydrological cycle. The warming climate is expected to evaporate more water from the world's oceans, thereby pumping more energy and moisture into the atmosphere — fodder for destructive storms. There is little doubt that the climate is getting warmer and that carbon dioxide from fossil fuel burning along with a variety of other so-called greenhouse gases are contributing to the temperature rise.

*Receding glaciers may be a sign of global warming due to greenhouse gas emissions.*

> Asian nations are expected to double their greenhouse gas emissions over the next 15 years, raising gas emissions from Asian nations from 28 to 45% of the world total.[M]
> — *Harvard International Review*

> During six out of the past nine winter-spring seasons in northern polar latitudes, the total ozone has declined in some months by 25-30% below the pre-1975 averages.[N]
> — *World Meteorological Organization*

More than 100 years ago, Swedish chemist Svente Arrhenius claimed that if fossil fuel burning doubled atmospheric $CO_2$, average global temperatures could rise by 10 degrees Fahrenheit (°F). During the last century, humans burning coal and oil have increased atmospheric carbon dioxide by 25%, and the concentration is still rising. Current computer models predict that average global temperatures will increase by 3 to 8 °F.[25]

Resistance to reducing greenhouse gas emissions by the international oil industry and some US politicians is fierce. With annual sales in excess of one trillion dollars and daily sales of more than two billion dollars, the oil industry drives the world economy. Cheaply produced oil supports the economies of the Middle East and large segments of the economies of Russia, Mexico, Venezuela, Nigeria, Indonesia, Norway, and Great Britain. Any serious disruption in the production and distribution of oil will impact the global economy and increase unemployment and depress the financial markets, and it could create a social crisis in certain countries. When economic and social costs are considered, it's no surprise that many of the world's politicians and the American media have been pushing the perception that the worries of climatic change are premature and inflated.

Yet the physical evidence is significant and increasing at an alarming rate. A 48-by-22-mile chunk of the Larsen Ice Shelf in Antarctic broke off in 1995, exposing rocks that had been buried for 20,000 years. A 70-square mile piece of ice shelf broke off in April 1998, and, more recently, on October 15, 1998, the US National Oceanic and Atmospheric Administration (NOAA) reported that an iceberg one-and-a-half times the size of Delaware had "calved" off from the Antarctica's Ronne Ice Shelf. NOAA scientists believe that the calving of icebergs plays an important role in the disintegration of ice shelves and is "a possible indicator of global warming."

Such weather changes are happening all over the world. Findings by a team of scientists at NOAA's National Climatic Data Center indicate that growing weather extremes in the United States are due, by a probability of 90%, to rising levels of greenhouse gases.²⁶ Ocean ecosystems are suffering, too. Researchers have discovered that warming surface water temperatures have generated a 70% decline in the population of zooplankton off the coast of southern California, jeopardizing the survival of several species of fish that feed on it. A British medical journal has linked changes in climate patterns to the spread of infectious diseases around the world. Ocean warming has triggered algae blooms linked to outbreaks of cholera in India, Bangladesh, and the west coast of South America, where, in 1991, the disease infected more than 400,000 people.

The executives who run the world's oil and coal industries know that the march of science and political action may be stymied by misinformation. Western Fuels Association, Inc., spent $250,000 in 1991 to produce and distribute a pro-energy video called *The Greening of Planet Earth: The Effects of Carbon Dioxide on the Biosphere*. On the video, scientists and researchers claim that doubling carbon dioxide concentrations over the next century could increase crop yields by 30% to 40% worldwide and triple forest productivity. Researchers on the video offer a barrage of optimistic hyperbole. One Ph.D. research scientist with the University of Florida claims, "For citrus, [global warming] will be a very, very positive thing."²⁷

Western Fuels Association is a $400 million consortium of coal suppliers and coal-fired utilities, a major source of human-generated carbon dioxide. In a letter of response to an article written by Ross Gelbspan and published in the April 1996 issue of *Harper's*, Frederick D. Palmer, Chief Executive Officer of Western Fuels wrote:

> "We do plead guilty to Gelbspan's observation that we are trying to focus discussion on whether there is a problem in the first place rather than what to do about it. To the extent climate has changed due to human presence, the change has occurred in the warming of winter nights. In this context, change has been beneficial by stimulating productivity of the biosphere through carbon dioxide enrichment of the atmosphere. We believe that renewable energy technologies have their place but that they should take their seat at the

> The bad news is that even if all greenhouse gas production stopped immediately, concentrations already in the atmosphere would cause enough warming (approximately 1 degree Celsius) for substantial ecological disruption.º
> — *Global Warming and Biological Diversity*

> ...nature has sent and is in a sense sending us a warning signal that human activity and technology are modifying our planet, and the potential for dramatic alteration of our climate system is present.ᵖ
> — *Global Climate Change and California*

> It will be another 20 years before the climate changes that are predicted to be associated with the greenhouse effect become large enough to be unambiguously differentiated from naturally occurring variations in climate. As a society we have the choice of ignoring the warning signs...or taking some action.[Q]
> — *American Scientist*

> The American Lung Association estimates that smog and fine soot account for 15,000 premature deaths, 400,000 asthma attacks, and 1 million various respiratory problems annually.[R]
> — *The Sacramento Bee*

table only where their cost is competitive with other fuels."[28]

That "chair at the table" is still empty because petroleum is currently incredibly inexpensive. As long as consumers are not willing to pay more money for clean and renewable energy sources, the energy companies will not provide it. Environmentalists call this "the price is more important than the air we breathe" approach.

Those who support the status quo have presented many interesting, albeit far-fetched, proposals. An example is one proposal by the Japan Gas Association to generate clouds by evapotranspiration from coastal mangroves and lagoons and to build man-made mountains to promote rainfall through orographic cooling. But increasing vegetation on the world's deserts will only make the surface darker, thereby absorbing considerably more sunlight. High evaporation rates will release water vapor, itself a greenhouse gas.[29]

Because "human-created problems interact in ways that can take decades to recognize, and longer than that to understand," many corporations and money moguls are taking advantage of this situation to line their own pockets.[30]

In a scientifically controlled environment, plants do seem to respond to elevated concentrations of carbon dioxide with more growth. But the world community must decide whether dumping 70 to 80 million tons of carbon dioxide into the atmosphere every year is really the best way to grow a bigger grapefruit.

Predicting the future of petroleum prices and politics is as difficult as prognosticating the weather. After observing the violent battles over water and mining rights in the 19$^{th}$ century American West, Mark Twain wrote, "Whiskey is for drinking; water is for fighting over." A similar aphorism can be applied to the petroleum politics of the 20$^{th}$ century. Conflict and military engagement over petroleum supplies and production played a significant role during the last 100 years. The geopolitics of oil will only increase in intensity as petroleum reserves diminish even further in the 21$^{st}$ century, and the strategic importance of the world's cheap oil supplies will have greater influence on the foreign politics of the industrialized nations. The reserves of many oil-producing nations are now either past their midpoint of depletion or close to it. The United States is far past its midpoint of oil depletion while five Middle East countries are still decades from their midpoint. Experts admit that the world is not running out of petroleum immediately, but it is quickly exhausting the supply of cheap oil that has fueled the world's wealthy industrial and mechanized economies. As major petroleum discoveries become increasingly scarce, known reserves will be controlled by a handful of countries in the Persian Gulf. Although oil will still be relatively inexpensive to produce and bring to market, it will become more expensive to buy.

In fact, Colin J. Campbell, an Oxford-educated and independent geologist, estimates that by the year 2010, the bulk of the planet's petroleum reserves will be owned and controlled by five Middle Eastern countries.[31] Will the ruling families in major OPEC producing countries, such as Saudi Arabia, still be pro-Western in the future? (Currently, about 70% of Persian Gulf exports to the United States come from Saudi Arabia. Nearly half of Persian Gulf oil exports to western Europe come from Saudi Arabia.)

A large portion of the population in these Middle East countries is comprised of young people under twenty years of age. No one can know with certainty how this upcoming generation will wield the economic power that abundant oil brings, but the day when OPEC oil production will exceed non-OPEC production is not far away. The US Department of Energy Information Administration predicts that the inevitable imbalance will occur sometime after 2015, but computer models formatted by Richard Duncan, Director of the Institute on Energy and Man, predict that as early as 2008, OPEC oil exports may exceed non-OPEC production.[32] The region's oil-based economies will become more powerful as Persian Gulf petroleum dominates global production in the 21st century. Because the countries in the Middle East have such large yet-to-produce reserves combined with a comparatively low depletion rate, they can behave as "swing producers," thus making

*Colin J. Campbell.*

up the deficit between world oil demand and what other countries can produce. Before the Persian Gulf nations reach their midpoint of depletion, they are likely to control both the level of world production and the price of oil.[33]

The long-range outlook for oil consumption and pricing is bleak for big oil consumers such as the United States and the European Community. In developing countries, oil consumption is expected to increase dramatically along with phenomenal growth in population. By year 2050, there will be nine billion people alive but with *three times less oil* per person available than in 1950. Even if the United States sent the marines to occupy Saudi Arabia to control its oil production, the diminishing supply of reserves would still drive up the cost of a barrel of oil. A permanent doubling in the price of oil will generate major economic and political upheaval in the world. Coal mining and burning would then be stepped up in some countries, such as China, and would have considerable negative impact on the environment. Nuclear power use will probably increase; however, finite quantities of uranium will limit its potential as a long-term solution.

Energy economics are predictable. Similar to fossil fuels, strategic minerals have also played a big part in the political arena of the 20th century. It has been suggested that Germany's dependence on Swedish iron ore might have been Germany's Achille's Heel. A disruption in Hitler's iron supply might have shortened World War II and saved countless lives. In 1978, during a civil war in Zaire, rebels invaded the Shaba Province, where 70% of the world's supply of cobalt is produced. Cobalt is perhaps the world's most strategic mineral as it is absolutely vital in jet engine manufacturing, and there is no substitute for cobalt in this use. After the rebels took over the cobalt-producing region, the price rose from $6 per pound to more than $40.[34] South Africa holds significant reserves of strategic metals such as platinum, manganese, vanadium, chromium, uranium and titanium. Without these vital metals, there can be no automobiles, airplanes, jet engines, satellites, or sophisticated weapons, not even many home appliances. Thus, it is no surprise that the United States treated South African apartheid with kid gloves, as

> Domestically, politicians compete to demonstrate their commitment to *saving the planet* while continuing to protect favored industries and groups.[5]
> — Cato Institute

disruption of that country's social, political, and economic institutions would be disastrous for US high-tech industrial production.

As the world begins to depend more and more on oil production out of the Persian Gulf, petroleum-hungry consumers, such as the United States, will have more and more at stake in the region. Small wonder that despite strong words of military action against Saddam Hussein in early 1998, the Clinton Administration found itself powerless against a world community that realizes the strategic importance and value of Iraqi oil reserves and production. The United States cannot afford to lose its few friends in the region, especially Saudi Arabia. US military and political support of King Fahd, leader of the Saudi royal family that rules Saudi Arabia, is almost exclusively based on the immense Saudi petroleum reserves. One possible theory (hinted at by some politicians and economists) of the 1991 Gulf War against Iraq holds that the whole invasion was set up to shut down Iraqi oil production in order to maintain a "strategic reserve" in the un-pumped oil fields of that country. (In 1998, 16% of Persian Gulf petroleum exports to the United States came from Iraq.)

*The Pentagon spends $50 billion a year safeguarding "cheap oil" from the Persian Gulf.*

When oil prices began to rise in the winter of 1995-96, the United Nations felt compelled to relax the Iraqi oil-embargo for "humanitarian" reasons. The $4 billion worth of petroleum that Iraq added to the world oil pipeline stabilized the market but has not markedly dropped prices, thus confirming the underlying growing scarcity in the world's oil supply. If Mark Twain were alive today, he might say, "Whiskey is for drinking; oil is for fighting over." He would be very close to the truth.

FOSSIL FUELS: THE WOLF IS AT THE DOOR    35

# Crude Oil / Petroleum Fact Sheet

**CRUDE OIL RESERVES** (*billion barrels*)

|  | 2000 |
|---|---|
| Saudi Arabia | 261.4 – 263.5* |
| Iraq | 100 – 112.5* |
| Kuwait | 94.7 – 96.5* |
| US (12th) | 21.8* |
| World | 981.4 – 1,016.8* (1,033.8*) |

**Analysis** (*using the higher estimates*): Saudi Arabia, with less than 0.4% of the world's population, may have over 25% of the world's crude oil reserves (over 12,000 barrels per capita). Iraq, also with less than 0.4% of the world's population, may have almost 11% of the world's crude oil reserves (almost 5,000 barrels per capita). And Kuwait, with less than 0.04% of the world's population, may have over 9% of the world's crude oil reserves (over 44,000 barrels per capita). For comparison's sake, the US, with less than 5% of the world's population, has a little over 2% of the world's crude oil reserves (about 79 barrels per capita). Kuwait has by far the most crude oil reserves per capita – over 3½ times more than Saudi Arabia, over 9 times that of Iraq, and over 557 times more than the US.

**PRODUCTION** (*million barrels/day*)

|  | 1998 | 1999 | | 2000 |
|---|---|---|---|---|
| **Crude Oil** | | | | |
| Saudi Arabia | 8.39 | 7.83 | (2,858/yr) | 8.4 |
| Russia | 5.85 | 6.08 | (2,219/yr) | 6.48 |
| US | 6.25 | 5.88 | (2,146/yr) | 5.84 (E) |
| World | 66.96 | 65.87 | (24,043/yr) | 67.99 |
| **Total Products** | | | | |
| US | 9.28 | 8.99 | (3,281/yr) | 9.12 |
| Saudi Arabia | 9.16 | 8.51 | (3,106/yr) | |
| Russia | 7.25 | 7.65 | (2,792/yr) | 7.87 |
| World | 75.15 | 74.18 | (27,076/yr) | 76.65 |

**Analysis** (*using 1999 figures*): Saudi Arabia, with less than 0.4% of the world's population, produced almost 12% of the world's crude oil (over 132 barrels/year per capita) and almost 12% of the world's petroleum products (almost 144 barrels/year per capita). Russia, with less than 2.5% of the world's population, produced over 9% of the world's crude oil (over 15 barrels/year per capita) and over 10% of the world's petroleum products (over 19 barrels/year per capita). With under 5% of the world's population, the US produced almost 9% of the world's crude oil (almost 8 barrels/year per capita) and over 12% of the world's petroleum products (almost 12 barrels/year per capita). Saudi Arabia produced by far the most crude oil per capita – over 8½ times that of Russia and about 17 times more than the US, as well as the most petroleum products per capita – almost 7½ times that of Russia and about 12 times more than the US.

**CONSUMPTION** (*million barrels/day*)

|  | 1998 | 1999 |
|---|---|---|
| US | 18.92 | 19.52 (7,125/yr) |
| Japan | 5.51 | 5.57 (2,033/yr) |
| China | 4.11 | 4.32 (1,577/yr) |
| World | 73.64 | 74.93 (27,349/yr) |

**Analysis**: With less than 5% of the world's population, the US consumed over 26% of the world's oil (almost 26 barrels/year per capita). Japan, with a little over 2% of the world's population, consumed almost 7.5% of the world's oil (about 16 barrels/year per capita). And China, with almost 21% of the world's population, consumed almost 6% of the world's oil (about 1¼ barrels/year per capita). The US was by far the most oil-gobbling country per capita of the three, consuming over 60% more than Japan and 21 times more than China.

**GROSS IMPORTS** (*million barrels/day*)

|  | 1998 | 1999 | 2000 |
|---|---|---|---|
| **Crude Oil** | | | |
| US | 8.71 (3,179/yr) | 8.73 | 8.93 |
| Japan | 4.29 (1,566/yr) | 5.50 | |
| S. Korea | 2.24 (818/yr) | | |
| World | 37.76 (13,782/yr) | | |
| **Total Products** | | | |
| US | 10.71 (3,909/yr) | 10.85 | 11.09 |
| Japan | 5.48 (2,000/yr) | | |
| Germany | 3.19 (1,164/yr) | | |
| World | 53.36 (19,476/yr) | | |

**Analysis** *(using 1998 figures)*: With less than 5% of the world's population, the US was responsible for about 23% of the world's crude oil imports (over 11½ barrels/year per capita) and about 20% of the world's petroleum products imports (over 14 barrels/year per capita). Japan, with a little over 2% of the world's population, was responsible for over 11% of oil imports (over 12 barrels/year per capita) and over 10% of petroleum products imports (almost 16 barrels/year per capita). South Korea, with less than 0.8% of the world's population, was responsible for almost 6% of oil imports (over 17 barrels/year per capita). And Germany, with less than 1.5% of the world's population, was responsible for about 6% of the world's petroleum products imports (over 14 barrels/year per capita). Per capita, South Korea's crude oil imports surpassed those of Japan by 40% and those of the US by 50%. And Japan's per-capita imports of petroleum products surpassed those of the US and Germany by over 11% each.

**EXPORTS** (*million barrels/day*)

| | 1998 | 1999 | 2000 |
|---|---|---|---|
| **Crude Oil** | | | |
| Saudi Arabia | 7.08 (2,584/yr) | | |
| Norway | 2.88 (1,051/yr) | | |
| Russia | 2.69 (982/yr) | | |
| US (34th) | 0.11 (40/yr) | | |
| World | 38.83 (14,173/yr) | | |
| **Total Products** | | | |
| Saudi Arabia | 7.91 (2,887/yr) | | |
| Russia | 3.73 (1,361/yr) | | |
| Norway | 3.15 (1,150/yr) | | |
| US (18th) | 0.95 (347/yr) | 0.94 | 1.04 |
| World | 54.8 (20,002/yr) | | |

**Analysis**: Saudi Arabia, with less than 0.4% of the world's population, exported over 18% of the world's crude oil (almost 120 barrels/year per capita) and almost 14.5% of the world's petroleum products (almost 134 barrels/year per capita). Norway, with a little over 0.07% of the world's population, exported almost 7.5% of the world's crude oil (over 234 barrels/year per capita) and almost 6% of the world's petroleum products (over 256 barrels/year per capita). Russia, with under 2.5% of the world's population, exported about 7% of the world's crude oil (almost 7 barrels/year per capita) and almost 7% of the world's petroleum products (over 9 barrels/year per capita). For comparison, the US, with under 5% of the world's population, exported almost 0.3% of the world's crude oil (almost one-seventh barrel/year per capita) and almost 2% of the world's petroleum products (about 1¼ barrels/year per capita). Per capita, Norway exported the most crude oil – almost 2 times that of Saudi Arabia; almost 35 times that of Russia; and 1,615 times that of the US. Norway also exported the most petroleum products per capita — about 92% more than Saudi Arabia, over 27 times that of Russia, and about 204 times that of the US.

**$CO_2$ EMISSIONS** (*million metric tons carbon equivalent*)

|       | 1998     | 1999     |
|-------|----------|----------|
| US    | 634.88   | 649.74   |
| Japan | 183.37   | 185.36   |
| China | 151.83   | 159.75   |
| World | 2,673.49 | 2,711.55 |

**Analysis**: The US, with under 5% of the world's population, emitted about 24% of the world's carbon dioxide emissions from oil (almost 2½ metric tons per capita). Japan, with a little over 2% of the world's population, emitted about 7% of the world's carbon dioxide emissions from oil (almost 1½ metric tons per capita). And China, with almost 21% of the world's population, emitted almost 6% of the world's carbon dioxide emissions from oil (about one-eighth metric ton per capita). The US exceeds the other two in per-capita emissions — over 60% more than Japan and almost 19 times more than China.

---

\* = as of January 1, 2000; (E) = estimate

*References:*
<http://www.eia.doe.gov/iea/>
> Each category leads to a table of various statistics links, plus a link to "other data," where most statistics reside.

<http://www.prb.org/pubs/wpds2000/wpds2000_Population2000-PopulationProjected.html>
> Table of population statistics for each country. The table claims that the figures are from mid-2000 year, but most sources actually come from 1998 to early 2000, and some numbers for more obscure locations actually come from earlier dates. In other words, the origin of this information probably coincides well with the 1998-2000 fuel statistics.

**Notes:** Total Products = crude oil, natural gas plant liquids, other liquids, refinery processing gain; Total Products Imports and Exports = crude oil and refined petroleum products; reserves variations due to numbers from two different publications: World Oil, and Oil and Gas Journal. Also, whereas one country exceeds the others in its category per capita, that does not mean that it exceeds every country in the world per capita; only the 3-5 countries with the highest overall totals were compared in this way.

# FOOTNOTES

[1] <www.iea.org/weo/index.htm> The International Energy Agency is an international organization based in Paris, France. Primary objectives include maintaining and improving systems for coping with oil supply disruptions as well as promoting rational energy policies in a global context through cooperative relations with non-member countries, industry, and international organizations. The IEA is assisting in the integration of environmental and energy policies and improving the world's energy supply and demand structure by developing alternative energy sources and increasing the efficiency of energy use.

[2] <www.hubbertpeak.com/summary.htm> Overview of geophysicist M. King Hubbert's theory on the depletion of finite resources. A chart reveals a selection of countries with suspect increases in reserves (usually due to political or economic motives). Colin J. Campbell's data also pinpoint a list of countries and their projected time to depletion midpoint in years. Another graph illustrates the relationship between Saudi Arabia petroleum exports and corresponding United States consumption. Loaded with hyperlinks to various aspects of the fossil fuel dilemma.

[3] <www.hubbertpeak.com/hubbert/index.html> Background on geophysicist M. King Hubbert and his amazingly prescient prediction regarding the patterns of discovery and depletion of the world's petroleum resources. Also offers information regarding "Hubbert's Prescription for Survival, a Steady State Economy."

[4] <www.altenergy.org/core/Fossil_Fuels_Futures/Joy_Ride/joy_ride.html> Pamphlet petroleum primer titled *When Will The Joy Ride End?* written by Randy Udall, Director of the Community Office Fore Resource Efficiency (CORE), with Steve Andrews, a Denver-based energy analyst. (Homepage under construction.) For more information, contact CORE at Box 9707, Aspen, CO, 81612 – (970) 544-9808.

[5] <www.spee.org/sld001.htm> Graph portraying the number of years that OPEC countries have left in "oil life" as compared to the rest of the world. It is one of fourteen informative "slides" that graph oil and natural gas reserves, production, field distribution, and so on. The Society of Petroleum Evaluation Engineers, or SPEE, "was formed in 1962 to answer a long-standing need for a professional organization which would bring together, for their mutual benefit, specialists in the evaluation of petroleum and natural gas properties."

[6] <www.bp.com/centres/energy/world_stat_rev/index.asp> Statistical review on world primary energy up to 2000. Covers global data, re: consumption, oil reserves, production, crude oil prices, refinery capacities, and trade movements. Data can be downloaded in Microsoft Excel spreadsheet form. The BP website also covers natural gas, coal, nuclear

energy, and hydroelectric energy data. Educational products are available with a discount for bona fide schools.

7   <www.geocities.com/RainForest/2958/oil.html> Weekly Foreign Oil Consumption Report dated February 12, 1998. A Congressional report presented by Senator Jesse Helms to President Clinton. Posted on the Renewable Energy Advocacy website, which encourages readers to send cyber-letters to their Representatives "asking them to join the Renewable Energy Caucus. The Caucus's mission statement is only to educate fellow members of Congress on the issues surrounding Renewable Energy—and does not take positions on any legislation."

8   <www.epa.gov/oar/oaqps/peg_caa/pegcaa01.html> *The Plain English Guide to the Clean Air Act* from the US Environmental Protection Agency. Brief summary that serves as a good introduction to the 1990 Clean Air Act. Touches on the role of the federal government and the states. Mentions interstate and international air pollution. Explains guidelines introduced by the 1990 legislation being implemented to clean up smog and other "criteria" pollutants.

9   Herbert Bormann and Stephen R. Kellert, *Ecology, Economics. Ethics: The Broken Circle* (New Haven, Connecticut: Yale University Press, 1991) pp.183–185. A group of environmental experts argue the importance of the interconnected circle of ecology, economics, and ethics. Addresses a wide range of concerns from global atmospheric degradation to loss of forests and loss of biodiversity.

10  <www.wri.org/health/note-ghg.html> World Resources Institute (WRI) document, "The Hidden Benefits of Climate Policy: Reducing Fossil Fuel Use Saves Lives Now." The benefits of reducing $CO_2$ emissions go substantially beyond averting potential disruptions of the Earth's climate. Even relatively small reductions in emissions worldwide could prevent 700,000 premature deaths a year by 2020. WRI's mission is to move human society to live in ways that protect Earth's environment and its capacity to provide for the needs and aspirations for current and future generations.

11  Thomas Gale Moore, *Climate of Fear: Why We Shouldn't Worry about Global Warming* (Washington, D.C.: Cato Institute, 1998) p.73. This book's subtitle reveals the writer's slant – Why We Shouldn't Worry about Global Warming. Moore is a senior fellow at the Hoover Institution at Stanford University and was a member of the President's Council of Economic Advisers in the 1980s. Moore's book offers a one-sided, optimistic outlook on the prospects of global warming. Endorsements by representatives of the US business sector agree that "happiness is a warm planet." This book is produced by a group with a specific pro-energy agenda and should be read cautiously as its premise suggests that "the rise in atmospheric carbon dioxide will benefit agriculture in particular and mankind in general." (Cato Institute – Conservative public policy research foundation, 1998)

[12] Gary Gardner, "When Cities Take Bicycles Seriously," *World Watch* (September/October 1998), p. 16–22. *World Watch* articles are written by the staff of the Worldwatch Institute. *World Watch* is a bimonthly, non-profit magazine that tracks key indicators of the Earth's health. *World Watch* is available at newsstands and bookstores or by subscription at <www.worldwatch.org>.

[13] Joseph B. Knox and Ann Foley Scheuring, *Global Climate Change and California: Potential Impacts and Responses* (Berkeley, California: University of California Press, 1991), pp. 112 – 114. Collection of essays by highly qualified scientists offers a glimpse of the environmental future of California. Topics range from greenhouse gases to the impact of climate change on California's vital water resources and water-supply systems. Discusses potential changes in various natural ecosystems.

[14] <www.monitor.net/rachel/r613.html> Rachel's Environment and Health Weekly is a website that posts a new article in its weekly newsletter. The site offers a free subscription to anyone who has an e-mail address. The article "Environmental Trends" includes information from an Index of Environmental Trends, written by Gar Alparovitz and others for the National Center for Economic Security and Alternatives. The authors' intent was to measure whether we are succeeding in curbing our environmental destruction. Nine countries were studied in the areas of air quality, water quality, chemical use, waste disposal, etc.

[15] <www.epa.gov/airmarkets/acidrain/effects/index.html> Fact sheet sponsored by the US Environmental Protection Agency. Focuses on the environmental effects of acid rain on surface waters, forests, and human health. It is short and to the point.

[16] <www.monitor.net/rachel/r476.html> Rachel's Environment and Health Weekly is a website that posts a new article in its weekly newsletter. Offers a free subscription to anyone who has an e-mail address. "The Dying of the Trees" is about a book of the same title by Charles Little. Whether the cause is acid rain, ozone depletion, or global warming, this article explains how the planet's trees, soils, and the "detritus food chain" (the "composting" micro-organisms) are jeopardized. Includes problems, surveys, and studies concerning these issues.

[17] <www.epa.gov/oar/oaqps/peg_caa/pegcaa05.html> This site is sponsored by the Environmental Protection Agency and the Office of Air Quality Planning and Standards. "The Plain English Guide to the Clean Air Act" is an overview of the current acid rain situation in the United States with a brief look at anticipated reductions in sulfur dioxide emissions due to its implementation.

[18] <www.ns.ec.gc.ca/aeb/ssd/acid/acidfaq.html> Eleven frequently asked questions about acid rain/deposition. The answers come from Environment Canada's "A Primer on Environmental Citizenship."

[19] <www.sft.no/english> Homepage for the Norwegian Pollution Control Authority (SFT), a directorate under the Ministry of the Environment, established in 1974. The SFT is working to ensure that pollution, hazardous substances, and waste do not cause health problems, affect people's well-being or harm nature's own powers of regeneration. This site makes the fish mortality and lake acidification in the Adirondacks and New England seem mild in comparison.

[20] <www.eei.org/issues/enviro/cleanair.htm> Edison Electric Institute (EEI) is the association of US shareholder-owned electric companies, international affiliates, and industry associates worldwide. US members serve over 90% of all customers of this shareholder-owned segment of the industry. They generate approximately three-quarters of all the electricity produced by electric companies in the United States and service about 70% of all customers in the nation. E.E.I.'s mission focuses on advocating public policy, expanding market opportunities, and providing strategic business information.

[21] <www.adirondackcouncil.org/> The Adirondack Council is a nonprofit environmental watchdog organization that monitors development on private lands around, and the environmental issues within, New York State's six-million acre Adirondack Park. The Council publishes regular newsletters, is involved in public education and research, and policy analysis.

[22] <www.skepticism.net/faq/environment/global_warming/index.html> The Skepticism.Net website offers a medley of articles and links to resources that argue against the perceived dangers of rising levels of greenhouse gas emissions or the threat of rapid climate change due to global warming.

[23] William K. Stevens, "New Evidence Finds This is the Warmest Century in 600 Years," *The New York Times*, April 28, 1998, p. C3.

[24] <www.newscientist.com/hottopics/climate/climatetimeline.jsp> A greenhouse timeline from the early 19th century until early 2000. Brief chronicle and overview of the increasing awareness of global warming by scientists over the last 175 years. Includes sidebars on the current effects from global warming on Alaska and Northern Europe. Covers the scientific debate and the international politics as well as further web resources. (2001)

[25] <www.wri.org/biodiv/b04-dnh.html> Article "*A Primer on Global Warming*" on the World Resources Institute (WRI) website. WRI is a non-profit educational organization and independent center for policy research and technical assistance on global environmental and development issues.

[26] Ross Gelbspan, "THE HEAT IS ON: The Warming of the world's climate sparks a blaze of denial," *Harper's*, (December 1995), pp. 31–37 Gelbspan, a thirty-year journalist with *The Philadelphia Bulletin*, *The Washington Post*, and *The Boston Globe*, is the author of *The Heat Is On: The Climate Crisis, the Cover-Up, the Prescription* (HarperCollins).

27 *The Greening of Planet Earth: The Effects of Carbon Dioxide on the Biosphere.* Free 1991 video produced by The Institute for Biospheric Research. Program funded by a grant from public utilities through Western Fuels Association, Inc. This 28-minute video presents optimistic future outlook about how increasing levels of $CO_2$ in the atmosphere will greatly enhance the growth of trees, plants, fruits, and vegetables. On the video, various scientists extol the virtues of carbon dioxide. Includes their forecast that cultivated crop yields will increase by 30% – 40%, the water-use efficiency of earth's vegetation could possibly double, and the productivity of the world's forests will triple.

28 Letter in *Harper's*, April 1996 pp. 85-86.

29 Ben Matthews, "Climate Engineering: A Critical Review of Proposals, their Scientific and Politic Contest, and Possible Impacts." A review of climate engineering proposals aims to provide a comprehensive resource of up to date information and ideas for people concerned about the development of large-scale technical fixes to counter the problem of global warming. In 1996 Ben Matthews was an academic with the University of East Anglia, UEA, Norwich, United Kingdom <www.uea.ac.uk>. This paper was compiled for *Scientists for Global Responsibility* SGR <www.sgr.org.uk> who promote the ethical practice and use of science and technology. SGR is affiliated to the International Network of Engineers and Scientists for Global Responsibility.

30 <www.monitor.net/rachel/r511.html> Rachel's Environment and Health Weekly is a website that posts a new article each week in its newsletter. A free subscription is offered to anyone who has e-mail. This August 1998 article "Environmental Trends" includes information from the *Index of Environmental Trends*, written by Gar Alparovitz and others for the National Center for Economic Security and Alternatives. The authors' intent was to measure whether we are succeeding in curbing our environmental destruction. Nine countries were studied in the areas of air quality, water quality, chemical use, waste disposal, etc.

31 <www.hubbertpeak.com/campbell/index.html> Exploration geologist Colin Campbell, a consultant who advises governments and industry experts regarding trends in the world petroleum market.

32 <www.dieoff.org/page224.htm> Richard C. Duncan, "The Peak of World Oil Production and the Road to the Olduvai Gorge." *Pardee Keynote Symposia*, Geological Society of America, Summit 2000, Reno, Nevada, November 13, 2000. Duncan's Olduvai theory has been called "unthinkable, preposterous, absurd, dangerous, self-fulfilling, and self defeating." The theory is defined by the ratio of world energy production (use) and world population. It states that the life expectancy of Industrial Civilization is less than or equal to 100 years: 1930 – 2030.

33 Colin J. Campbell, *The Coming Oil Crisis* (Essex, England: Multi-Science Publishing Company & Petroconsultants S.A., 1998), p. 176. This 210-page paperback discusses

how much conventional oil remains to be produced and its depletion pattern. It explains how to properly interpret published numbers, many of which are spurious or distorted by vested interests.

34 Walter Youngquist, *Geodestinies: The Inevitable Control of Earth Resources Over Nations and Individuals* (Portland, Oregon: National Book Company, 1997), pp. 321–322. Overview of the historical relationship between civilizations and minerals. This book also covers strategic minerals, free trade versus geologic provinces, earth resources, the future, and the potential of achieving a "sustainable" society.

## SIDEBAR FOOTNOTES

A  Colin J. Campbell, "Deepwater Oil: The End of the End Game." *Hubbert Center Newsletter* (April 1, 1999), p. 3. (M. King Hubbert Center for Petroleum Supply Studies, Colorado School of Mines, Golden, Colorado.)

B  Walter Youngquist, *GeoDestinies: The Inevitable Control of Earth Resources over Nations and Individuals.* (Portland, Oregon: National Book Company, 1997), p. 171.

C  Randy Udall and Steve Andrews, "When Will the Joy Ride End?" *Hubbert Center Newsletter #99/1,* January 1999, p. 3.

D  Colin J. Campbell, *The Coming Oil Crisis* (Essex, England: Multi-Science Publishing Company and Petroconsultants S.A., 1997), p. 1.

E  Bill Lindelof, "Citing smog and sprawl, groups to sue over road projects," *The Sacramento Bee*, September 10, 1999, p. 1.

F  Lester R. Brown, Michael Renner, and Christopher Flavin, eds., *Vital Signs 1998: The Environmental Trends That Are Shaping Our Future* (New York: W.W. Norton, 1998), p. 50.

G  Lisa Getter, "L.A. Study Finds Air Quality Better But Still Too Toxic," *Los Angeles Times*, March 1, 1999, p. 1.

H  Editorial, "Higher gas prices?" *The Sacramento Bee*, July 25, 1999, section I, page 4.

I  J. Raloff, "Climate Protection Saves Lives Now," *Science News* (November 8, 1997), p. 292.

J  F. Herbert Bormann and Stephen R. Kellert, eds., *Ecology, Economics, Ethics: The Broken Circle* (New Haven, Connecticut: Yale University Press, 1991), p. 140.

K  Ram M. Shrestha, S.C. Bhattachrya, and Sunil Malla, "Energy Use and Sulfur Dioxide Emissions in Asia," (*Journal of Environmental Management*, 1996), vol. 26, pp. 359–372.

L   Christopher Flavin, "Global Climate: The Last Tango," *World Watch* (Nov./Dec.1998), p.12.

M   William F. Martin, "Twin Challenges: Energy and the Environment in Asia," *Harvard International Review*, (Summer 1997), pp. 28-31.

N   "The 1998 Ozone Assessment," *World Climate News*, (January 1999), pp. 6-7. The *World Climate News* is issued by the World Meteorological Organization, based in Geneva, Switzerland.

O   Robert L. Peters and Thomas E. Lovejoy, eds., *Global Warming and Biological Diversity* (New Haven, Connecticut: Yale University Press, 1992), p. 11.

P   Joseph B. Knox and Ann Foley Scheuring, eds., *Global Climate Change and California* (Berkeley, California: University of California Press, 1991), p. 23.

Q   Kendrick Taylor, "Rapid Climate Change," *American Scientist* (July-August 1999), pp. 320-327.

R   H. Josef Hebert, "EPA targeting heaviest SUVs for pollution control," *The Sacramento Bee*, October 8, 1999, section D, p. 12.

S   Thomas Gale Moore, *Climate of Fear: Why We Shouldn't Worry About Global Warming* (Washington, D.C.: Cato Institute, 1998), p. 136.

# Coal

## 3

Coal is the most abundant of the carbon-based fossil fuels, but it is also a leading threat to human health and the environment. Oil replaced coal in the late 19th century as the fuel of choice in the United States, but rumors that the reign of coal is over are premature. Coal currently provides 24% of the world's primary energy requirements and, in 1999, generated 57% of all the electricity used in the United States. Burning coal accounts for 43% of annual global carbon emissions. Existing coal reserves may be large, but they won't last forever, and the health and environmental costs associated with this fuel limit its potential as an acceptable primary fuel in the future.

IMAGINE AN INEXPENSIVE ENERGY SOURCE that heats your home, cooks your food, and powers your appliances, but its health and environmental impacts are life- threatening. Would you risk getting emphysema, bronchitis, lung cancer, heart disease, arsenic poisoning, bone decay, or open sores in order to use this particular fuel? The answer is yes, because you already do.[1]

Carbon-based coal is the result of hundreds of millions of years of decomposition and compression of ancient plant matter. Some geologists call this form of ancient biomass "buried sunshine." It has been used as a source of fuel for centuries in many countries, and its extraction and processing have impacted the people whose land was once rich with it.

Often shortening a man's working life by 20 or 30 years, coal mining was — and still is — dirty, environmentally degrading, and dangerous. In the Industrial Age, many boys began working in the mines before they were 12, lying about their age in order to get a job. Despite the ever-present hazards — rock falls, gaseous explosions, accidents, pneumoconiosis ("black lung") due to particulate inhalation — mining became a necessary evil for the growing coal industry.[2]

In the 20th century, the world depended on "King Coal" for a large portion of its primary energy needs. In 1910, coal provided 62% of the world's energy and it is still a major producer of electricity. Coal is a very large energy source, but it must be mined and is not nearly as easy to handle and transport as is oil. It also has much less energy density than petroleum. Converting coal to a liquid to use as a transportation fuel is very expensive and doing it on a scale to significantly replace oil would require giant mining projects and vast environmental impact. The bottom line is that although the energy in coal reserves worldwide is greater than oil, it too is a finite fossil fuel.[3] In 1999, coal still generated 57% of the United States's electricity. Coal currently provides 73% of India's electricity and 97% of Poland's electricity. But coal's global energy contribution is diminishing as more countries recognize the "hidden costs" to human health and the environment. Today, coal accounts for about one quarter of the world's total energy production. The planet's vast but finite coal reserves will outlast those of oil and natural gas, but governments and energy providers should not rely on these resources. Coal reserves are extensive in certain regions of the world, but the high social cost exacted by converting them to energy is untenable. Miners are no longer the only ones at risk. Everyone on Earth is exposed to the damaging byproducts of coal, the consumption of which adversely impacts our health and environment. Air pollution, acid rain, and sulfur dioxide are just a few of the dangerous byproducts produced by coal combustion.

Burning coal has affected industrialized societies for more than a century. "Coal smogs" in London killed over

> The reign of King Coal has left a legacy of human and environmental damage that we have only begun to assess.[A]
>
> — *World Watch*

2,000 people in 1880, and the London fog killed some 4,000 in 1952. These killer smogs were comprised of dust, soot, and sulfur, which combined into a lethal atmospheric stew. Today, researchers know that prolonged breathing of soot and sulfur causes respiratory and cardiovascular diseases such as emphysema, asthma, lung cancer, and heart disease. In an effort to alleviate local air quality problems, tall smokestacks were built in order to thin the harmful nitrogen oxides and sulfur dioxide over a larger area. The result was atmospheric disruption, acid rain, and natural resource destruction. What was once a regional problem became a global concern as Norwegian fish, German forests, and Chinese sculptures were destroyed or altered. In fact, acidification may plague one-fifth of India's farmland, and, in 40 years, half of the 2,800 ponds and lakes in the Adirondacks may be too acidic to support living organisms. Meanwhile, billions of dollars have been lost in the US and China due to crop damage.

Everyday, the health and welfare of many people are at risk from the impacts of coal. Millions of people suffer bronchitis and respiratory illnesses, and lung cancer and death are not uncommon for those who depend on coal for indoor cooking and heating. The top ten most air-polluted cities in the world — nine in China, one in India — all use coal as a primary energy source. Atmospheric scientists have tracked large dust clouds of particulates and sulfur from Asia to the United States' west coast.

More stringent and costly air pollution mandates have inspired some coal-burning power plants to ignore the law in order to sell their electricity cheaper. According to a report published in *The Sacramento Bee,* many US power plants in the South and Midwest have been identified and prosecuted. In fact, the US Environmental Protection Agency recently took action against 32 plants to investigate their so-called "maintenance" expenses, which may have actually worsened the air quality.[4]

"Clean-coal" technologies that involve flue-gas desulfurization and nitrogen control equipment are the newest response to pollution regulations. The implementations of the clean-coal systems have lowered sulfide emissions,

> Coal is the most abundant and environmentally damaging of all fuels.[B]
> — *Discover*

> To mine enough coal [to replace oil] would involve the largest mining operation ever seen in the world.[c]
> — Walter Youngquist

but other resources are compromised. The less efficient, water-intensive technologies emit other compounds such as carbon dioxide, a greenhouse gas, and create sludge and other solid wastes that endanger people and the environment if they are not properly disposed.

In the United States, coal reserves surpass those of oil and natural gas by about 200 years and can be mined domestically, but using coal simply because there is plenty of it would be a serious mistake. Countries like India and China are also struggling with how to supply their growing populations with electricity but, at the same time, are scaling back using their reserves of coal to generate power. The air pollution, acid rain, greenhouse gas emissions, and other health dangers associated with processing vast amounts of coal into electricity would take its toll on countless people around the world.

Nonetheless, the coal industry and their lobbyists argue that the extraction and sale of coal is a deeply personal issue for those whose livelihoods depend on it. The World Coal Institute (WCI) assures politicians and consumers that coal is a viable source of energy for the future. The Institute urges governments and policymakers to consider the limitations in scientific understanding of possible climate change and the economic repercussions of inappropriate measures to control greenhouse gas emissions.[5] But John Slater, WCI's Chairman, recognizes that it may be too late to try to convince people that scientific studies may be erroneous. In a paper entitled "Global Overview of Coal," Slater concedes that the Institute's continued denial of the global warming theory is a lost cause because "the general public and the politicians believe it to be true and we can expect no sympathy from them." Slater believes that the coal industry may see a decline in the future rather than the growth that it enjoyed in the past, and he blames the Institute for failing "to make the case for coal to the general public and the political authorities in most countries."[6]

"Far from environmental pressure easing," says Slater, "the coal industry now faces the most extreme environmental threats it has ever seen." He admits that along with the low price and environmental credentials of natural gas, the Kyoto Protocol and efforts by the Environmental Protection Agency, future coal usage in the US could be limited. Slater points out that the Kyoto Protocol's potential halting of coal industries in Annex B countries (36 countries including the US, Canada, Japan, and most European countries) may not significantly cut greenhouse gas emissions if non-signatory nations increase their own coal consumption. Despite these dim prospects for the US coal industry, Slater insists that coal will continue to play a large role in the 21$^{st}$ century energy markets.

Many nations cannot afford to abandon the use of coal as a primary energy source. As a result, tests are now underway in various countries to separate carbon dioxide from fossil fuels and get rid of it before it can enter the atmosphere. Researchers hope to bury the $CO_2$ in the ocean, in geological formations, or in terrestrial ecosystems such as forests, soils, and crops. This short-term solution is somewhat understandable considering that the US Department of Energy (D.O.E.) reports that "concentrations of carbon dioxide in the earth's atmosphere could double by the middle of the 21$^{st}$ century and continue to build up even faster in later years, potentially creating a variety of serious environmental consequences."[7] Alternative long-term solutions may take more time and permit more harmful emissions in the short term, but a quick fix may cause unforeseen future hazards.

*Coal is a dirty business from extraction to consumption.*

Due to carbon sequestration research, forests have acquired value in their uncut form. As natural resources that have often been destroyed in favor of development and agriculture, forests are now being protected for their functional worth — they hold more value as carbon sinks. Their ability to absorb excess carbon dioxide is slowing deforestation and spurring reforestation in a worldwide effort to offset the greenhouse effect. The World Resources Institute currently monitors forest projects that are in progress around the world, from Costa Rica to Uganda.[8]

Other potential carbon sinks are geologic formations such as unused coal seams and depleted or abandoned oil and gas reservoirs. In the US, "nearly 90 percent of all coal is considered to be unmineable due to seam thickness, depth, and structural integrity," according to the D.O.E.'s National Energy Technology Lab (NETL).[9] Such seams have been shown to retain carbon dioxide that has been pumped into them. Because of the proximity of power plants to many of these coal seams, this approach may be cost-effective. The nearness of many coal-fired power plants to oil and gas producing regions makes the potential of storing $CO_2$ in depleted gas reservoirs or using the captured $CO_2$ in enhanced oil recovery operations economically appealing.

*The health and welfare of many people are at risk from the impacts of the coal industry.*

Using the ocean as a carbon sink has been a more cautious consideration. Technical and economic feasibility tests are being conducted by NETL in waters off the coast of Hawaii with the help of Japan and Norway. In an effort to cull even better ideas for capturing and sequestering carbon dioxide, NETL has offered monetary awards for novel plans that have been conceptualized up to the pilot stage. Workshops have also been sponsored to bring together the best from academia and industry to advance the most promising research and development that will make carbon sequestration a viable commercial technology.

Carbon sequestration is a favored solution by the coal industry because it provides a loophole for emission problems. Best of all, the federal government is picking up the tab to ensure the continued use of coal to generate electricity. But this type of technology is challenging. The expense of capturing, transporting, and injecting carbon dioxide in distant or difficult media could possibly double the retail cost of electricity. The additional energy used in the process could require 30% more power generation. Energy-intensive sequestration would offset the economic growth envisioned by the coal industry. Many institutions currently exude confidence in this new approach, but few seem concerned with the eventual harm this process could impose on the environment. Just as coal combustion has become more of a curse than a blessing, so might carbon sequestration — after the unintended impacts have already been made.

In April 1999, the D.O.E. released a 200-page document describing carbon sequestration and its potential for emissions reductions. It encourages government agencies and related industries to continue research and development projects that will determine future implementation of the technology. Renewable sources such as wind, solar, and hydroelectric power alone cannot meet the energy challenge, nor can a switch to natural gas significantly cut emissions in such a short time. Until more advanced alternative energy technologies are developed, carbon sequestration may be responsible for maintaining stable concentrations of greenhouse gases in the atmosphere while permitting worldwide economic growth.

If coal is to continue fueling the planet, as the coal industry desperately desires, our world community will have to endure health hazards and environmental degradation. The Chinese have decided that the hidden costs of this fossil fuel are not worth the energy produced. Informed citizens are educating the public and implementing policies to phase out coal while the government plans 40 coal-free zones. China and the United Kingdom are both shifting from coal-reliant industries like steel to the high-tech and tourism sectors. Considering that coal subsidies worldwide total around $63 billion, getting out of the coal business is good business.

> It's impossible to have strip mining or mountaintop removal and have adequate reclamation. [It's like] putting lipstick on a corpse.[D]
> — Ken Hechler, West Virginia Secretary of State

# Coal Fact Sheet

**RESERVES** (*billion short tons*)

|  | 1999 |
|---|---|
| **Demonstrated Reserve Base** | |
| US | 504.3* |
| **Estimated Recoverable** | |
| US | 275.57* |
| Russia | 173.07* |
| China | 126.22* |
| World | 1,088.6* |

**Analysis**: The US represents under 5% of the world's population and has over 25% of the world's coal reserves (almost 1,000 short tons per capita). Russia has under 2.5% of the population and almost 16% of the coal reserves (almost 1,200 short tons per capita). China represents almost 21% of the population and almost 12% of the reserves (almost 100 short tons per capita). Per capita, Russia has the most natural gas reserves of the three, most notably almost 12 times that of the US.

**PRODUCTION** (*million short tons*)

|  | 1998 | 1999 | 2000 |
|---|---|---|---|
| China | 1,479.64 | 1,118.11 | |
| US | 1,117.54 | 1,100.23 | 1,086.68 |
| India | 336.25 | 327.98 | |
| Australia | 313.67 | 320.59 | |
| World | 5,006.2 | 4,736.62 | |

**Analysis**: China represents almost 21% of the world's population and produced over 23% of the world's coal (less than 1 short ton per capita). The US, at under 5% of the world's population, also produced over 23% of the world's coal (almost 4 short tons per capita). India represents over 16% of the population and produced almost 7% of the world's coal (a bit over 0.3 million short ton per capita). And Australia, at a bit over 0.3% of the population, produced almost 7% of the world's coal (almost 17 million short tons per capita). Per capita, Australia produced the most coal of the four — over 4 times that of the US, almost 19 times that of China, and over 51 times that of India.

## CONSUMPTION (*million short tons*)

|  | 1998 | 1999 |
|---|---|---|
| China | 1,300.5 | 1,075 |
| US | 1,038.97 | 1,038.51 |
| India | 333.4 | 348.44 |
| World | 5,009.04 | 4,740.37 |

**Analysis**: At under 5% of the world's population, the US consumed almost 22% of the world's coal (almost 4 short tons per capita). While China, over four times the population of the US, consumed a little over 22% (almost 0.9 short tons per capita). At over 16% of the world's population, India consumed over 7% of the world's coal (over 0.3 short tons per capita). Per capita, the US far exceeded the coal consumption of the other two — almost 4½ times that of China and almost 11 times that of India.

## GROSS IMPORTS (*quadrillion btu/million short tons*)

|  | 1998 |
|---|---|
| Japan | 3.18 qbtu |
| S. Korea | 1.25 qbtu |
| Taiwan | 1.02 qbtu |
| US (14th) | 0.31 qbtu |
| World | 14.3 qbtu |

**Analysis**: Japan, whose population is a little over 2% of the world's total, imported over 22% of the world's coal (over 25 million btu per capita). South Korea, at less than 0.8% of the population, imported almost 9% of the world's coal (over 26 million btu per capita). Taiwan's population is almost 0.4% of the world's total, and it imported over 7% of the world's coal (almost 46 million btu per capita). The US, at almost 5% of the population, imported a bit over 2% of the world's coal (over 1 million btu per capita). Per capita, Taiwan imported by far the most coal of the four — over 73% more than South Korea, almost 2 times that of Japan, and almost 41 times that of the US.

## EXPORTS (quadrillion btu/million short tons)

|  | 1998 |
|---|---|
| Australia | 4.15 qbtu |
| US | 2.08 qbtu |
| South Africa | 1.62 qbtu |
| World | 14.43 qbtu |

**Analysis:** Australian exports accounted for almost 29% of total world coal exports, whereas Australia represents only a bit over 0.3% of the world's population (over 216 million btu per capita). At almost 5% of the population, the US exported over 14% of the world's coal (over 7.5 million btu per capita). At a bit over 0.7% of the world's population, South Africa exported over 11% of the world's coal (over 37 million btu per capita). Per capita, Australia exported by far the most coal of the three — almost 6 times that of South Africa and almost 29 times that of the US.

## $CO_2$ EMISSIONS (million metric tons carbon equivalent)

|  | 1998 | 1999 |
|---|---|---|
| US | 550 | 549.26 |
| China | 600.47 | 494.67 |
| India | 148.22 | 156.15 |
| World | 2,254.26 | 2,137.14 |

**Analysis:** The US was responsible for almost 26% of total world carbon dioxide emissions from coal but has only 5% of the world's population (about 2 metric tons emissions per capita). At almost 21% of the world's population, China emitted a bit over 23% of the world total (over one-third metric ton per capita). And India, the third most populous country in the world at over 16%, was responsible for a bit over 7% of world emissions (over one-fifth metric ton per capita). The average American was, therefore, responsible for about 5 times the emissions of the average Chinese, and almost 13 times the emissions of the average Indian.

---

\* = as of January 1, 1999

*References:*

<http://www.eia.doe.gov/iea/>

>Each category leads to a table of various statistics links, plus a link to "other data," where most statistics reside.

<http://www.prb.org/pubs/wpds2000/wpds2000_Population2000-PopulationProjected.html>

>Table of population statistics for each country. The table claims that the figures are from mid-2000 year, but most sources actually come from 1998 to early 2000, and some numbers for more obscure locations actually come from earlier dates. In other words, the origin of this information probably coincides well with the 1998-2000 fuel statistics.

**Note:** Reserves = anthracite (hard), bituminous (soft), lignite (brown), and subbituminous coal. Also, even though one country exceeds the others in its category per capita, that does not mean that it exceeds every country in the world per capita; only the 3-5 countries with the highest overall totals have been compared in this way.

FOOTNOTES:

1. Seth Dunn, "King Coal's Weakening Grip on Power," *World Watch* (September/October 1999), pp.10-19. A summary of historical views and the current status of coal, its unfavorable effects on human health and the environment, and the cutbacks of its use that China and other countries are instituting. Carbon sequestration and clean coal technologies are briefly mentioned. The Worldwatch Institute aims for "a sustainable society – one in which human needs are met in ways that do not threaten the health of the natural environment or the prospects of future generations."

2. <www.history.ohio-state.edu/projects/Lessons_US/Gilded_Age/Coal_Mining/default.htm> "Coal Mining in the Gilded Age and Progressive Era" is the title of a compilation of mining stories describing working conditions, disasters, child labor, and individual experiences from the mid-1800s to the early 1900s. Vivid historical photographs accompany the texts, which are posted on Ohio State University's Department of History website.

3. Walter Youngquist, "2000 Alternative Energy Sources: Water and Energy, the Basis of Human Society: Are They Globally Sustainable Through the 21st Century?" A pre-conference paper for the Kansas Geological Survey Open-File Repot 2000-51, p. 48.

4. "Dirty Utilities: EPA action against coal plants long overdue," Editorial in *The Sacramento Bee*, November 11, 1999. Opinion about the legal action the EPA has taken against 32 coal plants in 10 states that were operating poorly maintained utility services in order to undercut prices to gain a competitive edge.

5. <www.wci-coal.com> The World Coal Institute (WCI) is an industry-supported organization that endorses the use of coal primarily because of its abundance and cost competitiveness. The Institute argues that continued use of coal will sustain the international economy and protect people's interests, while assuring that clean coal technologies reduce greenhouse gas emissions. The WCI is a nonprofit, non-governmental association of coal producing and coal consuming enterprises and is the only international group working on behalf of producers and consumers. For more information on Clean Coal Technologies, visit <www.lanl.gov/projects/cctc/>.

6. <www.wci-coal.com/speeches/textonly/Spring_Coal_Forum_June_99.htm> "Global Overview of Coal," a paper presented by John Slater, Chairman of the World Coal Institute, at the Western Coal Council Spring Coal Forum in Colorado in June 1999. Slater explains the uncertain future of the coal industry based on environmental pressures relating to climate change, decreases in natural gas pricing, and measures that may soon be enforced by the Kyoto Protocol.

7 <www.fe.doe.gov/techline/tl_seqrpt.html> "DOE Reports on State of the Science of Carbon Sequestration" on the US Department of Energy's Fossil Energy website. This April 1999 article informs the public that the DOE has released a working draft that "provides a starting point for government, industry and academia to begin setting priorities and identifying specific directions for research and development activities that could extend over the next quarter century." Fossil Energy online is dedicated to "improving the fuels that power our world."

8 <www.igc.org/wri/climate/sequester.html> Climate Change: Carbon Sinks and Sequestration is a section of the World Resources Institute's website. It describes the major forest-related carbon sequestration projects throughout the world, implying that forests are now considered more valuable in their natural state due to their carbon dioxide filtering capabilities. The Institute strives "to move human society to live in ways that protect Earth's environment for current and future generations."

9 <www.fetc.doe.gov/publications/factsheets/carbon/carbon.html> "Carbon Sequestration Fact Sheet: Maintaining Economic Strength in a Carbon Constrained World" is posted on the United States D.O.E. Federal Energy Technology Center's website (FETC). This July 1999 fact sheet describes why we need to research sequestration, its potential cost challenges, possible injection sites, and R & D goals. The FETC, whose purpose is solving energy and environmental problems, was renamed the National Energy Technology Laboratory (NETL) in December 1999.

SIDEBAR FOOTNOTES:

[A] Seth Dunn, "King Coal's Weakening Grip on Power," *World Watch* (September/October 1999), pp. 10-11.

[B] Curtis Rist, "Why We'll Never Run Out of Oil," *Discover* (June 1999), p. 85.

[C] Walter Youngquist, *GeoDestinies: The Inevitable Control of Earth Resources over Nations and Individuals* (Portland, Oregon: National Book Company, 1997), p. 224.

[D] Maryanne Vollers, "Razing Appalachia," *Mother Jones* (July/August 1999), p. 42.

# Natural Gas

Natural gas is being touted by energy providers as an abundant, clean fuel for the 21st century. But this resource is non-renewable, and the D.O.E. states that the United States has only enough natural gas to last about 65 years. What then? US oil production peaked in 1970, and imported petroleum now costs the country $100 billion a year. The upcoming peak and subsequent plateau for our domestic natural gas supply will arrive two decades from now. After 2020, the bulk of the world's remaining supplies of oil and natural gas will be centered in the politically risky Persian Gulf and former Soviet Union. Nonrenewable natural gas cannot be considered civilization's long-term primary energy source.

NATURAL GAS (METHANE) has been getting a lot of press lately. Forecasted to be the fastest growing primary energy source in the 21st century, it burns cleaner than coal or oil and is not as controversial as nuclear power. According to projections by the US Department of Energy, natural gas (NG) is expected to be the fuel of choice for many countries in the future. In fact, the demand for NG will grow fastest in the developing countries of the world.[1]

Like other fossil fuels, natural gas is a product of organic matter that was trapped beneath the Earth's surface for tens of millions of years. Whereas coal is a solid and oil is liquid, NG was transformed by heat and pressure into a combustible gas. Coal, oil, and natural gas have similar origins, so it's no surprise that they are often found in geologic association. Oil is comprised primarily of ancient marine organisms, but natural gas can be formed both from land plants and from marine organic material. Oil and natural gas are frequently found together, in pockets or *traps* within sedimentary rock. Many fields produce both oil and gas, with the gas either dissolved within the petroleum itself or separated as a gas-cap above the oil.

NG is commonly extracted along with petroleum, but it also occurs in geological provinces by itself where oil doesn't exist. It is sometimes generated by deeply buried coal deposits, which have been subject to natural coking, or by sediments containing relatively woody-type plant materials, which are likely to be gas-prone, with little or no oil. Gas can also be formed in very young deposits, such as marsh gas found in swamps. At depths greater than 16,000 feet, the temperature is usually so hot that liquid oil cannot exist at all, and only gas survives.

Natural gas deposits are more widespread than oil fields, but the total endowment of conventional gas is probably less than that of petroleum because natural gas is comprised of smaller molecules than oil and is, therefore, more mobile and likely to escape through porous or cracked rock. NG deposits need an unpermeable geologic cap seal to trap and hold them. Most gas fields are found under salt caps, the most effective seal, or under the permafrost in Arctic regions, which is also an efficient cap. Less than 1% of the world's gas fields are considered *giants*, or *supergiants*, but 80% of the world's proven and produced reserves are contained in about 30 or 40 of these supergiant gas fields.[2]

Once located and extracted, raw natural gas is processed so that contaminants such as carbon dioxide, helium, nitrogen, hydrogen sulfide, and water are removed, leaving only low-moisture methane. Once impurities are removed, the gas is transported by pipeline to regional markets or is chilled to a liquid and shipped by tanker to distant ports. NG is characterized as either *wet gas* or *dry gas*. Wet gas contains higher hydrocarbons and is generally associated with oil accumulation. For every thousand cubic feet, wet gas can contain more than six gallons of recoverable hydrocarbons. Dry gas con-

> Over 70% of the oil and gas being produced still comes from fields that have been in production for over 30 years. Many of these fields are just beginning to experience depletion.[A]
> 
> — *Middle East Insight*

sists of methane and produces lesser amounts of hydrocarbons.

Until the 1890s, natural gas was primarily used to illuminate city streetlights. Providing it to individual households was not feasible at the time. After the turn of the century, electricity replaced NG for lighting. But the gas was not forgotten; it was just put on the backburner – namely the Bunsen burner – temporarily. Robert Bunsen's 1885 invention sparked a new use for gas – heat. Gas producers quickly shifted their focus to the thermal properties of natural gas, promoting it as a fuel for heating air and water as well as for cooking. But a combination of poor drilling and piping technologies, World War I, and the Great Depression delayed the development of an effective transportation network until the late 1940s. After World War II, there was a boom in construction technologies that advanced metallurgy, welding, and pipe-rolling processes. NG grew in importance as industries, electric utilities, and individual residences began using it as a primary fuel. In year 2000, there are more than 300,000 miles of pipe transporting gas throughout the US, not including local feeder lines.

Ironically, natural gas was originally considered a waste byproduct associated with oil production. As oil reaches the surface and the pressure around it diminishes, some light oils will vaporize and become gas. For many years, the gas was flared (burned off), but now the gas is either piped away to be sold on the market or pumped back into the oil reservoir to maintain the pressure needed to extract the oil.

As concerns over greenhouse gas emissions increase, energy providers predict that during the next several decades, natural gas will replace or supplement other fossil fuels because, when burned, it produces significantly less emissions than coal or oil. Burning methane does not release much $SO_2$ because the source of sulfur is removed before the gas is distributed. As opposed to burning coal, generating electricity by combusting NG in a state-of-the-art turbine can reduce carbon emissions by more than 60%. A standard boiler operating on natural gas emits, on average, only .0006 pounds of $SO_2$ per million BTU of fuel burned. Fuels like coal and oil can emit up to 6 pounds of $SO_2$ per million BTU.[3]

Natural gas costs four times as much as crude oil to transport through pipelines because it has a much lower energy density.[B]
— *Scientific American*

Of the approximately 620 gas processing plants in the US, most are located in Texas, Oklahoma, Louisiana, Kansas, New Mexico, and Wyoming. These six states account for about 71% of total US gas processing capacity, 72% of gas processed, and 87% of natural gas liquids production. The average US plant processes 77 million cubic feet of natural gas per day, with a potential capacity of 110 million cubic feet. According to the Energy Information Administration (EIA), natural gas provides 27% of the energy used today, second only to coal at 33%.[4] In 1997, the United States consumed 28% of the world's total gas production. Varying amounts of the natural gas supply in the US are consumed by the following sectors:

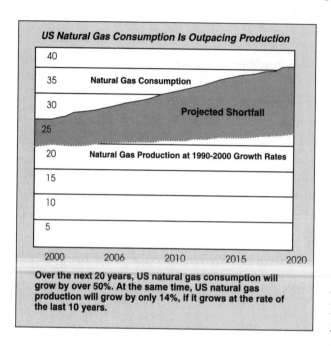

Over the next 20 years, US natural gas consumption will grow by over 50%. At the same time, US natural gas production will grow by only 14%, if it grows at the rate of the last 10 years.

Industrial ......... 49%
Residential ....... 21%
Commercial ..... 15%
Utility .............. 15%

Compared to the combustion of oil and coal, gas combustion is relatively benign as a contributor to air pollution. Oil and coal do not burn cleanly due to their complicated molecular structures and high ratio of carbon, sulfur, and nitrogen compounds. Toxic chemicals and ash are released into the air when coal and oil are combusted. In contrast, due to the simple composition of methane — one carbon atom and four hydrogen atoms per molecule — burning natural gas emits only carbon dioxide. Natural gas is currently considered the best choice among all fossil fuels. Consumption in the US and Canada is expected to grow 50% by 2020, due in part to the Clean Air Act Amendments of 1990, which are designed to curb acid rain, toxic emissions, and urban air pollution. Part of the intent of the amendments is to lessen oil dependency by substituting natural gas in various applications. As a result of these efforts, natural gas-powered vehicles, super-efficient boilers, and fuel cells are now reaching the market.

Annual US health costs due to exposure to gasoline and diesel engine exhaust are estimated at $20 to $50 billion. Vehicles burning compressed natural gas (NGVs), however, are environmentally friendly and are being integrated into municipal bus systems and corporate vehicle fleets. Over half a million NGVs are driven worldwide, with some 40,000 in the US alone. They typically generate about 10% of the carbon monoxide and 15% of hydrocarbon emissions that standard gasoline-powered vehicles produce. Although natural gas vehicles emit more methane than gasoline automobiles, the quantity is small as compared to $CO_2$ emissions. The biggest drawbacks to using compressed natural gas in private vehicles are limited driving range and greater fuel consumption. Due to the low energy density of NG, vehicles will burn 40% more fuel over their lifetime and will require heavy fuel storage systems, which currently limit mileage to about 250 miles per fill-up.

Facilities that use natural gas co-firing and reburning when generating electricity significantly reduce plant emissions. When a traditional fuel boiler is injected with natural gas during the burning cycle, $SO_2$ and NOx emissions are reduced without additional equipment or production costs. The reburning method is based on the same principle except that injection occurs just after the traditional fuel is burned. This process cuts $SO_2$ and $NO_X$ by 25% and 70%, respectively.

Fuel cell technology will be an important element in the changing global energy equation. Some hospitals and office buildings are already powered by stacked fuel cell systems. Fuel cells create electricity from the chemical reaction between the element hydrogen and electrolytes or metals such as gold and platinum. Methanol, a liquid derived from natu-

> The idea now is to encourage other companies operating large diesel fleets in Utah, California and Nevada to begin converting to LNG trucks and install public LNG refueling stations at their hubs.[c]
>
> — *The Sacramento Bee*

> Considerable CNG (compressed natural gas) storage technology development is necessary for cars with ranges equivalent to those of gasoline-fueled automobiles to be attractive.[D]
> — *Environmental Science & Technology*

ral gas, stores hydrogen efficiently. Once hydrogen is extracted from the methanol, it can be used to power a fuel cell. Hydrogen fuel cells produce virtually zero emissions. Because the extraction of pure hydrogen is expensive and uses significant energy, the first generation of fuel cells in cars will likely use a cell in which the methanol passes through a mechanism called a "reformer," which then extracts the hydrogen.[5]

Major American automakers like DaimlerChrysler and Ford are consulting with Ballard Power Systems, the world leader in fuel cell technology, to implement an aggressive program to market fuel-cell-powered automobiles by 2004. Cars running on methanol fuel cells will have little effect on a consumer's regular routine. A driver will simply stop at a pump and fill up with methanol rather than gasoline. In the future, we will rely on fuel cells to power everything from laptop computers to lawn mowers.

There is no doubt that natural gas will play an increasingly important role as a primary energy provider in the 21st century. It is imperative, however, that engineers address the problem of leaks in the natural gas operations, transmission, and distribution system. The gas industry emits about 315 billion cubic feet of methane per year due to equipment and pipe leaks, trap venting, and combustion.[6] If the leaks are not fixed, the environmental advantages of NG will be offset by methane releases to the atmosphere. The significant reduction in greenhouse gas emissions anticipated by converting from oil and coal to natural gas will be wiped out if only 3% to 4% of the projected methane production escapes into the atmosphere. Methane is 30 times more powerful as a greenhouse absorber than $CO_2$.

Some researchers believe that methane and other non-carbon dioxide gases are responsible for about 40% of global warming.[7] Although scientists working with the Intergov-

*Natural gas cannot match gasoline in versatility and ease of transport.*

ernmental Panel on Climate Change estimate that the entire natural gas industry was responsible for only 2% to 3% of total man-made methane emissions, environmentalists contend that natural gas production and transmission is one of the top five sources of human-produced methane in the United States. Landfills, coal mining, livestock, and manure are also primary culprits. In other regions of the world, rice cultivation and biomass burning contribute significantly to the problem.

Most energy experts believe that the global consumption of natural gas and non-conventional gas liquids will continue to increase rapidly in the next few decades as industrialized countries respond to environmental constraints and public demand for cleaner burning fuels. Some gas-rich countries that lack pipelines to major markets are considering building their own large-scale plants to convert natural gas into liquid form, which can then be transported as is oil. Technological developments are making it possible to produce synthetic liquid fuels from gas more efficiently and cheaply. These natural gas liquids (NGL) can be used in place of petroleum to generate energy, but their combustion emits greenhouse gases to the atmosphere. One trillion cubic feet of gas typically yields 13 million barrels of condensate and up to 40 million barrels of liquid by processing. Recently, researchers and chemists have converted methane to methanol directly using a catalyst. It is important to remember, however, that as tougher clean air legislation and new technological innovations encourage energy providers to use natural gas in a variety of applications, this non-renewable resource will experience accelerating depletion.

Some experts envision methane hydrates as the ultimate energy source of the future. Methane hydrates are a crystalline solid encased within an ice structure composed of natural gas and water.[8] They are found buried under permafrost and deep sea sediments below 300 meters — environments of low temperature and high pressure.[9] A recent United States Geological Survey (USGS) estimate that methane hydrates beneath US Gulf Coast waters hold some 200 trillion cubic feet of natural gas, enough to supply all the nation's energy needs for more than 2,000 years at current rates of use, has

> Unlike oil investments [in the Persian Gulf states], natural gas and petrochemical projects require massive amounts of capital.[F]
> — *World Oil*

> Many thousands of boreholes that have been drilled and cored in the oceanic seabed should have encountered ample indications of hydrates. But in fact, there are only three major known occurrences of massive hydrates.[G]
> — *Offshore*

piqued widespread interest. In April 1999, the US Senate passed the Methane Hydrate Research and Development Act by unanimous vote. In late October 1999, the House of Representatives passed a version of the bill that authorizes a five-year research and development program funded by grants and contracts.

Despite all the hoopla surrounding methane hydrates, there is no direct evidence that they exist in quantities sufficient to stave off the looming energy crisis. It is important to realize that methane in hydrates is a solid, not a gas or liquid. It cannot migrate and accumulate in large deposits as oil or natural gas do. Instead, methane hydrates are found in sediment laminates, and current estimates are extremely unreliable, most failing to indicate the areal extent, thickness, and concentration of the alleged deposits. The claims for widespread hydrate occurrence in thick oceanic deposits are suspect. Although some geologists believe that there are hydrate deposits in sea floor sediments hundreds of feet thick, after drilling thousands of boreholes, the thickest interval found so far is 39 inches.[10]

Since methane hydrate is deposited as dispersed grains or in thin laminates within seabed sediments, substantial technical obstacles are associated with its extraction. It will be necessary to melt the frozen matrix by depressurizing it or applying hot solvent before pumping to produce the methane. In fact, the DOE supported methane hydrate research from 1982 to 1991 and even developed production models for depressuring and thermal extraction. But other difficulties abound with mining hydrates. Hydrate under deep-sea pressure will expand 160 times its volume when brought to the surface and readily decomposes into water and methane, making it difficult to study in its original state. Even though the USGS has estimated that the amount of carbon bound in gas hydrates worldwide is twice the amount to be found in all known fossil fuels, some geologists find it difficult to believe that hydrates contained in the first 600 yards of deposits of oceanic sediments, representing less than 10 million years, could contain more carbon than coal and oil, which are buried deeper and have been for 500 million years.

> Our natural gas reserves won't last long if we burn them in both homes and cars.[E]
>
> — Randy Udall & Steve Andrews

It is possible that methane hydrate and the underlying free gas within the seafloor will become a commercially viable resource, but that day is probably decades away. Hydrate laminae at or near the seafloor are important to sediment stability, and environmentalists worry that mining hydrates could generate massive submarine landslides and lethal tsunamis. Rapidly dissociating hydrate may redistribute sediments away from continental margins by mass movement, releasing methane and disrupting wells beneath the drilling platforms. Entire oil rigs have already been lost due to hydrates that were gasified, causing surrounding sediments to collapse.[11]

Optimistic energy providers salivate at the thought of massive methane hydrate reserves ready to be commercially exploited, but exploration results have been very disappointing. In 1998, Chevron testified before a US Senate Committee that hydrates occur in low concentrations and have no commercial potential. There have been many thousands of boreholes drilled and cored in the oceanic seabed all around the world, but there is still no direct evidence for any massive hydrate deposits.[12]

Conventional natural gas has significant drawbacks. Existing problems include methane leakage from the production and distribution systems, the gas can be used conveniently only by countries within pipeline range of deposits, and it is not easily transported in bulk by other means. In addition, NG exploration, production, processing, and transmission are distinctly different from the equivalent processes for oil and require a different infrastructure.[13] Natural gas is a non-renewable resource, and predicted reserves are about 65 years' worth. As the world economy relies more and more on NG, it is hoped that research will lead to technological advances that will open up previously nonrecoverable reserves. On the other hand, if potential technological breakthroughs fail to deliver and global production and consumption of NG continue to soar, depletion may occur sooner than anticipated.

Top officials at the Natural Gas Supply Association (NGSA) believe that new technologies will almost certainly extend the production horizon and predict that the potential

> When and if supplies of natural gas begin to run out, the oil companies will focus on squeezing usable fuels out of even more difficult prospects.[H]
>
> — *Discover*

> While it is probably misleading to speak of a peak for [natural] gas, the high plateau will be centered around 2020.[1]
> — Colin Campbell

for growth in natural gas demand is excellent.[14] But the industry is also worried about falling prices and reservoir depletion. In the mid-1990s, many US natural gas wells depleted at average rates of 40% in the first year of production. In 1997, the 50 largest US exploration and production companies spent $28.6 billion on domestic oil and natural gas exploration and development, but production stayed flat. Regardless of whether the industry focuses on new developments or extracting known reserves, it is in danger of losing money.

Due to a collapse in domestic gas and world oil prices that began in 1998, the exploration and development of natural gas reserves have stalled as we enter the 21st century. Years of deficit spending have left many companies, both large and small, unable to borrow money. In fact, many are currently more focused on repaying debt than on expanding growth. As of February 1999, drilling had dropped to half of its activity prior to November 1997 — a major reduction for the industry. Compounding the situation is the estimation that 70% of the nation's natural gas supply comes from wells drilled after 1990, and the most recently drilled wells are running dry almost twice as fast as the wells brought on-line earlier in the decade.[15] Another threat to the rapid growth of the industry is a shortage of qualified employees. In 1983, there were 11,000 undergraduates enrolled in petroleum engineering programs in the United States. But by 1996, the number had fallen to 1,300, and personnel shortages reduced the industry's ability to increase drilling activity in 1996 and 1997.

Despite increasing consumption of NG, most regional reserves-to-production ratios have remained high. (A reserve-to-production ratio projects the longevity of a non-renewable resource based on predicted production.) But a problem looms on the distant horizon. More than 70% of the world's proved natural gas reserves are located in the Persian Gulf and former Soviet Union, both geologically oil-rich and politically volatile regions. In fact, Russia and Iran alone account for almost half of the world's total NG reserves. The whole Asia Pacific region contains less than 10%.

The United States has already seriously depleted its once vast oil and gas fields, and now holds less than 3.5% of the world's total NG reserves. If the lower-48-state per-well gas production continues to decline at the same rate it did in the 1980s, more than 20,000 wells will have to be drilled every year just to sustain the current US annual production for another decade. Gas drilling in the 1990s has averaged less than 13,000 wells per year and continues to decline.[16] Once again, the future energy scenario gives the Persian Gulf region the upper hand when it comes to non-renewable energy resources.

The US Energy Information Association estimates that America has only enough natural gas to supply this country's energy-hungry economy for about another 65 years, but even that assumption may be optimistic. Consider that in 1995, the USGS estimated that the proved US natural gas reserves were 135 trillion cubic feet, and those reserves would grow by another 322 trillion cubic feet due to improved production techniques and greater development of existing fields. Those projections failed miserably, however, when drilling in North America's largest gas field astride the Oklahoma-Kansas border did not produce as expected. Another blow to the projected yields by the USGS came from their inclusion of unconventional sources such as shale and tight sandstone, which are more expensive to produce and have a low recovery rate in mining extraction. Black shale beds located in the United States contain considerable amounts of natural gas, but the rock has low permeability and is currently not economic for large-scale commercial mining. Coal-bed methane from coal deposits is already in substantial production in North America, but the environmental costs are high.

It is important to remember that although natural gas appears abundant and relatively clean when compared to coal and oil as a primary energy source, it still produces dangerous emissions. Since the US Department of Energy estimates that the United States natural resource base will last about 65 more years, some experts see a high plateau for natural gas, centered around year 2020. Natural gas is not the economic or social equivalent of crude due to the inherent convenience, safety, and flexibility of oil. Because of an estimated production life span of less than a century, non-renewable natural gas cannot be considered civilization's long-term primary energy source.

> If [natural] gas production levels are to be maintained in the United States, new fields must be discovered and this is not happening to any significant extent.[j]
> — Walter Youngquist

## Natural Gas Fact Sheet

**RESERVES** (*trillion cubic feet*)

|  | 1998 | 1999 | 2000 |
|---|---|---|---|
| Russia |  |  | 1,700 – 1,705* |
| Iran |  |  | 790 – 812.3* |
| Qatar |  |  | 300 – 394* |
| US (6th) | 164 |  | 167.4* |
| World |  |  | 5,149.6 – 5,210.8* |

**Analysis** (using the higher estimates): Russia represents not even 2.5% of the world's population yet may have almost 33% of the world's natural gas reserves (almost 12 million cubic feet per capita). Iran has a bit over 1% of the world's population and over 15% of the natural gas reserves (over 12 million cubic feet per capita). Qatar, at under 0.01% of the world's population, may have over 7% of the reserves (almost 667 million cubic feet per capita). And the US, at under 5% of the population, may have a bit over 3% of the reserves (over 600,000 cubic feet per capita). Per capita, Qatar has the most natural gas reserves — over 55 times that of Iran, almost 57 times that of Russia, and almost 1100 times that of the US.

**PRODUCTION** (*trillion cubic feet*)

|  | 1998 | 1999 |
|---|---|---|
| **Gross** |  |  |
| US | 23.92 |  |
| Russia | 20.87 |  |
| Algeria | 5.3 |  |
| World | 101.89 |  |
| **Dry** |  |  |
| Russia | 20.87 | 20.83 |
| US | 18.71 | 18.62 |
| Canada | 6.05 | 6.26 |
| World | 82.79 | 84.69 |

**Analysis:** The US, with under 5% of the world's population, produced over 23% of the world's gross natural gas (over 85,000 cubic feet per capita) and almost 22% of the world's dry natural gas (over 75,000 cubic feet per capita). At under 2.5% of the world's population, Russia produced over 20% of the world's gross (almost 144,000 cubic feet per capita) and almost 25% of the world's dry natural gas (over 143,000 cubic feet per capita). Algeria, with a little over half of 1 percent of the world's population, produced over 5% of the world's gross natural gas (over 168,000 cubic feet per capita). And Canada, also with a little over half of 1 percent of the world's population, produced over 7% of the world's dry natural gas (over 203,000 cubic feet per capita). Per capita, Algeria produced the most natural gas — almost 20% more than Russia and almost twice that of the US. As for dry natural gas per capita, Canada produced the most — over 40% more than Russia and over 2½ times that of the US.

### CONSUMPTION — DRY (*trillion cubic feet*)

|  | 1998 | 1999 | 2000 |
|---|---|---|---|
| US | 21.26 | 21.69 | 22.68 (E) |
| Russia | 14.05 | 14.01 |  |
| UK | 3.07 | 3.26 |  |
| World | 81.9 | 84.2 |  |

**Analysis**: With under 5% of the world's population, the US consumed almost 26% of the world's dry natural gas (almost 79,000 cubic feet per capita). Russia, with under 2.5% of the world's population, consumed almost 17% of natural gas (over 96,000 cubic feet per capita). And the UK, with under 1% of the world's population, consumed almost 4% of natural gas (almost 55,000 cubic feet per capita). Of the three, Russia consumed the most natural gas per capita — over 23% more than the US and almost 77% more than the UK.

### IMPORTS — DRY (*trillion cubic feet*)

|  | 1998 | 1999 |
|---|---|---|
| US | 3.15 | 3.59 |
| Germany | 2.62 | 2.71 |
| Japan | 2.44 | 2.55 |
| World | 19.54 |  |

**Analysis**: The US, with under 5% of the world's population, was responsible for over 16% of the world's dry natural gas imports (over 11,000 cubic feet per capita). At less than 1.5% of the world's population, Germany was responsible for over 13% of the world's dry natural gas imports (almost 32,000 cubic feet per capita). At a little over 2% of the world's population, Japan was responsible for almost 12.5% of the world's dry natural gas imports (almost 20,000 cubic feet per capita). Germany exceeded the other two countries in natural gas imports per capita — over 65% more natural gas than Japan, and almost 3 times that of the US.

**EXPORTS — DRY** (*trillion cubic feet*)

|  | 1998 | 1999 |
|---|---|---|
| Russia | 7.14 | |
| Canada | 3.15 | 3.36 |
| Algeria | 1.87 | |
| US (14th) | 0.16 | 0.16 |
| World | 19.76 | |

**Analysis**: With under 2.5% of the world's population, Russia was responsible for over 36% of the world's dry natural gas exports (over 49,000 cubic feet per capita). Canada, with a little over half of 1 percent of the world's population, was responsible for almost 16% of the world's natural gas exports (over 102,000 cubic feet per capita). Algeria, also with a little over half of 1 percent of the world's population, was responsible for almost 9.5% of the world's natural gas exports (over 59,000 cubic feet per capita). Per capita, Canada exported more natural gas than the other two — over 72% more than Algeria and over 2 times that of Russia.

**$CO_2$ EMISSIONS** (*million metric tons carbon equivalent*)

|  | 1998 | 1999 |
|---|---|---|
| US | 319.26 | 320.89 |
| Russia | 203.83 | 203.37 |
| UK | 47.93 | 50.84 |
| World | 1,270.05 | 1,294.94 |

**Analysis**: The US, with less than 5% of the world's population, emitted almost 25% of total world carbon dioxide emissions (almost 1.2 metric ton per capita). With under 2.5% of the world's population, Russia was responsible for almost 16% of world's carbon dioxide emissions (over 1.4 metric tons per capita). The UK, with under 1% of the world's population, emitted almost 4% of world carbon dioxide emissions (almost 0.9 ton per capita). Of the three, Russia was responsible for the most emissions – over 20% more than the US and almost 65% more than the UK.

---

\* = as of January 1, 2000;   (E) = estimate

*References:*
<http://www.eia.doe.gov/iea/>
>   Each category leads to a table of various statistics links, plus a link to "other data," where most statistics reside.

<http://www.prb.org/pubs/wpds2000/wpds2000_Population2000-PopulationProjected.html>
>   Table of population statistics for each country. The table claims that the figures are from mid-2000 year, but most sources actually come from 1998 to early 2000, and some numbers for more obscure locations actually come from earlier dates. In other words, the origin of this information probably coincides well with the 1998-2000 fuel statistics.

Note: Reserves variations due to numbers from two different publications: *World Oil,* and *Oil and Gas Journal*. Although one country exceeds the others in its category per capita, that does not mean that it exceeds every country in the world per capita; only the 3-5 countries with the highest overall totals were compared in this way.

FOOTNOTES:

[1] <www.eia.doe.gov/oiaf/aeo/issues.html#natgas> US Department of Energy report on the prospects of natural gas supply and demand in the near future, and the current status of its production and consumption worldwide. It was last modified in April 1999.

[2] Robin Hill, Phil O'Keefe, and Colin Snape, *The Future of Energy Use*, (New York, NY: St. Martins Press, 1995), p. 90.

[3] <www.naturalgas.org/> Natural Gas Information and Educational Resources informs readers of facts, environmental issues, policy concerns, technology, and links associated with the fuel. "Designed to educate the general public on natural gas without company-specific overtones, the site consolidates general information on the US gas industry for use by students, teachers, journalists, legislators, foreign gas industry leaders and others." Sponsored by the Natural Gas Supply Association (NGSA), the Independent Petroleum Association of American (IPAA), and the National Ocean Industries Association (NOIA). Most information updated in 1998.

[4] <www.ngsa.org/Facts/Statistics/facts_stats.htm> Statistics on natural gas, oil, coal, overall electricity production, consumption, and emissions from 1997 and 1998. Originating from the Energy Information Administration, this page is posted on the Natural Gas Supply Association's website. The NGSA "encourages expanded use of natural gas and a regulatory climate that fosters competitive markets."

[5] <www.enn.com/enn-features-archive/1999/12/123099/featurefuel_6347.asp> "Fuel for the Future," an article by Michael Parrish dated December 30, 1999, posted on the Environmental News Network (ENN). It details the upcoming uses of fuel cells in commonplace items such as snow blowers and automobiles, and the innovations that will make them smaller, lighter, and more adaptable. ENN is recognized as one of the leading online producers of environmental news and information.

[6] <www.ngsa.org/News_Events/presentations/10_13_99STAR.htm> "Reducing Greenhouse Gas Emissions: Good Business for Natural Gas Producers" is a presentation by the Natural Gas Supply Association. The NGSA describes efficiency improvements and management practices that are reducing emissions. Based on content, the presentation appears to be from 1998 or 1999.

[7] <www.enn.com/enn-news-archive/1999/11/111099/methane_7197.asp> "Don't Forget Methane, Climate Experts Say," an article by Robinson Shaw dated November 10, 1999, posted on the Environmental News Network. It reports the conclusions that a team of atmospheric scientists, economists, and emissions experts reached upon assessing "the potential cost savings of introducing an additional greenhouse gas, methane, into a carbon dioxide emission-reduction strategy." They estimate that methane can reduce US

abatement costs by more than 25 percent compared to strategies involving carbon dioxide alone.

[8] <www.agiweb.org/gap/legis106/ch4106.html> "Update on Methane Hydrate Research and Development Act (12-10-99)" is posted on the American Geological Institute's website in the Government Affairs Program section. Included are background information on gas hydrate and descriptions of the sections taken by the House of Representatives, the Senate, and other government entities. The AGI is a non-profit federation of 35 geoscientific and professional associations that represent more than 100,000 geologists, geophysicists, and other Earth scientists.

[9] <www.ems.psu.edu/info/explore/MethanHydrat.html> Short article on methane hydrates dated January 2000 on the Earth & Mineral Sciences subsite of the Penn State website. What hydrates are, estimates on how much are present near the US, who is interested in them, and the dangers of their extraction are explained.

[10] Jean H. Laherrere, "Data shows oceanic methane hydrate resource overestimated," *Offshore* (September 1999), p. 158.

[11] M.D. Max and P.R. Miles, "Mapping natural gas hydrates with tuned detection tools," *Offshore* (August 1999), p. 136.

[12] Jean H. Laherrere, "Uncertain resource size enigma of oceanic methane hydrates," *Offshore* (August 1999), p. 160.

[13] *Global Energy Perspectives*, Edited by Nebojsa Nakicenovic, Arnulf Grubler, and Alan McDonald, (Cambridge: Cambridge University Press, 1998), p. 121.

[14] <www.ngsa.org/News_Events/presentations/4_22_99SP.htm> Statement given by Richard J. Sharples, President of Anadarko Energy Services and Company Chairman of Natural Gas Supply Association, before the Senate Energy and Natural Resources Committee on April 21, 1999. Mr. Sharples explains the obstacles that face oil and gas producers, such as decreases in prices, drilling activity, exploration, and qualified personnel.

[15] <www.eia.doe.gov/emeu/steo/pub/contents.html> The Energy Information Administration's January 2000 short-term energy outlook includes oil, coal, and natural gas supply and demand projections, and overall electricity demand. EIA, a statistical agency of the US Department of Energy, provides policy-independent data, forecasts, and analyses to promote sound policy making, efficient markets, and public understanding regarding energy and its interaction with the economy and the environment.

[16] J.P. Riva, Jr., "Domestic Natural Gas: A Fuel For the Future?" *The Professional Geologist* (March 1995), pp. 9-10.

SIDEBAR FOOTNOTES:

[A] Matthew R. Simmons, "Has Technology Created $10 Oil?" *Middle East Insight* (May-June 1999), p. 39.

[B] Safaa A. Fouda, "Liquid Fuels from Natural Gas," *Scientific American* (March 1998), p. 92.

[C] Kiley Russell, "Natural gas station opens on I-5 ranch," Associated Press article published in *The Sacramento Bee*, February 10, 2000, p. A-4.

[D] Heather L. Maclean and Lester B. Lave, "Environmental Implications of Alternative-Fueled Automobiles: Air Quality and Greenhouse Gas Tradeoffs," *Environmental Science & Technology* (January 15, 2000), p. 230.

[E] Randy Udall & Steve Andrews, "When Will the Joy Ride End?" Published in Newsletter #99/1 by the M. King Hubbert Center for Petroleum Supply Studies, January 1999, p. 3.

[F] A. F. Alhajji, "Will Gulf states open their upstream to foreign investment?" *World Oil* (December 1999), p. 21.

[G] Jean Laherrere, "Uncertain resource size enigma of oceanic methane hydrates," *Offshore* (August 1999), p. 160.

[H] Curtis Rist, "Why We'll Never Run Out of Oil," *Discover* (June 1999), p. 85.

[I] Colin Campbell, *The Future of Oil and Hydrocarbon Man*. Handbook that draws on the unique database of drilling, production, and reserves maintained by Petroconsultants' Global Energy Information Services. This booklet is derived from Dr. Campbell's *The Coming Oil Crisis*, MultiScience Publishing & Petroconsultants, 1997, p. 25.

[J] Walter Youngquist, *GeoDestinies: The Inevitable Control of Earth Resources Over Nations and Individuals* (Portland, Oregon: National Book Company, 1997), p. 195.

# Nuclear Energy

## 5

Twenty years ago, proponents of atomic energy proclaimed that by the 21$^{st}$ century, nuclear reactors would be producing abundant electricity cleanly, safely, and cheaply. Industry executives declared that electricity would be supplied at virtually no cost to the consumer and without creating air pollution. But splitting the atom to boil water is like using a chainsaw to cut butter. Carbon dioxide emissions from burning coal and oil seem like child's play in comparison to the tons of high-level radioactive nuclear waste piling up at reactors and weapon sites throughout North America. Many countries are now abandoning their nuclear programs, instead shifting to natural gas and renewable energy sources. Falling energy prices due to deregulation in the energy market may now be the final nail in the coffin for nuclear energy.

USING NUCLEAR ENERGY TO GENERATE ELECTRICITY presents similar problems as those associated with using nonrenewable fossil fuels. The planet's supply of uranium is limited, and mining the ore is hazardous to human health. Worse, the radioactive waste byproducts are a lethal and long-term danger to the environment. And worst of all is the unique and frightening scenario with nuclear energy that, in the wrong hands, the technology can be used to make terrible nuclear weapons. In 1999, nuclear energy provided about 17% of the world's electricity. The current global nuclear capacity of about 350 gigawatts is a historical high point, but it is a far cry from the expected 4,450 gigawatts predicted by the International Atomic Energy Agency back in 1974 — about 12 times less. Less than three decades ago, proponents of nuclear power were predicting virtually free electricity for everyone. What happened?

In the 1950s, 60s, and early 70's, the nuclear mania sweeping the United States was similar to today's excitement surrounding biotechnology, the Internet, and Dot Com stocks. In 1974, the United States boasted 200 reactors either operating, under construction, or on order. By 1988, there were 429 nuclear power plants in 25 countries producing about 17% of the world's electricity. The world's nuclear capacity had doubled in five years and quadrupled in ten, but then it stalled. In year 2000, the US has only about 100 operating nuclear power plants, all of them ordered before 1975. Technical engineering difficulties and unexpected equipment and system failures have dispelled the technological hubris that accompanied the dawn of the nuclear age. Today, there is little support among the American public for nuclear energy. Nobody wants a nuclear reactor in the backyard, and the frightening publicity regarding reactor meltdowns at Chernobyl and Three Mile Island have only enforced such notions regionally. The public's fear of the lethal radioactivity released in nuclear accidents is well founded. Health officials estimate that at least 4,365 people who took part in the Chernobyl cleanup have died in the Ukraine.

A major drawback to using atomic fission to generate electricity is the serious environmental impact of radioactive nuclear waste. Disposal of radioactive material has proven to be a much greater obstacle than originally estimated by industry experts. Nuclear power does not contribute to air pollution and greenhouse gas emissions, but a good solution to the challenge of how to store tons and tons of radioactive waste safely, a nuclear by-product that remains dangerous to all lifeforms for thousands of years, remains elusive. Other factors that have crushed the expected rapid growth of nuclear energy are increasing regulatory de-

> If by the mid-21$^{st}$ century it has become clear that nuclear fission is the only effective long-term source of energy, society will then have to consider the problem of accumulating waste on a longer time-scale.[A]
>
> — *Commonsense in Nuclear Energy*

mands, construction cost overruns, court challenges, and resistance by well-organized antinuclear activists. The final nail in the coffin for nuclear power will be the competitive prices brought about by deregulation in the energy sector.

The scale of atomic size has intrigued mankind since ancient Greek philosophers first recognized that all matter is composed of particles that are invisible to the human eye. The Greek word for these particles is *atomos*, which means indivisible. Modern scientists know that atoms can be modeled at the simplest level as miniature solar systems with electrons orbiting tiny nuclei like planets orbiting the Sun. Although the nucleus is about only one hundred thousandth of the size of the entire atom, it is so dense that it accounts for nearly all its mass. The nucleus is a cluster of nuclear particles, or nucleons, that cling tightly together. The particles are of two types, protons and neutrons. Each atom has the same number of orbiting electrons as it has protons. Since the electrons are negatively charged, they and the positively charged protons balance each other electrically.

Atoms of the same element always have the same number of protons in their nuclei, but the number of neutrons can vary. Nuclei with different numbers of neutrons but the same number of protons are categorized as *isotopes* of the same element. Many isotopes are unstable and transform themselves by shedding nuclear particles, a process known as *decay*. As each change occurs, energy is released and transmitted as radiation. The whole transformation process is called radioactivity. Radiation intensity varies according to the energy level and penetrating power of the expelled particles. For example, alpha particle radiation (nuclei of helium) won't penetrate a sheet of paper, let alone human skin; but if absorbed into the lungs or into an open wound, it can be especially damaging. Beta particles (electrons) can affect the outer layers of living tissue, and gamma radiation (electromagnetic radiation), which travels at the speed of light, can be stopped only by a thick wall of lead or concrete. Radiation is harmful to all

Megawatts by which the world's nuclear-powered capacity changed during 1998: -230.[B]

— *Harper's*

life forms, affecting the molecules of living matter. Even at low doses, it can lead to cancer or genetic damage.[1]

It was not until the last few centuries, however, that scientists began to question an atom's indivisibility. Only then did researchers discover the enormous power locked inside. The basic unit in nuclear energy is the atom. An atom's nucleus is held together by a strong force. When the atom is split, energy is released, resulting in heat. Electricity can be generated from this heat. When an atom of Uranium-235 splits, loose neutrons are absorbed, and the atom becomes unstable. Fission occurs again, and more neutrons and heat are released. When the neutrons hit other atoms, more fission occurs, additional heat is released, and the process repeats itself. In a conventional nuclear power plant, this continuous self-sustaining chain reaction heats and converts water into steam, which spins turbines to generate electricity. Unlike with coal, oil, or natural gas power plants, no combustion occurs in the nuclear process.[2]

> More disturbing is the elevation of nuclear science and nuclear scientists to the level of sacred icons, legitimizing in the process the militaristic, destructive, and non-productive functions of science.[c]
> — *Bulletin of Concerned Asian Scholars*

Much of the early atomic research in the US was conducted in utmost secrecy under the code name *Manhattan Project* as scientists developed an effective nuclear weapon to end the Second World War. On December 2, 1942, the world's first nuclear reactor was tested on the floor of an abandoned handball court beneath the University of Chicago. At 3:25 that afternoon, the fission chain reaction inside what was known as Chicago Pile-1 became self-sustaining, and the possibility of powering cities from the energy locked inside the atom became a reality. After the war, the US government began a "swords to plowshares" program, which encouraged researchers and engineers to focus on peaceful industrial applications. In 1946, Congress created the Atomic Energy Commission and authorized the construction of an experimental nuclear reactor. It took five more years, but, in December 1951, the first electricity generated by nuclear power made the dream a reality. Shortly after, optimistic executives representing utility industry interests assured the American public that nuclear power would be cheap, clean, and safe. The hyperbole and excitement eventually dissipated, however, as the occupational hazards, environmental costs, and biological dangers of splitting the uranium atom became apparent.

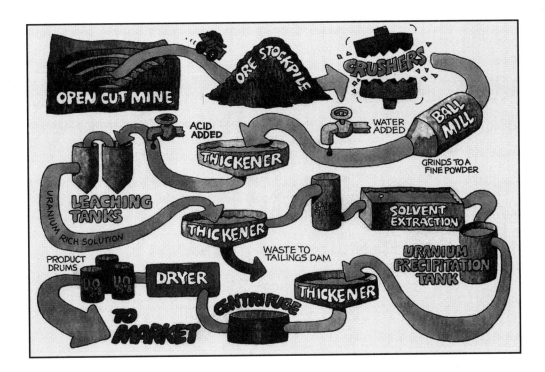

From start to finish, producing electricity with nuclear power is a hazardous enterprise. The power stations are just part of the nuclear fuel cycle. The mining and milling of uranium greatly increases the rates of cancer among miners. The processed and enriched uranium is used to fuel a nuclear reactor, which creates heat and drives steam generators to produce electricity. After use in the power stations, the irradiated fuel is sometimes reprocessed to recover uranium and plutonium. Eventually, the cycle will end with the disposal of the nuclear waste. Each stage in the nuclear cycle releases radioactive materials into the environment.

Uranium is a radioactive element. Primary uranium ores are derived from Precambrian (older than 570 million years) sources that have been buried and affected by high temperatures and pressure. Extracting the ore contaminates the air in

> Unless electric vehicles are invented that are efficient and used in quantity, uranium will not be a major substitute for gasoline or jet fuel.[D]
>
> — Walter Youngquist

> Health officials estimate 4,365 people who took part in the [Chernobyl] cleanup have died in Ukraine.[E]
>
> — *The Sacramento Bee*

the mineshafts, and, despite protection, miners are exposed to radioactive dust. These radioactive elements settle in the miners' lungs, delivering a strong dose of alpha radiation directly to the lung tissue. In a study published in 1980, the British Columbia Medical Association released a report entitled "The Health Dangers of Uranium Mining." The BCMA report cautioned that if Canada continued mining uranium, ever-increasing incidences of lung cancer among Canadian miners would be the result.[3] Beginning in the 1940s and continuing into the 1960s, thousands of men working the uranium mines of the US State of New Mexico were assured that breathing radon gas was not hazardous to their health. Only recently have researchers revealed the high frequency of radon-caused lung cancers among those miners.

To supply an average-sized reactor (1000 MWe) with one year's worth of uranium, between 45,000 to 90,000 tons of low-grade uranium ore must be dug from surface or underground mines. Once the rock containing the ore is brought to the surface, it is crushed, and the useful uranium is extracted. It takes an enormous amount of low-grade uranium ore to produce a meager 25 tons of enriched uranium, which is then used to fuel the nuclear reactor core. Uranium usually composes less than 1% of the total material mined, so the bulk rock that encases it ends up littering the landscape in the form of radioactive tailings.[4] The tailings are often stockpiled above ground, where they will continue to emit dangerous radon gas for millions of years. In the US, radioactive tailings make up more than 95% of the volume of all radioactive waste from all stages of the nuclear weapons and power production process.

There are currently millions of tons of radioactive waste tailings being stored at active mill sites in North America. As uranium decays, it creates a nasty stew of highly dangerous by-products: thorium-230, radium-226, radon-222, and the extremely dangerous elements lead-210 and polonium-210. A report delivered to the British Colombia Royal Commission of Inquiry declared that the failure to properly handle uranium tailings has led to internal lung doses calculated to be 100 rems per year to the local population. Calculations show that people living in areas near uranium mining deposits will receive 25% more radon radiation over the course of their lifetimes than populations living at a safe distance from the tailings.

The quantity of the planet's uranium reserves is as finite and limited as that of fossil fuels like coal, oil, and natural gas. The Uranium Institute estimates the total world recoverable resources of uranium at about 3,256,000 tons.[5] Existing nuclear reactors operating worldwide require 75,000 tons of uranium a year to produce less than 17% of the total world electricity requirements. These estimated resources will be sufficient to meet current and anticipated demand for another 45 years or so.

Nearly 80% of all the planet's uranium ore is located in just six countries, with only nine companies accounting for 82% of global uranium production. Australia holds 25% of the world's uranium reserves and Canada 15%. The US can claim only 4% of the planet's reserves. Of course, by the same economic principles that new petroleum reserves are discovered and developed, as the price of uranium rises, it will lead to increased exploration. But, as with oil, that will not diminish the serious economic and ecological problems associated with using nuclear fission to generate electricity.

Uranium mining tailings are responsible for much of the radioactive contamination that has plagued parts of the country over the last few decades. The Institute for

Energy and Environmental Research (IEER) reports that in the US, nearly one-third of all mill tailings from abandoned mill operations are on lands of the Navajo Nation. Many Native Americans have died of lung cancers linked to their work in uranium mines. Land and water contamination due to seepage and spills from radioactive mounds of discarded tailings is now endangering the water supply of millions of Americans in the US Southwest.

In January 2000, Energy Secretary Bill Richardson announced that the US government was returning 84,000 acres of land to the Northern Ute tribe as part of a deal to clean up 105 million tons of uranium tailings piled only 750 feet from the Colorado River. The radioactive rock and soil, a deadly legacy of uranium mining during the Cold War, is located just outside Arches National Park. A study by the US Fish and Wildlife Service reported that toxics such as arsenic and ammonia are leaching from the pile and contaminating the river, threatening endangered species of fish. California Representative George Miller, has raised the concern that the river contamination has tainted the drinking water for 25 million people in Arizona, Nevada, and Southern California.

The Denver-based company that operated the mine from 1962 to 1984 has declared bankruptcy, which shifts the problem to the Nuclear Regulatory Commission. The federal government is finally returning the land, after appropriating it from the Ute Nation in 1916 on the eve of World War I. The land contains oil-rich shale that the US government had wanted to use as an oil reserve for the US Navy during the war. Although the shale oil was never tapped, the Energy Department estimates that the section being returned holds six trillion cubic feet of natural gas, or about 30% of all the natural gas used in the US during 1998. The Ute tribe can open the land to oil and gas drilling but must pay a percentage of the royalties to the government. That money will help pay for some of the $300 million cost of relocating the uranium waste pile. The toxic pile is located about 50 miles south of the land the Utes are regaining.[6]

> Devising suitable storage canisters to confine permanently [high-level nuclear] waste is not possible since some of the fission products and remnant actinides have half-lives in the range of thousands to millions of years.[F]
> — Annual Review of Nuclear and Particle Science – 1998

The disposal of radioactive tailings is only one of the dangers associated with tapping uranium as an energy source. About half the world's uranium ore comes from open mines and an equal amount from underground tunnel excavation. The ore is taken to nearby mills for processing. Both mines and mills give off radioactive discharges to the environment. When enough uranium (it takes about 25 tons of high-grade ore to run a reactor for one year) has been extracted and the waste tailings discarded, it must be enriched for use in the reactor. The naturally occurring 0.7% ratio of the isotope uranium-235 must be increased to about 3.5%. First, it is converted into a gas. Then the isotope uranium-238 is extracted, wasting 87% of the material in order to retrieve the desired 13% of U-235-enriched uranium. This uranium dioxide is converted into powder and compacted into pellets, which are put into fuel rods and inserted into the reactor. The enriched 13% that enters the nuclear reactor core undergoes fission and causes a chain reaction. The heat generated by the nuclear reaction boils water and converts it into steam, which is used to drive a turbine and an electric generator. The average US nuclear power plant is capable of providing about 7 billion kilowatt-hours of electricity annually.

After the enriched uranium is used, 97% of it is recycled back into the reactor to be reprocessed with fresh uranium. About 200 tons of enriched uranium are needed to keep a plant going, but only 25 tons of new fuel are added each year. The remaining 3%, about 1,500 pounds, is high-level radioactive waste, which is extremely hazardous and must be isolated from the environment for a very long time. While in the reactor, some of the U-238 turns into plutonium and some fission products. These components are more dangerous and deadly than the original high-grade uranium that fueled the nuclear reaction.

Over the years, nuclear engineers have designed many different kinds of nuclear reactors. The most common types in use today are light-water pressurized water reactors and boiling water reactors. Originally developed in the United States, these two reactor-systems are now used all over the world, despite early difficulties with corrosion. There are also

> If plutonium or highly enriched uranium became available on the black market, virtually any state or well-organized terrorist group, might be able to make a nuclear bomb, and they could do so with virtually no warning to the international community.[G]
> — *Boston Globe*

gas-cooled reactors, pressurized heavy-water reactors, and light-water cooled graphite reactors. It was the availability of enriched uranium that allowed engineers to build reactors with compact cores, a greater flexibility of design, and better operating efficiencies. With a higher percentage of U-235, atoms could be lost without strangling the chain reaction.

The fast breeder reactor has been touted as the best hope for the future of the nuclear industry. The breeder is designed to create more fuel than it consumes by maximizing the process in which neutrons are absorbed by U-238 nuclei without causing them to split. The result is that the uranium atoms spontaneously transmute into plutonium. In theory, a breeder reactor can squeeze out more than 100 times as much energy from a pound of uranium than other reactors can.[7] But breeder reactors produce electricity less economically than other systems, and the recovery of plutonium from the highly radioactive fuel is complex and expensive. Another major drawback is that breeder reactors could make weapons-grade plutonium a common commodity. Governments worldwide are concerned that breeder reactors will increase the risk to non-proliferation of nuclear arms and terrorist theft. In addition, breeder reactors are the most dangerous of all the reactors in terms of environmental hazards, and the extra dangers imposed by the handling of extremely toxic plutonium make the breeder reactors very costly. The United States aborted its breeder reactor program, but other countries are still using them.

Breeder-reactor implementation has stalled in recent years due to high costs, safety concerns, and fears about proliferation in a worldwide plutonium-based nuclear energy economy. In 1998, France's breeder program stumbled when Prime Minister Lionel Jospin ordered engineers to close down the Superphenix, the world's largest fast-breeder reactor. As a key part of France's ambitious nuclear program, the Superphenix had an original cost estimate of $4.3 billion. Despite the hefty financial investment, Jospin pointed out that the fast-breeder reactor had run for a total of only 30 months in 12 years of operation. Over its lifetime, the reac-

> The United States has just over 100 operating nuclear plants, all of them ordered before 1975.[H]
> — Robert Pool

tor was shut down repeatedly in response to dangerous leaks in the cooling system.

Many European nations are backing away from nuclear energy completely. Austria banned it in 1978, and Italian voters approved a non-nuclear policy in 1987. In 1999, Germany's government announced plans to abandon nuclear power.

A recent report on nuclear power by the US Department of Energy states, "Both the Swedish and German governments have voted for an eventual phase-out of all nuclear power capacity."[8] In fact, despite intense resistance by its domestic energy utilities, Germany's Social Democrat/Green government intends to completely shut down its 19 nuclear stations even though they provide 30% of the nation's electricity. Sweden, which generates nearly half its electricity with 12 reactors, has also recently decided to phase out nuclear power.

In 1998, France, long recognized as the most pro-nuclear country, implemented a moratorium on new nuclear plant construction. The ban is telling because the French nuclear program has been one of the world's safest and most economical. Their plants are usually built and on-line within six years or less, are very reliable, and rank among the world's best in the percentage of time they are actually generating electricity. France spends $1 billion to $1.5 billion to build a domestic nuclear plant, which breaks down to about $1,000 per kilowatt capacity, about half the cost of an average US-built nuclear facility. France relies on fission to supply nearly 80% of its domestic electricity demand and leads the world in that category, but as other European countries retreat from nuclear power, cracks are starting to appear in France's pro-nuclear façade. Public support for the French nuclear program, which held firm even after the 1986 Chernobyl disaster, is now fading. The French are only shadowing the European trend to reduce or eliminate nuclear power altogether. Despite the optimistic hyperbole espoused by invested proponents of nuclear energy, and strong support for firing up more nuclear power plants in the United States by President George W. Bush in his administration's long-range energy

Disposal of spent fuel, the largest hurdle facing the US nuclear industry, is an issue Asian nations need to consider as well.[1]

— *Harvard International Review*

plan, Europeans have decided that nuclear power plants are too dangerous and too expensive. In Eastern Europe, Russia, and Ukraine, the level of electricity provided by nuclear reactors has dropped by 10% since the fall of the Berlin Wall.

In Asia, where 88 reactors are in operation and 26 under construction, nuclear power faces growing public opposition. China hopes to build 47 reactors over the next 20 years, but, as the country tries to make its power industry more competitive in an increasingly deregulated world market, it is not at all clear that the ambitious plan will reach fruition. Indonesia, Thailand, and Vietnam have all abandoned their nuclear programs.

The Nuclear Energy Institute claims that of all energy sources, nuclear energy has perhaps the lowest impact on the environment, including water, land, habitat, species, and air resources. Advocates of nuclear energy point out the Kyoto Protocol's limits on greenhouse gas emissions and argue that two million tons of coal must be burned to generate the equivalent of electricity produced from 21 tons of uranium fuel.[9] Proponents insist that nuclear energy is more ecologically efficient than fossil-fuel-derived energy because it produces more electricity in relation to its environmental impact.[10] The Institute gives credit to the industry for being the only energy provider established since the Industrial Revolution that has managed and accounted for all of its waste and attempted to prevent adverse impacts to the environment. Unfortunately, atomic energy is full of risks, including the fact that radioactive materials discarded by the nuclear fuel process cannot be filtered out of the environment for tens of thousands of years.

The transportation and storage of high-level radioactive waste are serious concerns for the nuclear industry, world governments, and the public in general. One option for storage is to heat the waste material and convert it into a dry powder that can be immobilized in Pyrex glass and stored in stainless steel canisters, a process called *vitrification*. Australian researchers are incorporating radioactive wastes in the crystal lattices of the naturally stable minerals within a synthetic rock. But neither of these techniques solves the problem of final disposal. Currently, the principle solution under consideration is deep geological disposal — the burial of radioactive waste in subterranean rock structures that inhibit groundwater movement. A few scientists are seeking ways to transmute radioactive waste into nonradioactive elements, thereby eliminating the radiological hazards and waste disposal problems. The transmutation of nuclear waste into less harmful elements has already been accomplished in the laboratory, but the technology needs more research and funding to become a viable solution to the radioactive waste problem.

Deep geologic burial of nuclear waste is not an acceptable solution to the radioactive waste problem. No known material can withstand the continuous assault of heat, helium, and hydrogen that the spent nuclear fuel produces. The nuclear industry frankly admits that subterranean storage risks contamination to underground water tables. Eventually, the unstable material will reenter the ground and poison the groundwater. Consider this — the longest any language has continuously been spoken is 4,000 years. This spent fuel will remain lethal for more than 10,000 years. What language can be used to make the signs warning people of the distant future to avoid the site?

Some countries are delaying deep burial of high-level radioactive wastes and spent fuel, hoping that a better solution will be devised soon. But in an effort to get this dangerous material out of sight and out of mind, the US nuclear industry has already identified possible dumpsites. Preliminary studies on the controversial Yucca Mountain High-Level Nuclear Waste Repository located 80 miles northwest of Las Vegas, Nevada, have already cost Ameri-

> Congress has struggled for six years over what to do with the more than 40,000 tons of radioactive used reactor fuel sitting at commercial power plants.[J]
> – H. Josef Hebert
> Associated Press

can taxpayers billions of dollars. The federal government is developing this site despite the huge price tag, intense opposition by Nevadans, and a warning by Environmental Protection Agency geologists that Yucca Mountain is riddled by 33 known earthquake faults and is the least stable of any site studied. The fact that the nuclear industry has asked the federal government for protection from liabilities from potential nuclear disasters and for help dealing with the constantly increasing amounts of high-level nuclear waste underscores the grave danger associated with this volatile environmental and political issue.

In 1987, Yucca Mountain, Nevada, was selected as a potential repository for the 77,000 tons of nuclear waste awaiting disposal from the United States' nuclear power plants.[11] Thousands of researchers and scientists have been testing the region's rock formations, climate, and groundwater flows. According to the Nuclear Waste Policy Act, "If, at any time, Yucca Mountain is found unsuitable, studies will be stopped immediately. If that happens, the site will be restored, and DOE will seek new direction from Congress."[12] The entire project from construction to closure — decommissioning is planned for 2116 — is estimated to cost $36.6 billion, almost double the original estimate. Testing and preliminary planning have cost almost $6 billion through fiscal year 1998, and 1999's budget was more than $307 million. Even if the DOE eventually decides that Yucca Mountain is not the best choice for the nation's nuclear depository, it won't be easy to walk away from those billions of dollars already invested in the project.

*For decades Americans have protested nuclear power.*

The 621 earthquakes of a 2.5 magnitude or greater that have occurred within a 50-mile radius of Yucca Mountain since 1976 have not shaken the government's resolve to develop the site. Hundreds of temblors have occurred during the Department of Energy's on-site evaluation, and according to Nevada's Nuclear Waste Project Office (NWPO):

Yucca Mountain itself is the result of millions of years of intense faulting and volcanism. Records of recent events indicate that faulting is an ongoing process in the vicinity of Yucca Mountain that is expected to continue long into the future. Thirty-three faults are known to exist within and adjacent to the Yucca Mountain site.[13] The NWPO, as well as two-thirds of all Nevadans, oppose the project.

If the lack of public support does not stop utility companies from building more nuclear power plants, perhaps the cost will. According to a recent study by the Worldwatch Institute, "Nuclear power generation has reached its peak and will begin a sustained decline in the year 2002 to its eventual demise."[14] Nicholas Lenssen, a Worldwatch energy analyst, who coauthored the study with Christopher Flavin, insists that nuclear energy cannot compete with natural-gas-fired plants, which will dominate the open, deregulated energy market. At $400 to $600 per kilowatt, gas-fired combined cycle plants are the obvious choice when compared to nuclear power's price tag of $3,000 to $4,000 per kilowatt. Even wind turbine installation is lower than $1,000 per kilowatt. This inability to compete successfully in energy markets is cited by the Washington International Energy Group and Wall Street analysts as the primary force in the expected long-term demise of nuclear power as an energy provider.

Splitting the atom offers energy, and the process generates no greenhouse gas emissions. But uranium mining, reactor accidents, nuclear proliferation, and the disposal of high-level radioactive material threaten the future well-being of the global community. The safe storage and transport of nuclear fuel waste have proven to be a serious obstacle to the industry and add significantly to the financial, social, and environmental cost of producing electricity with atomic energy. Underlying this reality is the fact that executives in the US nuclear industry have asked the federal government for protection from liabilities from

> There are many practices of the immediate postwar era that will not survive into the 21st century...the large-scale commercial use of nuclear power appears likely to join the list.[K]
> — *USA Today*

potential nuclear disasters and for help in disposing of increasing amounts of high-level nuclear waste. Scientists have few good ideas on how to safeguard future generations from radioactive material used to produce today's electricity. Yet we continue to rely on it to provide 20% of the United States' electricity demand.[15] The risk from nuclear power plants and nuclear waste disposal sites demands a vigilance and longevity of our social institutions that is unprecedented. It is a serious responsibility that will ultimately affect the health and safety of future generations.

# Uranium / Nuclear Power Fact Sheet

## RESERVES

|     | 1998   | 1999   |
|-----|--------|--------|
| US  | 276 mp | 274 mp |
|     | 106 tmt | 105 tmt |

## URANIUM PRODUCTION/ELECTRICITY GENERATION

|         | 1998      | 1999      | 2000      |
|---------|-----------|-----------|-----------|
| US      | 673.7 bkh | 728.3 bkh | 753.9 bkh |
| France  | 375 bkh   |           |           |
| Japan   | 306.9 bkh |           |           |
| Germany | 160.4 bkh |           |           |
| World   | 2366.7 bkh |          |           |

## CONSUMPTION

|        | 1999        |
|--------|-------------|
| US     | 197.7 mtoe  |
| France | 101.5 mtoe  |
| Japan  | 82 mtoe     |
| World  | 650.8 mtoe  |

## NET ELECTRIC POWER

|        | 1999        |
|--------|-------------|
| US     | 728.2 bkh   |
| France | 375.1 bkh   |
| Japan  | 308.7 bkh   |
| World  | 2,395.9 bkh |

## IMPORTS

|     | 1998     | 1999     |
|-----|----------|----------|
| US  | 43.7 mp  | 47.6 mp  |
|     | 16.8 tmt | 18.3 tmt |

## EXPORTS

|     | 1998    | 1999    |
|-----|---------|---------|
| US  | 15.1 mp | 8.5 mp  |
|     | 5.8 tmt | 3.3 tmt |

---

bkh = billion kilowatt hours
mp = million pounds concentrate
tmt = thousand metric tons
mtoe = million tons oil equivalent

*Reference:*
<http://www.eia.doe.gov/fueloverview.html>
   Alphabetical listing with nuclear and uranium statistics links, mostly for the US only.

FOOTNOTES:

[1] United Nations Environment Program, Radiation: Doses, Effects, Risks (Oxford, United Kingdom: Blackwell Publishers,1993), pp. 6,7.

[2] <http://nova.nuc.umr.edu/nuclear_facts/history/history.html> *The History of Nuclear Energy* from the US Department of Energy's Office of Nuclear Energy, Science & Technology. This Internet publication chronicles the development of the field of nuclear energy from its inception down to the present day. The article includes a nuclear power timeline and selected references.

[3] <www.ccnr.org/uranium_deadliest.html> "Uranium: The Deadliest Metal," article written by Dr. Gordon Edwards, President of the Canadian Coalition for Nuclear Responsibility (CCNR). This article, which appeared in *Perception*, vol. 10 no. 2, details the health risks faced by miners working uranium mines. The article covers the risks to workers and also the damage to the environment by uranium tailings — the end result of the mining cycle.

[4] <www.ieer.org/fctsheet/index.html> *IEER's* "Fabulous Factsheet File" is a list of links to information about uranium and plutonium, health and environmental dangers, etc. Most pages were revised in 1996 and are not date-sensitive. The Institute for Energy and Environmental Research (IEER) is dedicated to increasing public involvement in and control over environmental problems through the democratization of science.

[5] <www.uic.com.au/ne3.htm#3.3> "Nuclear Electricity," a publication by the Uranium Information Center Ltd. of Australia. This paper examines the role of uranium in the nuclear power cycle. Types of reactors are considered as well as grades of uranium ore. Several tables illustrate nuclear power's role in electricity production, types of plants in operation worldwide, and estimated world resources of uranium (table seven).

[6] Robert Gehrke, Associated Press: *The Sacramento Bee*, January 15, 2000, p. A-12.

[7] Robert Hill, Phil O'Keefe, and Colin Snape, *The Future of Energy Use* (New York, NY: St. Martin's Press, 1995), p. 165.

[8] <www.eia.doe.gov/cneaf/nuclear/page/nucforecast.html> US Department of Energy report on the prospects of nuclear energy supply and demand in the near future, and the current status of its production and consumption worldwide. It was last updated in April 1999. Located on the Energy Information Administration's website. EIA, a statistical agency of the US Department of Energy, provides policy-independent data, forecasts, and analyses to promote sound policymaking, efficient markets, and

public understanding regarding energy and its interaction with the economy and the environment.

9 "Effective Radioactive Waste Remediation," a paper presented by Paul M. Brown, at the First International Conference on Future Energy in April 1999 in Bethesda, Maryland. Brown, of International Fission Fuels, Inc., explains how an accelerator-driven reactor disintegrates radioactive waste and its significant advantages over waste burial. He also includes some facts about coal burning vs. nuclear energy and numbers of nuclear reactors and power plants worldwide.

10 <www.nei.org/doc.asp?catnum=2&catid=106> "Nuclear Facts and Quotes" is a page on the Nuclear Energy Institute's website. Operating plants, electricity production, economic performance, environmental protection, industrial safety, weapons conversion, license renewal, and other applications of radiation are some of the topics covered. Most information appears to be from 1998. The NEI "focuses the collective strength of the nuclear energy industry to shape policy that ensures the beneficial uses of nuclear energy and related technologies in the United States and around the world."

11 <www.state.nv.us/nucwaste/yucca/state01.htm> "Why Does the State Oppose Yucca Mountain?" is a page on Nevada's Nuclear Waste Project Office explaining the uncertainties regarding leakage, accidents, geological stability, and expenditure of billions of dollars to qualify a problematic site. The mission of the Nevada Agency for Nuclear Projects is to assure that the health, safety, and welfare of Nevada's citizens and the state's unique environment and economy are adequately protected with regard to any federal high-level nuclear waste disposal activities in the state.

12 <www.ymp.gov> The Yucca Mountain Project is fully outlined on this subsite of the US Department of Energy's website. They have listed most of the research activities that are currently taking place in order to evaluate the area's suitability for deep geologic radioactive waste disposal. Concentrates on the advantages of the area, unlike many other Yucca Mountain web pages.

13 <www.state.nv.us/nucwaste/yucca/seismo01.htm> "Earthquakes in the Vicinity of Yucca Mountain" is a short description of the high volume of seismic activity that has occurred near the proposed radioactive dumpsite. The page is a link off the State of Nevada's Nuclear Waste Project Office website, which is updated daily.

14 <www.cnn.com/NATURE/9903/09/nuclear.enn/> "Nuclear Power Can't Compete, Study Finds" is a March 1999 article originating from the Environmental News Network and posted on CNN's website under the Nature category. It describes nuclear's weakening grasp on its share of the energy industry due to high costs, both financial and environmental.

15 <www.uic.com.au/index.htm#uranium> The website of the Uranium Information Center of Melbourne, Australia. Of most use were the UIC Educational Resource Papers, linking to descriptions of uranium, the nuclear fuel cycle, waste management, and other related topics. Also quoted was *Nuclear Issues Briefing Paper* 41, "World Uranium Mining." Informative documents from late-1997 to mid-1999; most of them are slanted toward the benefits of nuclear power over other energy sources. The UIC was created in 1978, and its purpose is to increase Australian public understanding of uranium mining and nuclear electricity generation.

## SIDEBAR FOOTNOTES:

A Fred Hoyle and Geoffrey Hoyle, *Commonsense in Nuclear Energy* (San Francisco: W.H. Freeman and Company, 1980), p. 61.

B Harper's Index, *Harper's* (February 2000), pp. 15, 88. Index source credit: Worldwatch Institute (Washington).

C Hassan Gardezi and Hari Sharma, "Bulletin of Concerned Asian Scholars," (April-June 1999), p. 8.

D Walter Youngquist, *Geodestinies: The Inevitable Control of Earth Resources Over Nations and Individuals* (Portland, Oregon: National Book Company, 1997), p. 226.

E Associated Press Services, *The Sacramento Bee*, April 27, 1999, p. A-10.

F Peter Bunyard, "Nuclear Power — A Dead Loss," *The Ecologist* (November 1999), p. 412.

G John Donnelly and David Beard, "Russia's nuclear stockpiles pose big threat, report says," *Boston Globe* story published in *The Sacramento Bee*, February 26, 2000, p. A-12. Matthew Bunn is a nuclear nonproliferation specialist.

H Robert Pool, *Beyond Engineering: How Society Shapes Technology* (New York/Oxford: Oxford University Press, 1997), p. 118.

I William F. Martin, "Twin Challenges: Energy and the Environment in Asia," *Harvard International Review* (Summer 1997), p. 115. Martin is Chairman of Washington Policy and Analysis and Former US Deputy Secretary of Energy.

J H. Josef Hebert, "Senate Oks Nevada nuke dump; veto looms," Associated Press, *The Sacramento Bee*, February 11, 2000, p. A-6.

K Christopher Flavin and Nicholas Lenssen, "Nuclear Power Industry is at a standstill," *USA Today* (May 1993), p. 66.

# Section II
# Renewable Energy: A Traditional Bridge to an Alternative Future

WHAT IS RENEWABLE ENERGY, and how will it fit into the energy equation of the 21$^{st}$ century? Nonrenewable fuels like coal, oil, natural gas, and uranium exist in limited quantities, and their use threatens our health and environment. The following section focuses on traditional renewable energy sources like solar-, wind-, and water-power. Renewable energy systems will generate electricity cleanly for as long as the Sun continues to shine. Read on to learn why British Petroleum, Royal Dutch/Shell, and other oil companies are investing heavily in renewable sources of energy. Industry experts realize that these alternative energy systems not only help reduce greenhouse gas emissions, but they predict that over the next half century, renewables may grow to supply half the world's energy.

No doubt about it, renewable energy will play a major role in the energy industry of the 21$^{st}$ century and beyond. Successfully generating electricity by harnessing the perpetual power of the Sun and wind is not only technologically feasible, it is already a reality. The industrialized economies of modern times have relied on nonrenewable coal, oil, and natural gas, which are limited in supply and pollute the

*Wind farms require only small areas of land and have minimal impact on ranching and farming activities.*

> Only recently have the renewable energy units of national laboratories and academies of science begun to examine the commercialization of small-scale and decentralized technologies.[A]
>
> — *Environment*

environment when combusted. Nuclear energy won't carry the load because there is only so much uranium available, and widespread concerns about leaks and radioactive waste are well founded. The drama of the discovery, development, and exhaustion of non-renewable resources can be seen in the ghost towns of the American West and other mining regions around the world. Renewable energy will not solve the problem of increasing energy demand by a booming world population, but it does offer a bridge of hope until a replacement energy source for nonrenewable fossil fuels can be developed.

There are a handful of renewable energy resources that will help society survive the transition from the polluting, non-renewable energy used today to future technologies like hydrogen fuel cells and cold fusion. Similar to fossil fuel, which is hundreds of millions of years' worth of highly concentrated solar energy transformed into coal, oil, and gas, virtually all renewable resources also depend in one way or another on sunlight. Unfortunately, the electricity to be derived on a day-to-day basis from wind-, solar-, hydroelectric-, biomass-, and geothermal-powered technologies pales in comparison to the approximately 70 million barrels a day of the energy-dense, high-powered liquid fuel (crude oil) the world consumes every day.

Solar power relies on the energy produced by nuclear fusion in the Sun. This energy can be collected and converted in different ways, such as simply heating water for domestic use or by the direct conversion of sunlight to electrical energy using mirrors, boilers, or photovoltaic cells. The technology is improving, and the economics are getting more competitive. Photovoltaic panels don't generate electricity at night, but they can be used in the daytime to produce hydrogen, which can then be stored. Solar power will play an increasingly important part in the energy equation during the 21st century, but living off the daily sunshine will be vastly different than bingeing on a limited amount of cheap, energy-dense oil.

Scientists and engineers are continually improving the efficiency of renewable energy systems. Mankind has been harnessing the wind for thousands of years, and is now cleanly

> Power plants in the state [California] emit on average 71 tons of nitrogens of oxygen daily. Total carbon emissions from electricity generation this year [1999] will surpass 30 million tons.[B]
>
> — *The Sacramento Bee*

producing electricity with it. Air flowing through turbines or spinning blades generates power whenever the wind blows. Wind energy can be used to pump water or generate electricity, but the wind is inconsistent, and a "wind farm" requires extensive areal coverage to produce significant amounts of energy.

Hydroelectric power is a recent addition to the renewable energy field and a very successful one. The system is simple:
- Wind and Sun evaporate water, which is later released in the form of rain and snow.
- Hydroelectric dams catch the precipitation in reservoirs.
- The water is then channeled through turbines that turn generators to produce electricity.

Until recently, hydroelectric dams were considered environmentally clean sources of power, but fisheries and other wildlife habitat have been severely impacted on many dammed rivers. Most of the world's hydroelectric dams are historically recent, but all reservoirs eventually fill up and require very expensive excavation to become useful again. At this time, most of the available locations for hydroelectric dams in the United States are already developed.

Humans have been burning biomass materials since the dawn of time, and it is still the principal fuel used in many parts of the world. Incredibly, just 120 years ago, wood was the chief energy source in the United States. But today's economy runs on oil, and despite significant government support, converting wood to alcohol, or corn to ethanol has proven neither economical nor energy efficient. Researchers have recently discovered how to produce clean combustible gas from waste products such as sewage and crop residue, but biomass gas will not replace petroleum as the fuel of choice.

Hydrogen has been touted as the fuel of the future. It is the most abundant element known in the universe and can be burned cleanly as a fuel for vehicles with water as the main combustion byproduct. Hydrogen can also be fed into a fuel cell, a battery-like device that generates heat and electricity. Using hydrogen instead of gasoline or diesel will significantly reduce the health hazards and medical costs associated with

> One important fact highlighted by the [D.O.E.] studies is that no energy technology is completely environmentally benign.[c]
> — Scientific American

the exhaust from a conventional internal combustion engine. But the large-scale extraction of hydrogen from terrestrial resources such as water, coal, natural gas, or methane requires copious energy, which is currently produced by burning fossil fuels. Commercial hydrogen production is expensive and only shifts the pollution from vehicles back to the power plants. Producing hydrogen with solar power is the dream of environmentalists and renewable energy proponents. If done successfully, hydrogen and electricity will eventually become society's primary energy carriers in the 21$^{st}$ century.

Geothermal energy, left over from the original accretion of the planet and augmented by heat from radioactive decay, seeps out slowly everywhere, every day. In certain areas, the geothermal gradient (increase in temperature with depth) is high enough to exploit for the generation of electricity.

Another form of useful energy is "Earth energy," a result of the solar heat storage in the planet's surface. Soil maintains a relatively constant temperature throughout the year and can be used with heat pumps to warm a building in winter or cool a dwelling in summer. This form of energy can lessen the need for other power to maintain comfortable temperatures in buildings, but it cannot be used to produce electricity.

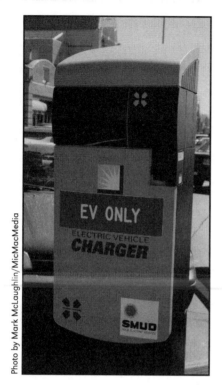

*In some cities, you can recharge your electric vehicle at the mall for no charge to your wallet.*

Tides, waves, and the heat differential within the world's tropical oceans are potent sources of clean energy. Various countries around the world are investing time and money into the technologies that may tap these renewable power producers, but overcoming the obstacles inherent in these systems will be difficult, and the energy produced limited in scope.

The media and industry claim that renewable energies are not yet economically competitive with fossil fuels. Perhaps not, but when one considers the health and environmental costs associated with burning coal and oil, the price of renewable energy becomes more attractive. No renewable energy system will single-handedly replace oil, but added together, they will become a very important part of the energy mix of the future. Traditional renewable systems are a logical step in the transition to advanced alternative energy sources such as cold fusion. Although scientists and engineers are working feverishly to overcome the various obstacles associated with "new energy" technologies, society should not stand by quietly while researchers wait for a breakthrough. Burning petroleum is polluting our air and water, and the bulk of the world's reserves of cheap oil are concentrated in the Persian Gulf, a region known for political volatility. Getting that oil will likely cost billions of dollars and possibly the lives of American soldiers.

Every year, American's consume 25% of all the energy produced in the world, but that conspicuous consumption can't last forever. To that end, the US Department of Energy (DOE) established the Renewable Energy Production Incentive (REPI) as part of an integrated strategy in the Energy Policy Act of 1992. This act promotes increases in the generation and utilization of electricity from renewable energy sources and furthers the advances of renewable energy technologies. In 1996, the Renewable Energy Policy Project released "The Environmental Imperative," a plan for the energy market to draw on renewable energy to avoid the severe environmental impacts of the fossil fuel cycle. This plan outlines the environmental imperative for accelerating the exploitation of renewable resources. It is important to realize that it usually takes about 50 years to significantly shift fuel patterns and that using electricity as an alternative to oil will require a major adjustment by the American public. The window of opportunity to make this energy transition without a major economic disruption will not be open for long.

Capital providers are very cautious about opportunities in the renewable energy arena because there are no major success stories yet.[D]

— *Solar Today*

It will, however, be several decades before renewables will make a significant impact on energy supply.[E]

— *Harvard International Review*

SIDEBAR FOOTNOTES:

[A] Daniel M. Kammen, "Bringing Power to the People: Promoting Appropriate Energy Technologies in the Developing World," *Environment* (June 1999), p. 15.

[B] Editorial, "Waiting for Davis: How PUC vacancies threaten to shift energy policy," *The Sacramento Bee*, May 30, 1999, p. B-6.

[C] H.H. Hubbard, "The Real Cost of Energy," *Scientific American* (April 1991), pp. 36 – 42.

[D] Richard T. Stuebi, "Attracting Equity for Solar Businesses," *Solar Today* (March/April 1999), p. 66.

[E] William F. Martin, "Twin Challenges: Energy and the Environment in Asia," *Harvard International Review* (Summer 1997), pp. 28 – 31.

# Solar Energy

Solar cells convert solar rays directly into electricity. Non-polluting photovoltaic cells use no fuel, mechanical turbine, or generator to produce electric current, and solar energy is renewable, clean, and abundant. The solar power industry has enjoyed double-digit growth in the last five years, but solar energy has historically suffered from inexpensive oil, which has been cheap and easy to produce. As air pollution worsens and global petroleum supplies get squeezed tighter in the future, the world's energy providers will look to the Sun for a clean, renewable, and decentralized energy source.

EVERY DAY, THE SURFACE OF PLANET EARTH is blasted with so much solar energy that, if harnessed, 60 seconds' worth could power the world's total energy requirements for one year.[1] The Sun is a colossal fusion reactor that has been burning for more than 4 billion years. In just one day, it provides more energy than the current human population would consume in 27 years.[2] By some estimates, the amount of solar radiation striking the earth every 72 hours is equivalent to all the energy stored in the planet's coal, oil, and natural gas reserves.

*Solar energy is replacing nuclear power.*

Solar radiation is a free and unlimited natural resource, yet converting it into an energy source is a relatively new idea. Using solar power for heat seems simple enough today, but it wasn't until 1767 that Swiss scientist Horace de Saussure built the first thermal solar collector. He used his solar collector to heat water and cook food. It wasn't until 1891 that the first commercial patent for a solar water heater was awarded to US inventor Clarence Kemp. The patent rights to this system were later purchased by two California executives who, by 1897, had installed the solar-powered water heaters in one-third of the homes in Pasadena, California.

In 1839, French physicist Edmund Becquerel determined that energy from the Sun could produce electricity through a "photovoltaic effect" ("photo" means light, and voltaic equals electrical potential). In the 1880s, selenium photovoltaic (PV) cells were developed that converted insolation into electricity with 1-2% efficiency.[3] (The efficiency of a solar cell is the percentage of available sunlight converted by the photovoltaic cell into electricity.) Because the principle behind the conversion of light into electricity was poorly understood and the efficiency rate of that conversion process so low, photovoltaic power remained just a curiosity for many years.[4] It wasn't until physicist Albert Einstein won a Nobel Prize for his explanation of the "photoelectric effect" in the early 1900s that scientists began to understand how photovoltaics really worked.

*Solar roof tiles use sunlight to generate household electricity.*

Solar technology advanced in 1908 when William J. Bailey of the Carnegie Steel Company invented a collector with an insulated box and copper coils. In the mid-1950s, Bell Telephone Laboratories utilized silicon PV cells to double the conversion efficiency to 4%, and then later to 11% efficiency. From then on, commercial interest in solar power intensified. During the late 1950s and 1960s, the United States' space program invested heavily in the development of photovoltaics. The cells were perfect sources of electric power for satellites because they were rugged and lightweight and could meet low power requirements reliably. The first solar-powered satellite, the Vanguard, needed only a *milliwatt* of electricity to run its tiny beeping signal. Solar cells today provide satellites with the *kilowatts* of electrical power required to operate complex on-board computers and communication equipment. Engineers have now miniaturized solar cells to the point that scientists at NASA believe that photovoltaic systems can produce enough energy to make solar-propelled space flight a reality.[5]

Solar energy has great potential for providing clean and unlimited electricity in many regions of the world. This renewable resource has largely been ignored by many US energy providers because there has been little economic motivation due to the abundance of cheap coal and oil. Corporate shareholders want their profits today, not sometime in the distant future. In the last few decades, however, global energy demand has surged, as have the environmental problems associated with burning coal and oil and the storage of nuclear-generated radioactive waste. In the late 1990s, more governments, utilities, and corporations were embracing renewable energy sources as environmentalists, consumers, and voters pressure them to do so. More importantly, many consumers are willing to pay for "green energy," so suppliers see future profit in non-polluting renewable energy production. Some governments and energy suppliers have been slow to recognize the potential of solar power. Historically, research and development in photovoltaics has progressed erratically, in short-lived bursts of interest. For example, the US Department of Energy (DOE) funded the installation and testing of over 3,000 PV cell systems during the 1973-1974 oil embargo. By the late 1970s, energy companies and government agencies were investing in the PV industry, and an acceleration in module development took place. But solar power remained far behind oil, coal, nuclear, and other non-renewable energy sources. Seri-

*Solar panels provide both power and shade at the zoo.*

Operating one solar water heater—instead of an electric water heater—saves the equivalent of nine barrels of oil every year and reduces carbon dioxide emissions by 1600 pounds and sulphur dioxide emissions by 12 pounds.[A]

— *Solar Today*

ous interest in photovoltaics increased again during the 1990s after several military conflicts in the oil-rich Persian Gulf.

Considering that the first practical solar cells were produced less than 30 years ago, the industry has come a long way. The biggest jumps in efficiency came with the advent of the transistor and its accompanying semiconductor technology. PV cell production costs have fallen to nearly 1/300 of what they were fifty years ago, and the purchase cost has gone from $200 per watt in the 1950s to less than $5 per watt in 1999.[6] The conversion efficiency of PV cells has increased dramatically, from 4% in the 1950s to 18.8% in 1999.[7]

Solar-powered energy systems are either thermal or photovoltaic. Thermal systems concentrate sunlight, convert it into heat, and then apply it to a steam generator or engine. A turbine converts it into electricity, which can be used to heat or cool buildings, or heat water. The advantage of a photovoltaic system over thermal generation is that it produces electricity directly, without any moving parts. PV systems convert the direct current (DC) generated by sunlight to an alternating current (AC), which home appliances run on. Today's photovoltaic systems are composed of cells made of silicon, the second most abundant element in the Earth's crust. Power is produced when sunlight strikes a wafer-thin semiconductor material and creates an electric current. The smallest unit of this system is a cell. Cells wired together form a module, and modules wired together form a panel. A group of panels is called an array, and several arrays are combined to form an array field.

There are advantages to photovoltaic solar power that make it one of the most promising renewable energy sources. The system is non-polluting, has no moving parts to break down, and requires little maintenance and no supervision. The average unit produces energy for 20–30 years with low operating costs. Solar energy systems are especially unique

*A vast array of reflectors powers the Weizman Institute in Israel.*

Environmental issues continue to be the driving force for the development of alternative energy technologies in much of Europe.[8]
— Solar Today

because they require no extra construction or developed land area, and function safely and quietly. Remote or underdeveloped communities can produce their own supply of electricity by constructing as small or as large a system as needed. When communities grow, more solar energy capacity can then be added as necessary. Photovoltaic-generated energy has a significant advantage over wind power, hydroelectric energy, and solar-thermal derived power. The latter three all require turbines with moving parts that are more expensive, are noisy, and require maintenance.

Solar energy is most popular in developing countries, which are the fastest growing segment of the photovoltaics market. Billions of people suffer a primitive existence without electricity while the Sun beats down on the land, making solar power the obvious energy choice for many populated regions in the world. The system's modular, decentralized character is ideal for filling the electric needs of remote villages in many nations. Independent solar systems are much more practical than extending expensive high-voltage power lines into remote areas, where most people do not have the money to pay for conventional electricity purchased off the grid.

India is one of the world's main producers of PV modules. Predicted to surpass China as the world's most populous nation, it plans to power 100,000 villages using solar energy and to install solar-powered telephones in half a million of them. Mexico hopes to electrify 60,000 villages with solar power. Zaire's Hospital Bulape serves 50,000 outpatients per year and is run completely on solar power, from the air conditioning to x-ray equipment. And in Moroccan bazaars, carpets, tinware, and solar panels lie side by side for sale. One of the most outstanding examples of a country's commitment to solar power is in sunny, arid Israel. More than half of all Israeli households (700,000) heat their water with solar energy systems. More than 50,000 new installations are added every year.

The island nation of Japan has no petroleum reserves and, therefore, is always looking for ways to produce energy without relying on imported oil. The Japanese government aggressively supports solar homes through subsidies and tax

> Architects who once shunned bulky blue solar panels are marveling at solar cells incorporated discreetly into the surface of roofing shingles, windows, skylights, and wall facades.[C]
>
> — *Discover*

> Israel is an outstanding example of how solar collectors are effectively used on virtually every rooftop in accordance with a law requiring solar water heaters to be installed in all new apartment buildings up to eight stories high.[D]
>
> — Madanjeet Singh (UNESCO)

incentives. Thousands of homes there have had solar systems installed in the last several years. Within a few years the Japanese government expects that more than 70,000 homes will be solar powered.⁰

Solar energy is as practical in urbanized and populated areas connected to the electric power grid as it is in remote areas. The average American home has more than adequate roof area for the solar panels that can produce much of the electricity needed to supply its domestic requirements. Solar electric roofing tiles are now available to accommodate custom as well as production housing. A modern solar home looks and operates very much like a home that is connected to a power line. Solar energy can provide up to 90% of an American's household electricity needs.

There are four broad categories that have been identified for the solar energy market: industrial, rural habitation, grid-connected, and consumer/indoor. Industrial use represents the largest application of solar power in the past 30 years. Telecommunications, oil companies, and highway safety equipment all rely on solar power for dependable electricity far from power lines. Roadside call boxes and lighted highway signs use the Sun's energy to provide service without buried cable connections or diesel generators. Navigational systems such as marine buoys and other unmanned installations in harsh remote areas are also ideal applications for solar power. Rural habitation includes cabins, homes, villages, clinics, schools, and farms. Grid-connected systems pair solar power with an existing grid network in order to supply a commercial site with enough energy to meet a high demand or to supplement a family's household supply. Consumer/indoor uses of PV cells include watches and calculators; PV modules power computers and radios.

That the practicality and environmentally safe nature of solar power is influencing people worldwide is evident in equipment sales. The production of PV cells and modules has increased threefold from 40 megawatts (MW) in 1990

*The power of the Sun can energize remote data collectors.*

Electricity generation remains a major source of pollution [in California]. Power plant carbon emissions will exceed 30 million tons this year.ᴱ
— *The Sacramento Bee*

to about 120 MW in 1998. Worldwide sales have been increasing at an average rate of about 15% a year during the last decade and leaped by 37% in 2001 as new applications hit the market. Unfortunately for American workers, it is European and Japanese companies that are reaping the harvest of that double-digit expansion. As recently as 1996, manufacturers in the US accounted for more than 40% of the world's photovoltaic shipments. Now Japan is the number one leading manufacturer of solar devices. It is a lost opportunity for the US economy because the United States pioneered the solar technologies finding commercial success today. Another obstacle for advocates of solar power is that of all photovoltaic panels produced in the United States, three-fourths go overseas, mostly to Germany and Japan, which have aggressive solar power programs. European and Japanese renewable energy companies have the advantage of citizen commitment and government subsidies more generous than those available to US firms. The most important factor in the technology shift, however, is that governments in Europe and Japan heavily tax oil, gas and coal to pay for some of the hidden cost of consuming fossil fuels. These taxes partially offset the billions of public dollars spent on health issues related to air pollution and also reduce traffic congestion and greenhouse gas emissions through conservation.

Many experts believe that there is a real possibility for the market to continue to grow at a lucrative clip into the next decade, making photovoltaics a multibillion industry. Oil companies are starting to believe it, too. According to the Worldwatch Institute, in 1997, British Petroleum revealed plans to increase sales of solar products 10-fold over the next decade. Shell Oil will invest more than $500 million in solar and other renewables over the next five years.[9]

Americans shouldn't have to wait another ten years for clean, renewable solar power. A professor of communications at Stanford University in California recently made a cogent economic argument for solar power now. In an April 25, 2001, letter to the *New York Times*, Dale Maharidge wrote: "The latest estimate is that California's failed experiment with deregulation will cost consumers $5 billion annually in increased rates. Watching the sun glint off my solar panels, I was moved

> We are seeing PV change from a cottage industry into a profitable business worldwide.[F]
> — Ken Zweibel (NREL)

> It's solar power for the fast-food generation. If we didn't tell you, you'd come in the house and never know we had solar. We have everything everyone else has.[G]
> —Katy Hundelt, off-grid customer near Sedona, Arizona.

to find out how many of them could be purchased for $1 billion. The answer is about one million 110 watt, 110 volt, utility-ready panels, the easiest kind for consumers to use. At peak power output on a California summer afternoon — when blackouts are most likely to occur — those one million panels would produce about 100 megawatts of electricity." Maharidge adds that if the state raised $30 billion by issuing bonds to buy solar panels, they would generate 3,000 megawatts — enough juice for 300,000 homes and the displacement of the emissions of three large coal-fired power plants. He notes that "If we rely solely on the free market to bring it to fruition, it could take decades for solar power to reach a level where it produces significant power. Utilities argue that panels don't work at night. But they produce power in precisely the hours when air conditioners are roaring and the most power is consumed."[10]

There are only two primary disadvantages to using solar power: a limited amount of sunlight and the cost of equipment. The value of sunlight a location receives varies greatly depending upon geographical location, time of day, season, and average cloud cover. The southwestern United States is one of the world's best areas for persistent sunshine. Globally, other areas receiving very high solar intensities include developing nations in Asia, Africa, and Latin America. Although solar energy technologies have made impressive cost improvements over the years, solar energy is currently still more expensive than traditional fossil fuel sources. However, solar energy is renewable and non-polluting, and the equipment will eventually pay for itself in 2 to 5 years, depending on how much Sun a particular location receives. Then the user will have a virtually free energy source until the end of the equipment's working life. Future improvements are projected to decrease the payback time down to 1–3 years.[11]

As the price of solar-generated electricity decreases and as the environmental and dollar costs of petroleum increases, photovoltaics will expand its international market. Solar power will be an excellent energy option, long after Hydrocarbon Man fades away into the sunset. Clean, renewable photovoltaic-generated power enjoys obvious advantages when compared to coal, oil, natural gas, or nuclear power.

FOOTNOTES

1. <www.ases.org/solarguide/factbase.html> The "Solar Guide Factbase" is located on the American Solar Energy Society (ASES) website. It links the reader to the history, terminology, trends, impacts, etc., of solar energy. Most of this reliable information dates from 1999. The ASES is "dedicated to advancing the use of solar energy for the benefit of US citizens and the global environment."

2. <www.nrel.gov/lab/pao/solar_energy.html> The US National Renewable Energy Laboratory (NREL) website about solar energy. It explains how much solar energy the planet receives, types of power systems in use, and the benefits of solar energy based on 1998 information. As the nation's leading center for renewable energy research, NREL is developing new energy technologies to benefit both the environment and the economy.

3. <www.nrel.gov/hot-stuff/press/999world.html> "World-Record Solar Cell a Step Closer to Cheap Solar Energy" a February 1999 press release from the National Renewable Energy Laboratory. The NREL has created a photovoltaic cell with an efficiency level of 18.8%, which has broad implications for both terrestrial and space technologies.

4. <www.solarpv.com/about_photovoltaics.html> Siemens Solar website dedicated to explanations of photovoltaics, history, various types of technology, strides made in the last decade, predictions of further progress, and so on. Describes the four categories of solar power use as well as production capacity worldwide. Web pages are undated but appear to be from 1997 to 1999. "The Siemens Solar Group is the world's leading manufacturer in the photovoltaic industry. We produce the highest quality solar cells and modules, and in conjunction with over 130 business partners worldwide have installed more photovoltaic generating capacity than any other manufacturer."

5. John Perlin, "Solar Propelled Space Travel," *Solar Today* (May/June 1999), pp. 16,18. The article describes NASA's efforts toward solar-powered space travel and the 1998 launch of Deep Space I, a solar-propelled spacecraft. John Perlin is co-author of *A Golden Thread* and author of *From Space to Earth: The Story of Solar Electricity*.

6. Sacramento Municipal Utility District (SMUD): PhotoVoltaic & Distributed Technology Department. Information flyer regarding SMUD's award-winning "PV Pioneer Program," which is a commercialization effort to bring the price of photovoltaic-generated power down. SMUD is taking an active role to stimulate the market and encourage mass production of the systems, the key to lower costs. "The attractive, lightweight, and durable PV panels are mounted in less than a day at the same pitch as your roof." In its 8$^{th}$ year, the project has succeeded in bringing the cost of the system down to less than $4.25/watt.

7. <www.si.edu/ndm/exhib/sun/2/obj_timeline.htm> This timeline of photovoltaic development is titled "Under the Sun: An Outdoor Exhibition of Light." The site is very informative and covers the benefits of using solar energy to the various designs of photovoltaic equipment that have been developed recently. The information is worth reviewing and is current up to 1998.

8. Suvendrini Kakuchi, "Environment-Japan: Solar Power vs. Global Warming," *InterPress Service*, November 29, 1997.

9. Lester R. Brown, Michael Renner, and Christopher Flavin, *Vital Signs*, (New York, NY: W.W.Norton & Company, 1998), p. 60.

10. Dale Maharidge, "Look to the Sun," *The New York Times*, April 25, 2001.

11. <www.pvpower.com> The "PV Power Resource Site" contains 13 links to information about past and present research and development in photovoltaics. Information included is current — 1998 and 1999. The site was developed and is maintained by Mark Fitzgerald "for the coordination and dissemination of information of global photovoltaic (PV) technologies, applications, history, and resources."

## SIDEBAR FOOTNOTES

A. Tom Lane, "The Southern Star—From Catastrophe to Inspiration," *Solar Today* (May/June 1999), pp. 46-49.

B. Pamela Murphy Kunz, "Solar Energy in Europe," *Solar Today* (January/February 1999), pp. 28-31.

C. Kathryn Brown, "Invisible Energy," *Discover* (October 1999), p. 36.

D. Madanjeet Singh, *The Timeless Energy of the Sun for Life and Peace with Nature* (Published by the United Nations Educational, Scientific, Cultural and Communication Organization—UNESCO, 1998), p. 152.

E. Editorial, "Waiting for Davis," *The Sacramento Bee*, May 30, 1999.

F. Christopher Flavin, "Solar Power Markets Boom," *Worldwatch* (September/October 1998), pp. 23-27. Ken Zweibel is with the US National Renewable Energy Laboratory, a federally funded lab based in Golden, Colorado.

G. Fred Bernstein, "O spacious skies, with nary a power pole in sight," *The New York Times* (September 23, 1999), p. F-4.

# Wind Energy

**7** In the 1960s, popular American folk musician Bob Dylan wrote, "You don't need a weatherman to know which way the wind blows." In year 2001, the wind represents more than weather; it means money and energy. High-tech wind turbines as tall as the Statue of Liberty are now producing *mega*watts of electricity. Wind is the oldest source of renewable energy, and it will play an important role as a nonpolluting energy provider in the 21st century.

HUMANS HAVE TAKEN ADVANTAGE of wind power for thousands of years. The first known use occurred in 5000 BCE when people used sails to navigate the Nile River.[1] By 900 ACE, Persians were using windmills to pump water and grind grain.[2] The Chinese have written documentation of windmill use by 1219 but had probably acquired the technology centuries before. Agrarian communities on the island of Crete depended on hundreds of sail-rotor windmills to pump water for crops and livestock. By the 16th century, some 10,000 windmills were in use in the Netherlands. The Dutch were responsible for many refinements of the windmill, which was used primarily for pumping excess water off flooded land. As early as 1390, they had connected the mill to a multi-story tower with separate floors devoted to grinding grain kernels, removing chaff, and storing the processed grain. Living quarters for the windsmith and his family were

*Early windmills were used to pump water and grind grain.*

> Denmark now generates 8% of its electricity from wind power, and another fraction from the combustion of agricultural wastes.[A]
> — World Watch

arranged on the lowest floor. The system was so efficient that eventually there were thousands of such windmills operating in England.

But it took almost 500 years to perfect the windmill's efficiency to the point that it had the major features recognized by contemporary designers as being crucial to the performance of modern wind turbine blades. By then, applications ranged from saw-milling timber to processing spices, tobacco, cocoa, paints, and dyes.

The windmill was further refined in the United States during the late 19th century. Heavy, inefficient wooden blades were replaced by lighter, faster steel blades around 1870. Over the next century, more than six million small windmills were erected in the western US, where they pumped ground water for livestock and provided the domestic supply for families living on remote ranches. The first large windmill to produce electricity was a multiblade design built in 1888. Modern 70–100 kilowatt wind turbines blow away its meager 12-kilowatt capabilities.

Today, individuals and corporations are realizing that wind power is a promising clean energy resource that can serve as an alternative to fossil-fuel-generated electricity.[3] In 1999, global wind-generated electricity exceeded 10,000 megawatts, which is approximately 16 billion kilowatt-hours of electricity. That's more than enough to serve five cities the size of Miami. In a world where more than 2 billion people live without electricity, decentralized wind power is projected to be one of the developing world's most important sources of electricity. Wind-generated energy also has a bright future in industrialized nations like the United States and Europe. According to the American Wind Energy Association, wind energy could provide 20% of America's electricity with turbines installed on less than 1% of the nation's land area. Within that area, less than 5% of the land would be occupied by wind

*Among all renewable energies, wind power is the closest to market competitiveness in year 2000.*

equipment—the remaining 95% could continue to be used for farming or ranching.[4] By the year 2010, 10 million American homes may be supplied by wind power, preventing 100 million metric tons worth of $CO_2$ emissions every year.

Lessening our dependence on fossil fuels is critical to the health of all living things, and wind energy can help in that effort. The 3 billion kWh of electricity produced by America's wind-powered turbines annually displace the energy equivalent of 6.4 million barrels of oil and avoid 1.67 million tons of carbon emissions, as well as sulfur and nitrogen oxide emissions that cause smog and acid rain. More wind power means less smog and soot, less acid rain, and reduced greenhouse gas emissions.

Windmills may have been around for almost 1500 years, but no one ever imagined that clean wind power would become affordable enough to compete with inexpensive coal and oil. But that time has come. In fact, many utility services around the world offer wind-generated electricity at a premium of 2 to 3 cents per kWh. If a household used wind power for 25% of its needs, it would spend only $4 or $5 dollars per month for it, and the price is still dropping. Compare this to 4.8 to 5.5 cents per kWh for coal or 11.1 to 14.5 cents per kWh for nuclear power, both of which threaten the environment. Clean wind energy is, therefore, cheaper than any other electric generation system except natural gas, which emits one pound of greenhouse gases for every kilowatt-hour of electricity it generates. The success of wind power is due to better turbine technology, which has helped reduce costs by more than 80% since the early 1980s. Many industry analysts see the cost of wind power dropping an additional 20% to 40% by 2005.

The global wind energy industry is growing by leaps and bounds. The United States, Germany, Spain, Great Britain, Denmark, and India are large producers of wind-generated power. In 1998, the city of Navarro, Spain, was already utilizing wind power to generate 23% of its electricity needs. Denmark now produces 8% of its electricity from wind power. According to *Worldwatch Institute Online*:

"The world added 2,100 megawatts of new wind energy generating capacity in 1998, a new all-time

It is believed that electricity produced by a "farm" of large-sized turbines would be as cheap as conventional electricity and, in time, cheaper.[B]
— *Canada Today*

A law has been passed by the German Parliament forcing the utility companies to pay 90 percent of the electricity rates charged to consumers for electricity supplied to the grid from wind energy.[C]
— *Energy and the Environment: The Policy Challenge*

Local authorities who fail to find sites for wind farms in the Netherlands will have to face the wrath of not one, but two government ministers.[D]
— *Windpower Monthly*

> Offshore wind has the potential to deliver substantial quantities of energy at a price that is cheaper than most of the other renewable energies.[E]
>
> — *Solar Today*

record, and 35% more than was added in 1997. Wind power is now the world's fastest growing energy source and has also become one of the most rapidly expanding industries, with sales of roughly $2 billion in 1998."[3] Extensive offshore developments are likely in northern European waters in the next decade. These developments will be a major step for this technology and will result in a dramatic increase in decentralized electricity generation. Offshore wind has the potential to deliver substantial quantities of energy at a price that is cheaper than most of the other renewable energies because wind speeds are generally higher offshore than on land.

Worldwide, wind energy capacity has expanded at an annual rate of nearly 30% during the 1990s. Christophe Bourillon, executive director of the European Wind Energy Association, recently stated that Europe has emerged as a world leader in wind energy development in the 1990s. Since 1998, wind power has been the fastest-growing new source of electricity in the world and the Europeans are cashing in on the trend. Similar to the photovoltaic energy market, the US was a pioneer in key wind technology improvements, but now seriously lags behind countries like Denmark. Denmark's government used to subsidize the installation of wind turbines but abolished the program in 1989, when wind power was regarded as fully competitive with electricity produced from heavily taxed fossil fuels. In 2001 Danish firms are supplying 60% of the wind turbines being installed in the rapidly expanding American market.

As far as the US wind industry is concerned, 1998 was the best year yet. The American Wind Energy Association attributes this "wind rush" to progressive state policies and growing consumer demand for "green" (low-environmental-impact) power. More states require that a portion of their energy production must come from renewable sources, and utilities are now offering customers the choice of buying green power at a premium over energy generated from conventional, envi-

*In contrast to the modest traditional windmill, modern wind turbines stand taller than the Statue of Liberty.*

ronmentally-damaging sources, such as coal and oil. As one of the least expensive renewable energy resources, wind is often the primary beneficiary of the green energy market. Utilities as well as policymakers are continuously surprised by the public's positive response to the availability of this environmentally friendly power supply.

Although wind energy is now more affordable, more available, and still pollution-free, it does hold some drawbacks. Wind power suffers from the same lack of energy density as direct solar radiation. The fact that it is a very diffuse source means that large numbers of wind generators (and thus large land areas) are required to produce useful amounts of heat or electricity.[6] Wind turbines cannot be erected everywhere simply because many places are not breezy enough for suitable power generation. When an appropriate place is found, building and maintaining a wind farm can be costly. It is a highly capital-intensive technology. If the interest rates charged for manufacturing equipment and constructing a plant are high, then a consumer will have to pay more for that energy. One study found that if wind plants were financed on the same terms as natural gas power plants, their cost would drop by nearly 40%. Fortunately, as more facilities are built, the cheaper wind energy will become. The cost of wind-generated electricity has dropped by 15% with each doubling of installed capacity worldwide, and capacity has doubled three times during the 1990s.

Bird fatalities on wind farms are also a concern. A study in the Altamont Pass Wind Resource Area in California found 182 dead birds, 119 of which were raptors. In response, the wind industry is committed to modifying the equipment in order to make the area safer for birds. Ideas include reducing the number of perches on turbines, spacing turbines far apart and in the direction of migration, painting patterns on blades that contrast with landscape colors, and even broadcasting a radio frequency to keep birds away altogether. At the same time the wind industry increases its efforts to take responsibility on this issue, it also points out how many millions of species are killed every year due to the acquisition and distribution of conventional sources of energy.

> The installation is part of a global trend toward larger wind turbines.[F]
> — Colin Taylor, SMUD's director of power generation

> Wind turbines will generate a projected 21 billion kilowatt-hours of electricity in 1999 — enough power for 3.5 million suburban homes.[G]
> — Christopher Flavin

> The Advanced Alternative Energy Corporation is developing a concept for combining renewable biomass and solid waste along with wind power alone for producing combined heat and power.[H]
> — *Environmental Technology*

Overall, the advantages of wind power heavily outweigh the disadvantages. Although it can only supplement other sources of energy for now, it provides skilled jobs for people in rural communities, replaces environmentally harmful energy sources, and is inexhaustible. Wind energy will never be subject to embargoes or price shocks caused by international conflicts, and, unlike oil energy, clean wind power is renewable year after year.

## FOOTNOTES

[1] <http://solstice.crest.org/renewables/re-kiosk/wind/history/index.shtml> "History of Wind Power" is a brief description of the ways in which wind energy has been acquired and used for various tasks since 5000 BCE.

[2] <www.telosnet.com/wind/early.html> "Illustrated History of Wind Power Development" is a detailed explanation of the various types of windmills and their construction and use since 1000 BCE.

[3] <www.awea.org/news/index.html> Index of news releases on the American Wind Energy Association's website. A 1999 article entitled "US Wind Industry Finishes Best Year Ever, By Far" explains how wind power is catching on quickly as consumers and policymakers restructure their energy priorities. "World Wind Capacity Tops 10,000 Megawatt Mark" from April 1999 describes the increasing use of wind energy in Europe and the US, and the social and environmental benefits of such energy. "People count on AWEA to provide up-to-date, accurate information about the domestic and international wind energy industry. The Association is an industry clearinghouse, utilizing [various] vehicles for communicating facts and statistics."

[4] <www.awea.org/pubs/factsheets.html> Links to Wind Energy Fact Sheets in PDF format on the American Wind Energy Association's website that gives information on the affordability, cleanliness, and sensibility of using wind as a source of electrical generation.

[5] <www.worldwatch.org/alerts/981229.html> "Wind Power Sets New Record" is a December 1999 press release written by Christopher Flavin of the Worldwatch Institute. Wind power's viability is being realized in the US and even more so abroad, and is the fastest growing renewable energy source. The Worldwatch Institute "is a nonprofit public policy research organization dedicated to informing policymakers and the public about emerging global problems and trends and the complex links between the world economy and its environmental support systems."

# Hydrogen & Fuel Cells

## 8

Hydrogen burns clean and lean and is the most abundant element in the Universe. Ever since "Big Bang" gave birth to the Cosmos 15 billion years ago, hydrogen has fed the fusion reactions in stars. Scientists and engineers are working hard to harness the energy potential of hydrogen to propel trains, planes, and automobiles without harmful emissions. But hydrogen extraction has a dark side: it requires significant energy to free the element from terrestrial resources such as water, coal, and methane before it can be used in fuel cells to create electricity. When hydrogen is obtained by using power from fossil fuels, it only transfers air pollution from the transportation sector back to the power plant.

HYDROGEN IS THE MOST ABUNDANT ELEMENT known to man, and the supply is virtually limitless. Hydrogen can be made from fresh or salt water by electrolysis, which uses electricity to split the water molecule into its elemental components of hydrogen and oxygen. Because hydrogen energy has long been considered the ultimate universal fuel source, it is no wonder scientists have been working on hydrogen energy cells for more than 150 years. The technology hasn't slipped by NASA either, where such fuel cells have supplied power in all manned space missions since Project Gemini in 1965. Fuel cells provide astronauts with heat, electricity, and drinking water.

Hydrogen fuel cells are making big headlines in the new millennium, but this technology has actually been around for a long time. Ever since the British physicist and judge Sir William Grove constructed a prototype fuel cell in 1839, researchers have been improving the system. Grove demonstrated that hydrogen and oxygen could be compelled to bind together to form water while, at the same time, producing an electric current.[1] Later in the 19th century, a hydrogen-rich gas was extracted from coal and used in the gas lamps and heaters of European and American towns and cities. Known as "town gas," it consisted of 50% hydrogen and 50% carbon monoxide, and its distribution helped lay the foundation for the safe use of hydrogen as an energy source.

In 1932, Francis Bacon, a descendant of the famed English scientist and philosopher Sir Francis Bacon, developed the first modern fuel cell. The cell was later refined and upgraded until a five-kilowatt system was successfully demonstrated in 1952. A few years later, scientists and engineers involved in America's nascent space program were attracted to fuel cell technology because it was compact and lighter than any type of battery and infinitely less dangerous than any known nuclear application. NASA employs hydrogen fuel cells in all its modern efforts, such as the US Space Shuttle and Space Lab projects. Using hydrogen today for commercial energy production and transportation makes good economic, political, and environmental sense.

Although much of the recent excitement regarding fuel cells as a clean, unlimited, and decentralized energy source will play out in some future decade, it is no secret that the hydrogen industry is already well established in the United States. About 50 million pounds are used every day in production plants across the country. Hydrogen is used in various industrial applications such as welding and the refinement of heating oil and gasoline from crude. Hydrogen is a key component in the manufacture of chemicals, especially ammonia (fertilizer) and methanol. Even everyday items like soap, vitamins, margarine, cosmetics, peanut butter, and vegetable oil are processed using hydrogen.[2]

The real news is that hydrogen fuel cells are being designed that will replace gasoline and diesel to power your vehicle and your home, hospital, or community center. The advantages to the typical internal combustion engine are significant. Hydrogen fuel cells power vehicles cleanly and will

**Hydrogen H**
Atomic Number: 1
Atomic Weight: 1.00794
Hydrogen is a gaseous element discovered by Henry Cavendish in 1766. It is the first element on the Periodic Table.
Hydrogen is:
- Colorless
- Tasteless
- Odorless
- Slightly soluble in water
- Highly explosive

dramatically reduce the health hazards and the billions of dollars of medical costs associated with conventional automobile exhaust. Additionally, fuel cells and the hydrogen that energizes them can be produced in the United States, as opposed to importing politically risky Persian Gulf oil.

Hydrogen is the only common fuel that is not bound to carbon; it burns cleanly, producing only heat energy, water, and trace amounts of oxides of nitrogen. It is nontoxic and abundant. If you include ocean water, hydrogen can be considered inexhaustible since it returns to water and eventually to the ocean after use. Hydrogen is normally a gas that can be compressed and stored in cylinders. It can also be kept as a liquid, but the gas does not liquefy until it is chilled to 423.2 degrees below zero Fahrenheit (-253 Celsius). Most hydrogen today is derived from light hydrocarbons such as natural gas. It is distributed by pipeline, over-the-road trailers, rail, and barge.

Like electricity, hydrogen does not occur naturally in a usable form — it must be generated or produced by burning fossil fuels or other forms of energy. One way of obtaining hydrogen is by cracking a water molecule into the components hydrogen and oxygen. The hydrogen is then fed into a fuel cell, a battery-like device that generates DC current. It supplies electricity by combining hydrogen and oxygen electrochemically without combustion. Unlike a battery, however, a fuel cell does not run down or require lengthy recharging. It will produce electricity and heat as long as hydrogen and oxygen are supplied. The oxygen is typically derived from ambient air, but the hydrogen comes from a system called a reformer, which produces the gas by breaking down a fossil fuel. Reformers do release pollutants as they break down the hydrocarbons to release hydrogen. Fuel cells are basically electrochemical engines that produce electricity by harnessing the reaction of hydrogen and oxygen. The only byproducts of the cell itself are clean water and useful heat.[3]

All fuel cells have the same basic operating principle. An input fuel is catalytically reacted in the fuel cell to create an electric current. Fuel cells consist of an electrolyte material, which is sandwiched in between two thin electrodes, an anode and a cathode. A fuel containing hydrogen flows to the

> Fuel cells are not new devices, although the intense recent interest in them is.[A]
> — *The New York Times*

anode, where the hydrogen electrons are freed, leaving positively charged ions. To speed the reaction, a catalyst, such as platinum, is frequently used. The electrons travel through an external circuit while the ions diffuse through the electrolyte. At the cathode, the electrons combine with the hydrogen ions and oxygen to form water, a harmless by-product. Fuel cells and batteries are similar in that both rely on electrochemistry, but the cell must be fed a hydrogen-rich fuel, whereas in a conventional battery, the reactants are the materials used in the electrodes.[4]

In order to produce significant amounts of electricity, hydrogen fuel cells are often arranged in multiple sets or stacked. Fuel cells produce electricity with 40–60% efficiency, compared to the 15–18% efficiency obtained by the internal combustion engine. (One reason for the superior performance of fuel cells is that they do not have to idle when a vehicle is stationary.) Cell designs vary from small cells to power your sports car all the way to larger industrial-size cells, which may someday supply electricity to entire cities.[5] However, experts stress that, although hydrogen fuel cells have tremendous multiuse potential, they still have technological and cost hurdles to overcome before they become a practical alternative to gasoline engines.

> Industry sources estimate that sales of the smaller fuel cells for residences and small businesses could reach $50 billion a year in the US by 2030.[8]
> — Scientific American

The race to develop an affordable fuel cell system remains an expensive marathon for the auto industry, but dramatic progress has been made. In the late 1960s, each kilowatt of fuel cell energy for the space program cost up to half a million dollars to produce. That expensive price tag has been slashed by a thousandfold to about $500 kilowatt of power for an automobile today. Automakers are spending hundreds of millions of dollars on fuel cell research as government pressure for zero-emission vehicles increases. In 1990, the California Air Resources Board challenged the Detroit-based auto industry with legislation mandating that 2% of the automakers' fleets sold in the Golden State would have to be zero-emission vehicles by 1998, increasing to 10% by 2003. There are compelling reasons for California's aggressive approach to emission-free vehicles. Auto exhaust accounts for 90% of the state's carbon monoxide, 77% of its nitrous oxides, and 55% of its reactive organic gases. Ozone levels in

Southern California occasionally exceed federal air quality limits by 300%.

To prove that emission-free transportation was feasible, the Schatz Energy Research Center built the first "neighborhood electric vehicle" (NEV) at Humboldt State University. The US Department of Energy, the Southcoast Air Quality Management District, and others funded the $3.9 million transportation project. The energy-efficient, non-polluting NEV was designed as a small city car, suitable for short driving errands. Although the vehicle worked well in its designed role, both the public and industry were indifferent, and sales were negligible.

Automotive manufacturers began experimenting with hybrid fuel mixtures and vehicles powered by a combination of electric batteries and a conventional internal combustion engine. Hybrid vehicles drastically reduce exhaust pollution but are not emission-free. There are two types of hybrid electric car systems, series or parallel. In a series hybrid, a small gas or diesel engine generates power to drive an electric motor and recharge a small battery pack. The engine basically serves as a mobile charging station. Parallel hybrids have two power systems, gas and electric. Either can propel the vehicle, but the two are designed to combine for quick acceleration. Hybrids have modified transmissions, cutting down on energy lost during shifting, and regenerative braking, thereby absorbing energy lost in braking and returning it to the engine. American consumers had no interest in the early models, which were small and equipped with low-range lead-acid batteries. In year 2001, hybrid vehicles have improved in speed, range, fuel economy, and style to the point that the auto industry is fairly confident that they can be successfully niche-marketed.

In the mid-1990s, however, opinion polls that had initially indicated strong public support for electric vehicles began to reveal an undercurrent of concern about subsidies, battery technologies, and range. These fears were bolstered

*California is pressuring Detroit with legislation mandating that more zero emission vehicles be sold in the Golden State.*

by an extensive multimillion dollar misinformation campaign launched by the auto industry against the electric vehicle mandates. In 1996, the California Air Resources Board responded to the inflated concerns by dropping the 2% state requirement of vehicles sold for 1998 but retaining the 10% in 2003 mandate. In order to meet that target, the automotive industry is working feverishly on fuel-cell powered vehicles that are emission-free. These futuristic vehicles are expensive but will soon be coming to a dealer showroom near you.

Commercial fuel cell buses and automobiles are already on the road, with more on the way. In Reykjavik, Iceland, the city's commercial bus service is being converted to non-polluting electric motors driven by hydrogen fuel cells.[6] The cities of Chicago (USA) and Vancouver (Canada) are both examining using fuel-cell-powered buses as part of their transportation systems. DaimlerChrysler has developed the New Electric Car, or NECAR 4, a zero-emission vehicle. The NECAR 4 has an electric motor powered by hydrogen fuel cells. It can carry five adults, has a range of 280 miles, and reaches speeds of 90 mph. DaimlerChrysler intends to introduce the car to the US market by 2004.[7]

*Fuel cell buses transport passengers without polluting the air.*

Fuel cells are based on a family of technologies that use a variety of electrolytes and operate at different temperatures. Each family member tends to be best suited for certain applications. Fuel cell design varies according to the power demands of a given system and the operating temperatures that best suit a particular application. Higher operating temperatures allow the use of less pure hydrogen sources because fuel cells are capable of chemically extracting hydrogen from a variety of fuels such as methanol and other fossil fuels like oil or gasoline.

NASA uses alkaline fuel cells in its space shuttle missions. The electrolyte, or membrane material, is potassium hydroxide, and the catalyst for the electrode reactions is often platinum, which contributes significantly to a fuel cell's high cost. (Fuel cells that run at temperatures low enough for mobile use rely on a catalyst, generally platinum, to speed the reaction and increase power generation. It has been estimated that the world's annual production of platinum is only enough to build six million 50-kilowatt electrochemical engines a year, about 15% of current automobile production.) Alkaline cells reach power generating efficiencies of up to 70%, meaning that of the hydrogen consumed in the reaction within the fuel cell, only 30% of its potential heat energy is lost in the chemical reaction converting hydrogen into electricity.

Phosphoric acid fuel cells are used where stationary power plants are appropriate, such as in hospitals, nursing homes, office buildings, and utility power plants. Proton exchange membrane (PEM) cells are the smallest and lightest, making them the engineer's choice for transportation applications such as automobiles and trucks. All of the automobiles that are being readied for the consumer marketplace by major manufactures worldwide, including GM, Ford, DaimlerChrysler, Honda, and Nissan, are using stacks of PEM fuel cells for their power plants. PEM cells also have the lowest operating temperatures, around 200 degrees Fahrenheit, making them ideal for cars and trucks, as lower operating temperatures mean faster start-up times for the internal chemical reactions that power the fuel cells. At these lower temperatures, however, the PEM cell requires pure hydrogen to operate effectively.

Solid oxide fuel cells in Japan and Europe are supplying power for large installations such as electricity generation for utility companies and industrial production facilities. Finally, molten carbonate fuel cells are in competition with phosphoric acid cells for commercial applications, such as in hotels and airport terminals. Using a molten mixture of carbonate salts to screen the ions, this fuel cell can process less pure fuels, such as coal- and oil-based hydrocarbon fuel sources. Cost is still a major handicap in the United States.

*Fuel cells have taken astronauts to the moon.*

Electricity from fuel cells now costs $3,000 to $4,000 per kilowatt, as opposed to $500 to $1,000 per kilowatt produced by a typical utility gas-fired combustion turbine.

Other promising developments in fuel-cell technology are the direct methanol fuel cell, which extracts hydrogen from methanol within the cell, and regenerative fuel cells, where the fuel cell is a closed system, a perfect technology for space flight applications. Another major thrust in hydrogen technology is the use of the element as a directly combustible fuel. No major technical innovations are necessary for using hydrogen in today's existing fossil fuel internal combustion engines. Similar to the conversion that allows internal combustion engines to run on natural gas, the adaptation to hydrogen is minor and affordable, presenting many advantages over the use of nonrenewable hydrocarbon fossil fuels.

> It's not generally realized that the energy cost of going to the Moon is less than a hundred dollars in terms of kilowatt hours of electricity.[c]
> — Arthur C. Clarke

Ford Motor Company has built a modified version of a 2-liter four-cylinder gasoline engine that runs on hydrogen. It promises about 30% better fuel efficiency and produces no hydrocarbons, carbon monoxide, or carbon dioxide. Emissions of nitrogen oxide would meet proposed federal clean air standards. In 1999, Ford unveiled a $1.5 million hydrogen filling station at its North American research complex in Dearborn. The company believes that until engineers refine the fuel cell powertrain, the hydrogen-powered engine could be a good alternative to the conventional fossil-fuel-burning internal combustion engine.

Vehicles powered by an internal combustion engine fueled with hydrogen are not completely emission-free, but, compared to an engine running on hydrocarbon fuel or even cleaner natural gas, the contrast is striking. Fossil fuels produce carbon monoxide, carbon dioxide, particulate pollution, and hydrocarbon chemicals that interfere with the Earth's ozone layer. Engines burning hydrogen produce only water vapor and small amounts of nitrogen oxide. Russian engineers are trying to run jet aircraft engines on hydrogen whereas Japanese, German, and American companies are competing to build more efficient hydrogen-fueled

*Sign of the times.*

engines. Recent research suggests that adding hydrogen to natural gas and burning the combination fuel increases the efficiency of the engine while reducing exhaust emissions significantly.

But hydrogen is not the panacea that the media and futurists portray. According to the US Department of Energy Office of Power Technology, the most daunting problem associated with hydrogen production is that it takes twice as much energy to produce the hydrogen than it will return in power. This energy input/output imbalance has been a major stumbling block in the development of new hydrogen-based energy technologies. The economics of industry always focus more on short-term profit and loss than on environmental and social concerns.

It is clear that significant technological and economic barriers must be overcome before hydrogen can become the energy solution for our global future. Several alternative production methods are presently being explored in hopes of bringing down the financial and environmental cost of manufacturing hydrogen. The most promising utilizes the clean, renewable power of the Sun to produce and collect hydrogen from sea water, from recycled water, and even from the garbage that threatens to overflow landfills worldwide. The US Department of Defense and the Xerox Corporation recently joined forces to build a solar-powered photovoltaic (PV) hydrogen fuel generating facility. The demonstration unit used the Sun to extract enough hydrogen to service five vehicles, and cost was $1.4 million.[8]

A solar hydrogen system would alleviate air pollution and greenhouse gas emissions while allowing remaining hydrocarbon fuel supplies to be used for vital purposes other than energy use, such as the manufacture of plastics, synthetic fibers like nylon and polyester, and other durable goods. Fuel cells also are well suited to using hydrogen derived from trees and plants (biomass), from waste gases generated at landfills and water pollution control plants, as well as from sources such as wind and tidal wave energy.

The distribution and storage of hydrogen represent another challenge. Hydrogen can be stored safely as a compressed gas. Storage tanks have been reinforced with composite car-

> Things that five years ago seemed to be 20 years away may be only 5 or 10 years away today.[D]
> — David Cole
> University of Michigan

> A basement fuel cell working in tandem with solar cells on a house's roof could churn out all the necessary local energy, rain or shine.[E]
> — Discover

> If solar energy is used in the daytime to produce electricity which in turn is used to produce hydrogen, the hydrogen can be stored.[F]
> 
> —Walter Youngquist

bon fibers, making them ten times stronger than steel. Journalist Roy McAlister reports that composite fiber tanks can "...readily resist the impact of a 100-MPH collision, an attack with a .357 magnum pistol, or a bonfire test in which the tank's surface reaches 1,500 degrees Fahrenheit."[9] One drawback to the advantages of hydrogen is that, as a compressed gas, it requires six to ten times more storage space than gasoline. This requirement creates a problem for engineers designing fuel tanks for the hydrogen-powered automobile.[10] Hydrogen gas is usually shipped as a super-cooled liquid at temperatures approaching absolute zero. An equivalent energy content of gasoline, liquid hydrogen, and the required refrigeration system needed to keep the fuel cooled to such low temperatures requires six to eight times more storage space than gasoline.

Natural gas pipelines are capable of delivering hydrogen gas. In fact, the two gases can be transported together and separated at the point of use. Natural gas is methane, which has a greater density than hydrogen. It takes three times the volume of hydrogen to equal the energy in a given amount of natural gas. But at its lower density, hydrogen can be pumped through a pipeline at three times the flow rate of methane, balancing the energy equation between the two gases.

A significant obstacle to developing a true hydrogen economy is the energy needed to produce it and to provide for energy losses in the hydrogen-to-application chain. Hydrogen requires at least twice as much energy as electricity — twice the tonnage of coal, twice the number of nuclear plants, or twice the field of photovoltaic panels — to perform an equivalent unit of work. Economics dictate that most of today's hydrogen is produced from natural gas. It is an inefficient process, which discards 30% of the energy in one nonrenewable and polluting fossil fuel (natural gas) to obtain 70% of another (hydrogen).

> The problems of storing and refueling associated with hydrogen can be avoided by using hydrocarbon fuels like methane.[G]
> 
> — Chemical & Engineering News

Despite current technological challenges, the race to tap the pure power of hydrogen is very real. The US Dept. of Energy's Office of Power Technologies has launched a program that supports hydrogen produced by renewable energy power systems — including those generated by wind, solar, hydropower, and gasification of biomass. To achieve that mission, the program has a multipoint strategy to expand the use of renewable energy to produce hydrogen by working with fuel cell manufacturers and national laboratories, and to demonstrate safe and cost-effective fueling systems for hydrogen vehicles.[11]

Researchers are optimistic that hydrogen-fuel technology will play a big part in our future energy needs. Burning hydrogen contributes nothing to the greenhouse effect, ozone depletion, or acid rain. A sustained solar-hydrogen economy will go a long way towards solving some of the world's energy and environmental problems. To achieve commercial success with future consumers, however, a fuel cell vehicle must have clear economic advantages over the internal combustion engine — just having negligible emissions won't be sufficient. No matter how virtuous the clean-burning hydrogen economy may seem to a world choking on polluted air and water, it will involve very large investments in new technology and equipment. No single energy source can slake the thirst of a world consuming about 70 million barrels of oil a day; but considering the hidden costs to our health and environment from fossil fuel use, the conversion to renewable hydrogen will ultimately seem like a real bargain.

> ...customers may find themselves living in a home whose electricity comes not from a generating plant, tens, hundreds, or even thousands of kilometers away, but rather from a refrigerator-size power station right in their own basements or backyards.[H]
> — *Scientific American*

## FOOTNOTES

[1] <www.ch.ic.ac.uk/kucernak/Research/fuelcells.html> Biography on Sir William Grove from fuel cell page by Dr. Anthony Kucernak, Imperial College of Science, Technology and Medicine, London, England.

[2] <www.ttcorp.com/nha/hhs_2.htm> National Hydrogen Association: a nonprofit membership association whose primary objective is to organize and support the industrial development and use of hydrogen as an energy carrier and to accomplish the US transition to a hydrogen economy. The NHA represents the hydrogen industry and facilitates cooperation between industry and government.

[3] <www.fuelcells.org> Fuel Cells 2000 website is maintained by the Breakthrough Technologies Institute, a nonprofit educational organization formed to promote the development and commercialization of hydrogen fuel cells. The site contains a description of how a fuel cell works, FAQ's about fuel cells, images of different types of fuel cells, press releases, and technology updates.

[4] <www.dodfuelcell.com> US Department of Defense (DoD) Fuel Cell Demonstration Program. The purposes of the program are to stimulate growth and economies of scale in the fuel cell industry and to determine the role of fuel cells in the DoD's long-term energy strategy.

[5] <www.anuvu.com/NewFiles/Carbon-X%20Features.html> Anuvu (A New View) is a California corporation established in 1994. Its mission is research and development in the field of advanced consumer technologies. The primary focus of Anuvu, Inc. is to provide production-ready designs to solve critical issues for a sustainable future and to provide real answers to the environmental problems of today and the future. Anuvu, Inc. is designing fuel cell systems to be used in cars, trucks, buses, trains, and boats to provide transportation systems that will have superior performance and clean air technology.

[6] <www.hfcletter.com/letter/march99/feature.html> *Hydrogen and Fuel Cell Letter.* Edited by Peter Hoffman, a former international news correspondent, *Hydrogen & Fuel Cell Letter* has reported on developments worldwide involving hydrogen as an energy source since 1986.

[7] Jim Motovalli, "Your next car? A new generation of clean cars is coming to a showroom near you," *Sierra* (July/August 1999), p. 40.

[8] <www.energy.ca.gov/afvs/ev/hybrids.html> California Energy Commission page on hybrid- and hydrogen-powered automobiles. The CEC defines a hybrid vehicle as a vehicle that runs on something other than battery power. This report outlines vehicles in the hybrid-, hydrogen-, fuel cell-, and solar-powered categories and specifically cites the Xerox project.

[9] Roy McAlister, "The Energy Reformation: Part One – Solar Hydrogen: Power the New Millennium," *Natural Science* (January 1999), pp. 168, 169.

[10] <www.tgifdirectory.com/clean/cleanair/hydrogen.html> Document from the hydrogen section of the Clean Air Transportation website, sponsored by the TGIF Travel Group of California. It lists examples of projects undertaken in the area of hydrogen fuels, outlines the benefits and the drawbacks of hydrogen fuel technology.

[11] <www.eren.doe.gov/power/hydrogen.html> The mission of the Hydrogen Research Program is to "enhance and support the development of cost-competitive hydrogen technologies and systems that will reduce the environmental impacts of energy use and enable the penetration of renewable energy into the US energy mix." This program is part of the US D.O.E.'s Office of Power Technologies: Clean Power for the 21st Century.

## SIDEBAR FOOTNOTES

[A] Ann Eisenberg, "Fuel Cell May Be the Future Battery," *The New York Times* (October 21, 1999), p. G-13.

[B] Alan C. Lloyd, "The Power Plant In Your Basement," *Scientific American* (July 1999), p. 82.

[C] Arthur C. Clarke, "2001: The Coming Age of Hydrogen Power," *Infinite Energy* (October/November 1998), p. 16.

[D] David Cole, Director of the University of Michigan's Office for the Study of Automotive Transportation. Quoted by Matt Nauman, "Green cars are hot at auto show," *The Sacramento Bee* (January 14, 2000), p. C-12.

[E] Katherine Brown, "Invisible Energy," *Discover* (October 1999), p. 36.

[F] Walter Youngquist, *GeoDestinies: The inevitable control of Earth's resources over nations and individuals* (Portland, Oregon: National Book Company, 1997), p. 258.

[G] Mitch Jacoby, "New fuel cells run directly on methane," *Chemical & Engineering News* (August 16, 1999), p. 7.

[H] Alan C. Lloyd, "The Power Plant In Your Basement," *Scientific American* (July 1999), p. 80.

# Hydropower

**9** Hydroelectric energy produces significant power generation for many countries around the world. But this renewable energy resource often exacts a heavy environmental cost. Huge dams built for flood control, upstream water storage, recreational use, and hydroelectric power plants alter or destroy important and complex ecological systems.

WATER HAS ALWAYS BEEN civilization's most precious natural resource; it is critical for the survival of every life form on the planet. Irrigated water enables gardens to flourish in the desert, and, when harnessed for hydroelectric energy, water can power the lighting for entire cities. Thousands of years ago, the Greeks devised water wheels, which picked up water in buckets around a wheel. The water's weight caused the wheel to turn, converting kinetic energy into mechanical energy for grinding grain and pumping water.[1] In the 1800s, the water wheel was often used to power machines such as timber-cutting saws in European and American factories. Engineers eventually realized that the force of water falling from a height would turn a turbine connected to a generator and produce electricity. Niagara Falls, a natural waterfall in New York State, powered the first hydroelectric plant completed in 1879.

*Hydroelectric power is a leading source of renewable energy, but the environmental impacts are costly.*

When archaeologists from some other planet sift through the bleached bones of our civilization, they may well conclude that our temples were dams.[A]

— *Cadillac Desert*

Tens of thousands of spillways and dams were constructed in the US during the 20th century in order to maximize this important renewable source of energy. Aside from a plant for electricity production, a hydropower facility consists of a water reservoir enclosed by a dam with gates that can open or close depending on how much water is needed to produce a particular amount of electricity. Once electricity is produced, it is transported via transmission lines to an electric utility company.[2] Hydroelectric power can provide vital electricity during peak demand, such as on hot summer afternoons when it is needed to power air conditioners.

Over the past 100 years, the United States has led the world in building dams. Secretary of the Interior Bruce Babbitt recently observed that "On average, we have constructed one dam every day since the signing of the Declaration of Independence."[3] Of the more than 75,000 dams in the US, less than 3% of them are used to produce 10–12% of the nation's electricity. With more than 2,000 facilities, the US is the world's second largest producer of hydropower, behind Canada. The dams that do not produce electricity are used for irrigation or flood control. Many people believe these pre-existing sites could contribute to the country's power supply in a cost-effective manner if hydroelectric facilities were constructed.

There are several favorable features of hydropower. Anywhere sufficient rain falls, there will be rivers. If a particular section of river has the right terrain to form a reservoir, it may be suitable for dam construction. Once the system has been installed, no fossil fuels are required to produce the electricity, and the earth's hydrologic cycle naturally replenishes the "fuel" supply. Therefore, no particulates or greenhouse gases are released into the atmosphere, and no waste that requires special containment is produced. Although water is a naturally recurring domestic product and is not subject to the whims of foreign suppliers, it is susceptible to changing weather patterns and regional drought.[4]

Photo by Thomas McLaughlin

*The US is one of the world's largest producers of hydropower, second only to Canada.*

Hydropower is very convenient because it can respond quickly to fluctuations in demand. A dam's gates can be opened or closed on command, depending on daily use or gradual economic growth in the community. The production of hydroelectricity is often slowed in the nighttime when people use less energy. When a facility is functioning, no water is wasted or released in a contaminated state; it simply returns unchanged to continue the hydrologic cycle. The reservoir of water resulting from dam construction, which is essentially stored energy, can support fisheries and preserves, and provide various forms of water-based recreation for locals and tourists. Land owned by the hydroelectric company is often open to the public for hiking, hunting, and skiing. Therefore, hydropower reservoirs can contribute to local economies. A study of one medium-sized hydropower project in Wisconsin showed that the recreational value to residents and visitors exceeded $6.5 million annually, not to mention the economic stimulation provided by employment.

Hydroelectric power is also very efficient and relatively inexpensive. Modern hydroelectric turbines can convert as much as 90% of the available energy into electricity. The best fossil fuel plants are only about 50% efficient. In the US, hydropower is produced for an average of 0.7 cents per kilowatt-hour (kWh). If the costs for removing the dam and the silt it traps are excluded, hydro-generated electricity costs about 30% the price of using coal, oil, or nuclear space power at 20% the cost of using natural gas. Refurbishing dated hydroelectric equipment could further increase efficiency. An improvement of only 1% in the US system would supply electricity to an additional 300,000 households.

Hydropower has become the leading source of renewable energy. It provides more than 97% of all electricity generated by renewable sources worldwide. Other sources, which include solar, geothermal, wind, and biomass gasification, account for less than 3% of renewable electricity production. In the US, 81% of the electricity produced by renewable sources comes from hydropower. Worldwide, about 20% of all electricity is generated by hydropower. Some regions depend on it more than others. For example, 75% of the elec-

> First-world nations from Australia to Norway depend on governments for storage and delivery systems rationalized on the basis of economic development with almost no consideration given to fiscal or environmental impacts.[B]
> — *Water Markets: Priming The Invisible Pump*

tricity produced in New Zealand and over 99% of the electricity produced in Norway come from hydropower.

The use of hydropower in the US avoids burning 22 billion gallons of oil or 120 million tons of coal each year. The carbon emissions prevented by the nation's hydroelectric industry are the equivalent of an additional 67 million passenger cars on the road. The advantages of hydropower are quite compelling, but there are some serious drawbacks that should cause people to reconsider its overall benefit.

Since the most feasible sites for dams are in hilly or mountainous areas, the seismic faults that created the rugged topography can pose a serious danger to the dams and, therefore, to the occupied land below them for thousands of years, long after they have become useless for generating power. In fact, dam failures do occur regularly due to these terrain conditions, and the effects can be devastating.[5]

Dammed water floods the countryside behind the project, forcing people who live in the affected area to move and relinquish their former lifestyles. This is very stressful and often bitter and controversial, especially if a community has maintained a particular way of life on the same land for generations. Such is the case in Chile, where the indigenous Pehuenche are currently fighting construction of the 570MW, $500,000,000 Ralco Dam on the Biobío River. Eight families continue to refuse to negotiate land exchanges with the utility company, and wish to remain on their lands.[6] If the project succeeds, a 13-square-mile reservoir would inundate the land and force 600 people out of their homes, 400 of whom are Pehuenche whose ancestral home is the upper Biobío. A total of five dams have been planned, which will force the relocation of 1,000 Pehuenches, 20% of the survivors of this ancient culture.

The Chinese are building a mammoth hydroelectric and flood-control dam on the Yangtze River. The Three Gorges Dam project will help control the region's devastating floods and will generate up to 8% of China's current electricity demands, the equivalent of 18 nuclear power plants. The 15-year construction effort began in 1994 and will last until 2009. Although the massive dam will alleviate downstream

> ...the Sriram Sager Dam in Andhra Pradesh [India], completed in 1970, lost a third of its capacity in two years because of silt building up on the reservoir bottom.[c]
>
> — Donald Worster - Environmental Historian

flooding and produce up to 84.6 terawatts of electricity per year, the social and environmental cost will be high.

More than 1.2 million people will be forced to move from the region, and rising waters behind the dam will drown many sites of archaeological importance. The Yangtze River is one of the world's most heavily silted rivers, and Chinese engineers are unsure how quickly sedimentary deposits will accumulate within the dam's reservoir. Geologists indicate that the dam site is located in a seismically active area, and environmentalists are concerned that air and water pollution will increase with projected industrial growth in the region. The Three Gorges Dam represents China's transition from an agrarian-based economy to one of industrialization.

*Salmon depend on steady water flows to flush them down river when young and then guide them upstream years later to spawn.*

The construction of a dam not only affects the people nearby, it can severely alter a river's natural functions. By diverting water for power, dams remove water needed for healthy in-stream ecosystems. Stretches below dams often dry up, and birds that have migrated to a specific riparian environment for thousands of years no longer have enough insects on which to prey when the water level drops. If they have few migration alternatives, that could mean the endangerment of species that once flourished. Fish, such as salmon, depend on steady flows to flush them down river early in their life and guide them upstream years later to spawn. Stagnant reservoir pools disorient migrating fish and significantly increase the duration of their migration. Native populations of fish may decrease or disappear altogether due to temperature changes caused by dams. Slower water flow means warmer temperatures, and bottom-release of cold water means cooler temperatures. Several of hydropower's disadvantages focus on fish. It is easy to forget the importance of fish and other aquatic life, some of which reside at the bottom of the food chain.

The environmental changes caused by hydroelectric projects may be obvious to the local biologist but often elude

In the case of Butte Creek, removing one antiquated 12-foot hydroelectric dam and modifying a handful of small but obstructive waterfalls could more than double the river's spring-run spawning habitat.[D]

— *The Sacramento Bee*

the average person. Most people will more readily notice a smoggy haze developing in an area where a coal plant is operating than a smaller population of a particular bird species where a hydropower facility functions. Such oversights lead people to erroneously believe that hydropower has little or no negative impact on the environment.

Utility companies often emphasize that hydro-generated electricity is "clean and green" but neglect to mention the long-term environmental hazards. Dams hold back silt, debris, and vital nutrients. Silt collects behind the dam on the river bottom, accumulating toxic heavy metals and other pollutants. Eventually, this silt will render the dam inoperable, leaving an expensive and dangerous mess for future generations, who will have to remove the collected debris or live with a potentially catastrophic mudflow poised to inundate the area below the dam.

There is also a debate between preserving rivers for their aesthetic value versus meeting the energy needs of thousands of people. In 2001, there are 600,000 river miles impounded behind dams in the US In contrast, only 10,000 river miles (less than half of 1%) are permanently protected under the National Wild and Scenic Rivers System. The only free and undammed river in the US that is longer than 600 miles is the Yellowstone.

Hydropower may be better on the environment than fossil-fuel sources, but its future is so uncertain that we need to focus on other alternatives. An increasing array of statutes, regulations, agency policies, and court decisions have made the hydroelectric licensing process costly, arbitrary, and time-consuming. A typical hydropower project takes 8 to 10 years to push its way through the licensing process. By comparison, a natural-gas fired plant, which emits significant carbon dioxide ($CO_2$) gases, can typically be sited and licensed in 18 months. Given this uncertain licensing climate, few investors are willing to risk capital on new hydropower development.

Relicensing is a complex process in which private dams are reevaluated every 30 to 50 years. The Federal Energy Regulatory Committee (FERC) considers committing public river resources for private power generation. When deciding whether or not to issue a license, FERC is now required

> Hydropower — and raw political power — won out over tradition and treaty-promised fishing rights.[E]
> — *Searching out the Headwaters*

to consider not only the power generation potential of a river but also to give equal weight to energy conservation, protection of fish and wildlife, protection of recreational opportunities, and preservation of other aspects of environmental quality. Relicensing was infrequent until 1993 when hundreds of licenses began to expire. The Hydropower Reform Coalition formed in 1992 to take advantage of this once-in-a-lifetime opportunity to restore river ecosystems through the relicensing process. To the Coalition's dismay, a new bill is being considered called the Hydroelectric Licensing Process Improvement Act, which, if passed, would limit the ability of federal agencies to protect natural resources, making relicensing much easier for dam operators.

Some people favor dam removal so that healthy rivers and riverside communities can be restored, but most of the larger dams in the US are not likely candidates for removal. If a dam is not removed, it is wasteful not to use them to their full potential as long as they are still sturdy. A hydropower assessment conducted by the US Department of Energy found that 4,087 sites could be developed without constructing a new dam. The assessment considered such values as wild/scenic protection, threatened or endangered species, cultural values, and other nonpower issues. If all of this hydroelectric potential were developed, 22.7 million metric tons of carbon could be avoided, but this savings in carbon emissions pales when compared to the tonnage of silt and other material that must be handled if the river is to be restored to a free-flowing state. All dams will eventually be silted up their rivers. Future generations will have the choice to keep the obsolete and dangerous dam or pay to remove it. Keeping the poorly consolidated silt and mud behind the dam is potentially hazardous and removal costs may actually exceed the value of energy produced over the dam's lifetime.

*Hydropower remains a highly controversial issue.*

> "Water flows uphill to money."
> — *A Sense of the American West*

Unlike other renewables such as wind and solar power which receive more praise than criticism, hydropower remains a highly controversial issue. While hydroelectric power generation does have much merit, it is not a panacea. Despite the pros and cons, hydroelectricity is just one part of our contemporary arsenal of renewable energies. It will be up to future generations to decide if the damage dams cause to our natural resources is worth the price of "cheap" electricity.

## FOOTNOTES

1. &lt;lsa.colorado.edu/essence/texts/hydropower.htm&gt; Brief but informative webpage describing what hydropower is, its history, how modern generating plants work, and some of the pros and cons of its use.

2. &lt;www.hydro.org/hydrofacts.htm&gt; Hydro Facts is a list of five links to quick facts about the benefits of hydropower, including FAQ's and 8 hydropower myths. The National Hydropower Association is a "trade association dedicated exclusively to representing the interests of the hydropower industry." Its latest sources are from 1998.

3. &lt;www.americanrivers.org/issues/default.htm&gt; This section of the American Rivers site is about dams in general, hydroelectric dams, and dam removal. Included are "Ten Reasons Why Dams Damage Rivers," information about the relicensing of hydropower dams, and an article entitled "Removing Dams that Don't Make Sense." American Rivers was founded in 1973 in a protective effort to give more rivers National Wild and Scenic status. It is a national conservation organization dedicated to protecting and restoring America's river systems and to fostering a river stewardship ethic. The information for this article is from 1997 to 1999.

4. &lt;www.wvic.com/hydro-facts.htm&gt; Facts About Hydropower lists statistics and related graphs that include the top producers of hydropower, its cost, emissions avoided, its share of electricity production compared to other renewables, recreation opportunities, etc.

5. &lt;www.ecy.wa.gov/programs/wr/dams/failure.html&gt; A Washington State Dam Safety program analysis of the 857 dams in the state entitled "Status of Dams in Washington State and Notable Dam Failures." Graphs show data from safety inspections of the dams whose failure would pose a threat to populations. Several dam failures are listed, with photographs and statistics of the damage. The top two problems responsible for dam failures in the US are overtopping and defects in the foundation. Information is undated but not time sensitive.

6. &lt;www.irn.org/programs/biobio/action990818.html&gt; A plea for people to show their opposition to the Ralco Dam project on the Biobío River in Chile, this document states that the Pehuenche people along the river will lose their ancestral home if a dam is constructed. It urges readers to write a letter to the utility company, furnishing a sample letter and listing the effects of the proposed project. It was posted August 18, 1999.

## SIDEBAR FOOTNOTES

A  Marc Reisner, *Cadillac Desert* (New York: Penguin Books, 1993-revised ed.), p. 104.

B  Terry L. Anderson and Pamela Snyder, *Water Markets: Priming The Invisible Pump* (Washington, D.C.: Cato Institute, 1997), p. 204.

C  Donald Worster, *An Unsettled Country: Changing Landscapes of the American West* (Albuquerque: University of New Mexico Press, 1994), p. 50.

D  Marc Reisner, "State's policy of water consensus is a dry well," *The Sacramento Bee*, April 11, 1999, p. H-2.

E  Sarah F. Bates, *et al.*, *Searching Out the Headwaters* (Washington, D.C.: Island Press, 1993), p. 98.

F  James E. Sherow, *A Sense of the American West* (Albuquerque: University of New Mexico Press, 1998), p. 178.

# Geothermal Power

## 10

Planet Earth formed in a fiery consolidation of cosmic dust and gas 4.6 billion years ago, but it is still seething with primordial heat. The use of geothermal energy for electricity has grown to about 7,000 megawatts in 21 countries around the world. In the United States, the Department of Energy has set ambitious goals that call for geothermal energy to provide for up to 10% of the West's electricity by 2020. Renewable geothermal energy is clean and reliable and has great potential in many parts of the world.

ENERGY FROM THE EARTH ITSELF will play an important part in the renewable energy equation of the 21$^{st}$ century. Ever since the world's first geothermal-generated electricity was produced at Larderello, Italy, in 1904, humans have tapped this primordial power source. Geothermal energy is derived from the heat contained within the planet, heat that in some places is so intense that it melts mantle rock to create molten magma. Experts believe that the ultimate source of geothermal energy is radioactive decay occurring deep within the Earth. Geothermal heat is a renewable energy source primarily produced when ground water descending from the Earth's surface meets

*Geysers are spectacular evidence of geothermal activity.*

molten magma rising toward it. Some of this geothermal water circulates back up through faults and cracks and reaches the Earth's surface as hot springs or geysers, but most of it stays deep underground, trapped in cracks and porous rock. In most regions of the world, this heat reaches the surface in a very diffuse state; however, in some areas, including substantial portions of the western United States, geothermal reservoirs exist close to the surface and are easily tapped for power generation.

Archeological evidence suggests that the first use of geothermal energy in North America occurred more than 10,000 years ago with the settlement of Paleo-Indians around hot springs. The mineral springs served as a source of warmth, cleansing, and healing. The value and importance of hot springs ran so deep in American Indian society that they were considered neutral zones, where members of warring nations could bathe together in peace.[1] The Romans used geothermal water to treat eye and skin disease and to heat buildings at Pompeii.

*Humans have enjoyed the comfort of natural hot springs for more than 10,000 years.*

Today, Americans benefit in a much different way from this important natural resource. Tapping the heat emanating from beneath the Earth's crust can generate electricity without harmful fossil fuel emissions. In geothermal power plants, steam, heat, or hot water provides the physical force that spins turbine blades to generate electricity. Engineers have developed several methods for converting geothermal energy into electricity, primarily dry steam, hot water, and binary systems. "Dry steam" reservoirs produce steam but little water. The steam is piped to where it can spin turbine generators that produce electricity. Hot water reservoirs form where magma flowing relatively close to the surface directly heats groundwater. Naturally pressurized, hot water flows to the surface via the production well, where a separator flashes the water into steam and turns turbines. When geothermal-heated water is not hot enough to flash into steam, it can still produce electricity in a "binary" power plant. In a binary system, the geothermal water is passed through a heat exchanger, where its heat is transferred into a second liquid, which boils at a lower temperature than water. When heated, the binary liquid flashes to vapor that expands across the turbine blades. The vapor is then recondensed into a liquid and reused repeatedly. In this closed-loop cycle, there are no emissions released into the air.[2]

Another implementation is to use hot water directly. Direct use involves utilizing heated water (without a heat pump or power plant) for industrial processes, to heat buildings and greenhouses, to farm fish, or to supply heated mineral water for resorts. In 1999, the US capacity of direct-use systems totaled more than 470 MW, or enough energy to heat 40,000 average-sized houses.

Sometimes, magma pockets far below the surface heat subterranean rock, which, in turn, warms deeply circulating groundwater.[3] In order to maximize the energy gleaned from these so-called "hot dry rocks," geothermal facilities can fracture the hot rocks and pump more water through them for greater power generation.[4]

The concentration of geothermal energy must be very high in order to make heat extraction economical, and not all geothermal sites are equal in their power potential. Geographical regions that possess well-developed geothermal systems are located in geologically active areas. These favored regions have continuous, concentrated heat flow to the surface. In the U. S., the western states have the best geothermal potential. Iceland, New Zealand, Japan, the Philippines, and South America are also prominent global hot spots. Since the 1960s, France has been heating up to 200,000 homes using geothermal water. In Iceland, geothermal energy caused by the constant movement of geologic plates coupled with the volcanic nature of the island, is used to heat virtually all homes.

Indonesia, with numerous volcanoes, possesses some of the most geologically active territory in the world. This densely

> Geothermal power is one resource that calls out for greater utilization.[A]
> — US Energy Secretary Bill Richardson

*In Iceland, geothermal energy heats 95% of all homes.*

populated, energy-hungry island-nation has a wealth of geothermal power potential. In 1997, Texaco and Chevron's joint Indonesia representative drilled an exploratory geothermal well in the oil and gas producing Darjat area. The well had the highest dry steam rate in the Eastern Hemisphere, enough to power a 20-megawatt electricity generator.[5]

The world's largest geothermal energy facility, The Geysers, is located in northern California. Its 14 geothermal power plants boast a combined capacity of approximately 700 megawatts of electricity. In some ways, The Geysers is a working model on how *not* to approach a "renewable" geothermal resource. It was initially built as a health resort in the 1850s on an active steam field, but over the years, more extensive use of the thermal activity was established.[6] Electricity production at The Geysers began in 1960, but, by the late 1980s, power generation had diminished because steam was being extracted faster than it could be naturally replaced. Because of declining steam supplies, the geothermal units at The Geysers Power Plant are now operating at reduced capacities. Plant operators and steam suppliers are seeking new strategies to maximize future power generation.

*Tapping the heat from beneath the Earth's crust can generate electricity without harmful fossil fuel emissions.*

Engineers know that to prevent rapid depletion, geothermal facilities should operate on a closed-loop system. This procedure, which re-injects water back into the system for constant steam generation, has now been implemented at The Geysers. In 1996, The Geysers' power plants were producing enough electricity to supply a population the size of San Francisco, but despite increased efficiency and conservation, it is projected that the massive steam field will be defunct in about 50 years.[7]

Another commercial geothermal-energy-extraction technique utilizes heat pumps. Ground-source heat pumps use the Earth or groundwater as a heat source in winter and a heat sink in summer. Heat pumps move warmth from one place to another, transferring heat from the soil to the house in winter and from the dwelling into the ground in summer.

Similar to the temperature range in a cave, the temperature within the ground maintains a constant average in contrast to the constantly changing air above. The geothermal heat pump is one of the most efficient and nonpolluting home cooling systems available. This method reduces reliance on the electrical grid, resulting in significant environmental benefits and reduced energy costs. It is estimated that between 10,000 to 40,000 heat pump systems are installed every year.

Geothermal energy is a reliable, decentralized power source for some regions, but, like all renewable energy sources, inexpensive oil supplies undermine its potential. Despite the fact that this energy source is clean and renewable and can reduce our dependence on imported fuels, the fact remains that fields of sufficient quality to produce dependable electricity economically are relatively rare in the US. Environmental concerns also cloud the implementation of geothermal facilities. Many of the most highly active areas are located in protected wilderness zones that environmentalists want to preserve.[8] And although no combustion occurs, some systems produce carbon dioxide and hydrogen sulfide emissions, require the cooling of as much as 100,000 gallons of water per megawatt per day, and must dispose of toxic waste and dissolved solids.

Geothermal energy alone won't solve the energy problem, but it does help reduce reliance on fossil fuels. In 1999, geothermal-generated electricity saved the US 60 million barrels of oil. Considering the health and environmental costs from burning that much oil, this natural hotbed of energy should not be overlooked. US geothermal electric power generation was approximately 2200 MW in 1999, or about the same as four large nuclear power plants but without the radioactive waste.[9] Geothermal energy currently ranks third among renewables, following hydroelectricity and biomass gasification, and ahead of rapidly expanding solar and wind power. The heat of the Earth contributes to our arsenal of clean and renewable energy sources, but it still can't come close to replacing our reliance on petroleum.

> US geothermal power plants currently have a total electricity generation capacity of 2,700 megawatts, enough electricity to meet the residential needs of more than 3.5 million people.[B]
>
> — *Mother Earth News*

## FOOTNOTES

[1] <www.eren.doe.gov/geothermal/geohistory.html> "A History of Geothermal Energy in the United States," published by the Geothermal Energy Program under the Office of Power Technologies (OPT). The OPT is part of the Office of Energy Efficiency and Renewable Energy at the US Department of Energy. Their mission is to develop clean, competitive power technologies for the 21$^{st}$ century, including renewable energy (solar, wind, geothermal, and biomass), energy storage, hydrogen, and superconductors.

[2] <http://geothermal.marin.org/pwrheat.html> Excellent geothermal energy tutorial by the Geothermal Education Office (GEO). The purpose of the GEO is to ascertain that citizens of both today and tomorrow understand what geothermal energy is, what it can do, and its growing place in providing the clean energy necessary to protect our environment while providing needed power. The GEO office assists schools, industry, and energy and environmental educators. GEO is partially funded by the Office of Geothermal Technologies of the US Department of Energy.

[3] <http://solstice.crest.org/renewables/re-kiosk/geothermal/theory/index.shtml> The Center for Renewable Energy and Sustainable Technology introduction to geothermal power. A brief explanation of the following topics: Origin of Heat in the Earth, How Geothermal Systems Form, Geothermal Occurrences Today, Potential of Geothermal Energy, Geothermal Use Today, Resource Base of Geothermal Sites, Reliability of Geothermal Systems, Plant Construction, and Domestic Resources.

[4] <www.eia.doe.gov/cneaf/solar.renewables/renewable.energy.annual/chap04.html U. S. Energy Information Agency profile of the geothermal industry written by Jim Disbrow, Operations Research Analyst for the Energy Information Administration. This profile outlines the activity of the geothermal power industry through 1996, assesses the availability of resources in the United States, and details the corporate changes taking place within the industry. A table shows the temperatures of geothermal fluids required for various uses.

[5] <www.eia.doe.gov/emeu/cabs/indoe.html> Energy evaluation of Indonesia and Environmental Issues. Published in December 1999 by the United States Energy Information Administration (D.O.E.), this analysis covers the various energy sources and environmental problems associated with Indonesia.

[6] <www.unocal.com/geopower/evolution/index.htm> "A Continuing Evolution" written by Paul Atkinson of Unocal about The Geysers. He includes a historical description, the current status of the facility, new developments, and other information.

[7] <http://vulcan.wr.usgs.gov/LivingWith/PlusSide/geothermal.html> "The Plus Side of Volcanoes — Geothermal Energy." Subsection of "Living With Volcanoes" by the US Geological Survey at the Cascades Volcano Observatory in Washington.

8   <www.public-policy.org/~ncpa/studies/renew/renew6.html> National Center for Policy Analysis study by Robert L. Bradley, Jr., President of the Institute for Energy Research in Houston. Author of the two-volume *Oil, Gas, and Government: The US Experience*, Bradley is an adjunct scholar at the Cato Institute. This report details the economic problems associated with geothermal energy, examines the site-specific nature of geothermal resources, and outlines the difficulties of current geothermal power plants in the United States.

9   <www.oit.osshe.edu/~geoheat/whatgeo.htm> Geo-Heat Center, Oregon Institute of Technology. Research at the Center is supported by the US Department of Energy's Geothermal Division under a grant by the Assistant Secretary for Conservation and Renewable Energy. The Center offers the transfer of technological information to consultants, developers, potential users, and the general public as an important element in the development of direct heat utilization of geothermal energy. The Geo-Heat Center's resources are available to the public through the US Department of Energy.

SIDEBAR FOOTNOTES

A   *Los Angeles Times*, "Energy chief wants to spur geothermal power in West," article published in *The Sacramento Bee*, January 24, 2000, p. A-4.

B   "The State of US Renewable Power," *Mother Earth News* (February 1999), p. 16.

# Biomass Energy

## 11

Plants and trees are natural storehouses of solar energy. The solar energy that drives photosynthesis is retained within the chemical bonds of the structural components of biomass. Special processing of biomass materials, such as corn and sugar cane, can produce liquid fuel like ethanol. Burning organic matter dissolved in water releases a clean combustible gas, which can generate electricity for homes or even run an internal combustion engine.

THE TERM "BIOMASS" REFERS TO ORGANIC MATTER that has stored solar energy through the process of photosynthesis. It exists in one form as plants and may be transferred through the food chain to animals' bodies and their wastes, all of which can be converted for energy use through simple processes, such as combustion and decomposition, which release the carbon dioxide stored in the plant matter.[1] Burning fossil fuels uses "old" biomass and converts it into "new" carbon dioxide that contributes to the greenhouse effect and depletes a nonrenewable resource. Burning "new" biomass contributes no new carbon dioxide to the atmosphere because if we replant harvested biomass, carbon dioxide is returned to the cycle of new growth.

*Biomass is a renewable resource that consists of common plants and trees.*

Most of the biomass fuels in use today come from wood products, dried vegetation, crop residues, and aquatic plants. Biomass has become one of the most commonly used renewable sources of energy in the last two decades, second only to hydropower in the generation of electricity.[2] Due to its low cost and renewable nature, biomass now accounts for almost 15% of the world's total energy supply and as much as 35% in developing countries, where it is mostly used for cooking and heating.[3]

In 1997, energy generated from municipal solid waste (MSW) facilities in the United States reached 449 trillion Btu. That year, there were 107 waste-to-energy (WTE) facilities in the US, with the capacity to process over 100,000 tons per day. (More than 80% of those operations generated electricity as an energy product.) Ironically, the growth in new MSW projects has slowed dramatically in the US during the 1990s. Analysts point out that recycling has reduced the quality and quantity of waste streams to WTE facilities. Other factors include costly pollution controls, a federal tax policy that no longer favors investments in capital-intensive products, and the competitive pressures of cheap oil and a deregulated market where electricity prices are dropping.[4]

There are real environmental benefits to cofiring biomass in existing electric utility facilities. Adding wood to a coal fuel-stream reduces the amount of coal burned to generate electricity. Typical biomass fuel feedstocks contain almost no sulfur ($SO_2$), have about 50% of the nitrogen content of coal ($NO_x$), and reduce greenhouse gases. Despite the reduced emissions achieved by blending biomass with fossil fuels, there has been only one US project developed to grow energy crops for electricity production. In 1992, the Comprehensive National Energy Act created a tax credit for electricity produced and sold from wind and "closed-loop" biomass. (Closed-loop biomass is defined as any organic matter derived from a plant that is planted for the exclusive purpose of being used to produce electricity.) According to the Internal Revenue Service, though the tax credit has been used extensively for wind power projects, no biomass project has yet claimed this tax credit.[5]

> Approximately 16% of all municipal solid waste tonnage generated in the US is disposed of by direct combustion.[A]
> — Government Advisory Associates, Inc.

> In Western Germany, about a quarter of garbage landfills produce power from biomass energy. Switzerland reuses 80% of its waste, and France about 50%.[B]
> — Madanjeet Singh (UNESCO)

Despite the moribund state of biomass development in the industrialized US, the World Bank is counting on a pilot demonstration program of a biomass gasification facility to generate megawatts of electricity in northeastern Brazil. The World Bank recently conducted an economic and risk evaluation to assess whether biomass gasification/combined cycle technology will eventually become competitive with other new and evolving electricity supply resources available in northeastern Brazil. The World Bank's short-term objective is to establish a globally replicable prototype unit on a commercial scale for the cogeneration of electricity based on the gasification of wood chips or sugarcane fiber. No native forest will be used in fulfilling this objective. The long-term objective is to reduce global warming by lowering $CO_2$ emissions from fossil fuels.[6]

One drawback with current biomass waste-to-energy technology is that commercial facilities that generate electricity by burning MSW or landfill gas are still causing air pollution by burning coal, oil, or natural gas for start-up and fuel stabilization (or even as a primary fuel). A new and unique biomass gasification process has been developed that efficiently converts organic feedstock solutions such as agricultural and industrial byproducts and waste into a hydrogen-based gas ($COH_2$). The Carbo-hydrogen™ gas is a chemical raw material, an industrial fuel, and a potential transportation fuel. Developed by researchers Wil Dammann and W. David Wallman at DW Energy Research, LLC., in Colorado Springs, Colorado, their patented process uses forced, rapid oxidation by means of a direct current (DC) carbon-arc immersed in a water-based organic solution. This energy conversion process (greater than 90%) provides a simple way of using biomass in solution to produce a stable, versatile, and clean-burning gas. The gas has potential applications in fuel cells, at remote site power generation, in metal ore processing, and in the synthetics industry as well as others. The process is being researched by the National Renewable Energy Laboratory (NREL) in Golden, Colorado, for

*The gasification of organic waste is a carbon-neutral system.*

Photo courtesy of Alternative Energy Institute, Inc.

the production of biodegradable plastics. Best of all, this gasification process is a carbon-neutral system since it uses plant materials that would have decayed and produced $CO_2$ anyway. The gas is a hydrocarbon but not a fossil-fuel product.[7]

In California, a "biorefinery" is being built by Boston-based BC International Corporation (BCI) near Oroville, which will convert rice stubble, orchard prunings, leaves, fruit pits, and other biomass products into ethanol. Mixed with gasoline, the fuel is called "gasohol." Ethanol is a familiar product in the American Midwest, where factories convert cornstarch into about 1.6 billion gallons annually. Ethanol as a transportation fuel is very clean, adds octane, and is a benefit to air quality, a very serious problem in California's inland valleys. Midwestern ethanol growers also see California — and its more than 20 million licensed drivers — as a huge new market. BCI's factory, expected to open by 2001, will save California's rice farmers money because current state policy mandates that most leftover rice straw be plowed back into the ground, not burned — consuming about a fourth of the grower's profits. California environmentalists hope the ethanol will replace the additive MTBE in gasoline, which has been linked to lake and groundwater contamination.

Unfortunately, ethanol is no pot of gold when it comes to its use as an alternative transportation fuel. It is not economically competitive and is subject to shifting political winds. The federal government heavily subsidizes the ethanol fuel industry, giving 60 cents for every gallon brought to market. State subsidies add another 20 cents. Without this state and federal government financial support, charged to the taxpayer, the price of oil would have to be at least $40 a barrel before ethanol would be competitive even with corn at the low price of $2 a bushel. The US government also waives the federal gasoline tax on gasohol, a tax that pays for the maintenance of the federal highway system. This gas tax exemption for ethanol currently costs the federal Highway Trust Fund about $475 million dollars a year and is another form of subsidy to ethanol producers.

More importantly, it takes about 80% more energy to produce a gallon of ethanol than can be obtained in net fuel. Therefore, not only does the nation have to import oil to

> We see [the biorefinery near Oroville] as an efficient waste-based ethanol plant that takes advantage of low-cost and no-cost materials.[c]
> — *The Sacramento Bee*

fuel this corn/alcohol system, but its production adds to environmental degradation of land, water, energy, and biological resources.[8] Growing corn saps soil fertility and increases erosion; furthermore, it is estimated to take about 3,600 gallons of water to produce a bushel of corn. And although ethanol has been touted as reducing air pollution when mixed with gasoline or burned as the only fuel, there is no reduction when the entire production system is considered. The production and use of ethanol as fuel contribute to the increase in atmospheric carbon dioxide because twice as much fossil energy is burned in ethanol production than is produced as ethanol. T. Stauffer, a research associate at Harvard, stated in a lecture on *Economics of Energy* that "The bottom line is that using alcohol to stretch gasoline is like using filet mignon to stretch hamburger."

When evaluating biomass as a viable solution to the global energy problem, one must consider that although emissions associated with the factory conversion of biomass into usable energy or electricity are reduced, they do exist. Another consideration must be where will all the wood and other plant material for fuel come from? Some worry that it may cause accelerated deforestation practices by developing nations. The problems associated with widespread clear cutting can lead to groundwater contamination, floods, and irreversible erosion patterns that could literally change the structure of the world ecology.[9]

Biomass will succeed at some level as an alternative source of renewable energy because it is capable of being implemented at all levels of society. Tree and grass plantations show considerable promise in supplying a biomassed-based energy source, but the commercial use of plantation-grown fuels for power generation in the US has been stifled by an IRS tax code that denies a tax credit for electricity produced from scrap wood and agricultural waste or from standing timber planted before 1992. One European company, Bical, helped by Europe's aggressive renewable energy policy, is flourishing by promoting and planting Miscanthus, a tall grass that is also a short rotation crop that can be grown almost anywhere. First established by 20 English farmers interested in diversifying into profitable alternative crops, Bical is now a multi-

> There does not appear to be any crop which can be raised and processed efficiently enough into a liquid fuel to have an energy positive result.[D]
> — Walter Youngquist

national company with interests in the US, Caribbean, and Europe. When combusted with coal, Miscanthus reduces $CO_2$ emissions significantly. Miscanthus is also excellent for roof thatching and for chicken and animal beddings.[10]

Supplying the United States' enormous energy needs would require a planted area of one million square miles. That's roughly one-third of the area of the 48 contiguous states. There is no way that plantations could be implemented at this scale, not to mention the soil exhaustion that would eventually occur. Biomass alone cannot replace our current dependence on coal, oil, and natural gas; but it can complement other renewables such as solar and wind energy.

Chris Flavin, Senior Vice President of the Worldwatch Institute, states, "If the contribution of biomass to the world energy economy is to grow, technological innovations will be needed, so that biomass can be converted to usable energy in ways that are more efficient, less polluting, and at least as economical as today's practices."[11] Biomass needs more government support and financial incentives in order to compete with cheap but polluting fossil fuels. Only after biomass-based fuel crops are well established might we have a successful form of alternative energy. But as long as worldwide prices of coal, oil, and gas are relatively low, the establishment of plantations dedicated to supplying electric power or other higher forms of energy will occur only where financial subsidies or incentives exist or where other sources of energy are not available.

The US Department of Energy has supported the Biomass Resource Assessment Task, which has created a Biomass Resource Information Clearinghouse.[12] They hope to develop biomass resource assessment methodologies that will provide the energy community with critical data. They are developing high-quality biomass resource databases that show quantity, quality, opportunities, and obstacles. The goal is to provide one-stop shopping for publicly available biomass resource data and information. Although it retains potential as better processing technologies emerge, biomass energy is still somewhat limited and clearly not capable of sustaining the world's increasing energy needs on its own.

> The ability to combine biomass and solid waste with wind energy in a combined heat and power system...constitutes a substantial energy production breakthrough because when the wind doesn't blow, you can dispatch the solid fuels.[E]
> — *Environmental Technology*

## FOOTNOTES

1. <www.seda.nsw.gov.au/ren_biomass.asp> The Sustainable Energy Development Authority (SEDA) is a government agency in New South Wales dedicated to the investment in and use of sustainable energy technologies. A look at the site map shows their interest in solar, thermal, wind, and biomass energy as implemented in the home, at work, and in the community. This page gives a brief description of what biomass is and what processes are used to convert it into different forms of energy. NSW hopes to include biomass into their renewable energy-generation system.

2. <www.eren.doe.gov/RE/bioenergy.html> Energy Efficiency and Renewable Energy Network (EREN), a department of the US Department of Energy. Bioenergy technologies use renewable biomass resources to produce an array of energy related products including electricity, liquid, sold, and gaseous fuels, heat, chemicals, and other materials. Bioenergy ranks second (to hydropower) in renewable US primary energy production and accounts for 3% of the primary energy production in the United States. This website offers basic information on biomass resources, biopower, biofuels, biobased chemicals and materials and more.

3. <www.newafrica.com/energy/biomass/> Africa is the world's largest consumer of biomass energy (firewood, agricultural residuals, animal wastes, and charcoal), calculated as a percentage of overall energy consumption.

4. <http://tonto.eia.doe.gov/FTPROOT/renewables/renewablespubs.htm> *Renewable Energy Annual – 1998*, an annual report released by the Energy Information Administration (EIA), US Department of Energy, Washington, D.C. The EIA is an independent statistical and analytical agency within the US Department of Energy. Report was released in December 1998.

5. <www.geocities.com/RainForest/2958/graham.html> Background on Section 45 of the Internal Revenue Code (Energy Crops) and the "20% Rule." According to the IRS, in order to qualify a facility built prior to 1992 (i.e., used property) under Section 45, a 20% Rule must be satisfied. This means that the fair market value of the existing facility built before 1992 cannot exceed 20% of the fair market value of the modified facility after its retrofit to use biomass fuels. This information is contained in a sample letter addressed to the US Senate. The letter demands that in order for Section 45 to be useful to encourage project development, there needs to be no restriction in the definition of a qualifying facility between a pre- and post-1992 power generation facility.

6. <www.worldbank.org/html/fpd/em/biomass/biomass_turbine.htm> *Economic & Risk Evaluation of the Brazil Biomass-Gasification/Gas-turbine Demonstration Project.* Analysis of the biomass gasification power-generation facility suggests a high probability that it

will deliver electricity to northeastern Brazil at a cost lower to or competitive with the cost of all other electricity supply options after year 2010.

[7] <www.altenergy.org/4/cofe/cofe.html> Profile on W. David Wallman and Wil Dammann, partners at DW Energy Research, who have developed a unique Biomass Gasification Process to produce a clean-burning gas from biomass feedstock solutions. The profile is part of the Alternative Energy Institute's review of the 1999 Conference on Future Energy.

[8] David Pimentel, "1993 Environmental and Economic Benefits of Sustainable Agriculture: Socio-economic and Policy Issues for Sustainable Farming Systems," Cooperativa Amicizia, Padova, Italy, pp. 5-20.

[9] <www.worldgame.org/recall/deforest.html> World Game Institute web page on deforestation. This overview of deforestation discusses the causes, associated problems, and statistics on deforestation worldwide. Accompanying graphs show deforestation in tropical areas and in Norway, present a global comparison of forest cover in 1980 and 1990, give statistics on the tropical timber trade from 1987, and calculate the percentage of fuelwood's share in the total energy consumption worldwide.

[10] <www.bical.net> Bical has contracts to supply baled or chipped Miscanthus to power stations under the Non-Fossil Fuel Obligation, which provides premium prices for "green" power. Bical can supply Miscanthus rhizome, as well as provide support and advice on financing the crop. They also help with maintenance, harvesting, and sale of Miscanthus cane products.

[11] <www.worldwatch.org> Christopher Flavin is Senior Vice President at the Worldwatch Institute, where he directs the Institute's research programs and heads its climate and energy team, which analyses energy resource, technology, and policy trends.

[12] <rredc.nrel.gov/biomass/welcome.html> Biomass Resources Information Clearinghouse (BRIC). The purpose of this clearinghouse is to provide high-quality biomass resource data for the United States to the energy community. They are creating an atlas of US biomass resources, which will show how much biomass is available, county-by-county, from a wide variety of sources.

## SIDEBAR FOOTNOTES

A  Governmental Advisory Associates, Inc., *The Municipal Waste Combustion Industry in the United States – 1997-98 Resource Recovery Yearbook and Directory* (Westport, CT 1997).

B  Madanjeet Singh, *The Timeless Energy of the Sun for Life and Peace with Nature* (Published by the United Nations Educational, Scientific, Cultural and Communication Organization — UNESCO, 1998), p. 37.

C  John Howard, "Plant will turn rice stubble into ethanol," *The Sacramento Bee*, August 14, 1999, p. B-5.

D  Walter Youngquist, *GeoDestinies: The inevitable control of Earth's resources over nations and individuals* (Portland, Oregon: National Book Company, 1997), p. 246.

E  Editorial, "New Tech Combines Biomass/Wind Power Generation," *Environmental Technology* (Sept./Oct. 1999), p. 8.

# Tidal & Ocean Thermal

## 12

Tides, waves, and the heat differential within the world's tropical oceans are potent sources of clean energy. The trick is how to harness these renewable resources effectively and economically. Engineers have come up with several mechanisms to capture these natural power producers.

HUMANS HAVE ALWAYS BEEN IMPRESSED BY THE POWER of the world's oceans, but engineers have only recently tried to tap the energy potential exhibited by tides, waves, and seawater temperature differentials. Tidal energy utilizes the gravitational energy of the Sun, Earth and Moon. Wave power converts the energy released in crashing waves, driven onshore by wind. Ocean thermal systems exploit the greatest solar collector on Earth — the sea. Each of these energy forms has its own advantages and disadvantages, but none of them is the answer to the looming energy crunch.

Tidal energy works on the same fundamental principal as the water wheel. In the case of tidal energy, however, the difference in water elevation is caused by the fluctuation between low and high tides. Engineers build a dam or barricade across an estu-

*The oceans are the world's largest solar energy collector and energy storage system.*

ary to block the incoming tide, the outgoing tide, or both. When the water level on one side of the dam is higher than the level on the other side due to a tidal change, the pressure of the higher water increases. The water is then channeled through a turbine in the dam, which produces electricity by turning an electric generator.[1]

Tidal energy is being harnessed in several countries around the world, from facilities in Russia to France with 400 kW to 240 MW capacities.[2] Some proposed sites, however, exhibit extraordinary potential. Britain's Severn Estuary and Canada's Bay of Fundy have potential capacities of as much as 8,000 and 30,000 MW, respectively.[3] The Severn Estuary average tidal range is 26 feet, and the Bay of Fundy boasts a 32-foot tidal range, ideal for substantial electricity generation.[4] But exceptionally high tides are rare and found only at high latitudes in relatively remote areas, a major limitation to this energy source. A tidal range of at least 21 feet is needed to build a sufficient head of water for the turbines. There are few places in the world that make such a facility economically worthwhile.

> On an average day, 23 million square miles of tropical seas absorb an amount of solar radiation equal in heat content to about 250 billion barrels of oil.[A]
> – National Renewable Energy Laboratory

Another important obstacle to the tidal system is the sheer amount of time that passes in which little electricity can be generated between the rising and falling tides. During these times, the turbines may be used to pump extra water into the basin to prepare for periods of high electricity demand, but not much else can be done in the interim to generate more electricity. By its very nature, a tidal-based energy facility can generate a maximum of only ten hours of electricity per 24-hour day. Because tides change in time from day to day, there is a problem of integrating this varying power production into a large regular power grid. It cannot supply power at a steady rate or increase generation during peak times.

*Power can be produced as long as the temperature between the warm surface water and the cold deep water differs by 36 degrees Fahrenheit.*

Although the operation and maintenance of a tidal power plant is low, the cost of the initial construction of the facility is not, making the overall price of the electricity generated economically prohibitive. It is estimated that the Severn tidal project with a proposed capacity of 8,640 MW, will cost $1,600 per kW, or over $13.8 billion.

Tidal energy may be physically limited in scope and expensive to produce, but in contrast to the combustion of fossil fuels, using the tides to produce electricity is renewable and avoids air pollution. But tidal energy facilities do have an environmental impact. The alteration of the natural cycle of the tides may affect shoreline as well as aquatic ecosystems. Pollution that enters a river upstream from the plant may be trapped in the basin, where the natural erosion and sedimentation pattern of the estuary may be altered. Local tides could decrease by more than a foot in some areas, and mixing the water could stimulate the overgrowth of organisms that paralyze shellfish. Little is known about the potential environmental harm of a tidal energy facility, and some people fear that its impact won't be understood until after installation. With such uncertainty, tidal power is still an unproven alternative energy candidate.

*If less than 1/10$^{th}$ of 1% of the ocean's stored solar energy could be converted into electric power, it would supply more than 20 times the total amount of electricity consumed in the US on any given day.*

Even if the high cost and environmental issues were circumvented, distributing the electricity generated by tidal facilities is a logistical problem. There are less than two dozen optimal tidal energy sites in the world, and nearly all of them are in remote locations. Since the collection sites are limited, the electricity produced at the tidal plant must be distributed inland by an expensive transmission system.

It has been estimated that worldwide, approximately 3000 gigawatts (1 gigawatt = 1 GW = 1 billion watts) of energy are continuously available from the action of tides, but only 2% or 60 GW can potentially be recovered for electricity generation. Despite tidal power's inability to replace conventional energy sources, it will not be dismissed in the near future. Britain, India, and North Korea have planned to supplement their grid with this renewable energy source.

Wave energy potential is greatest in regions with strong on-shore winds. The best areas are on the eastern edges of the oceans (western side of the continents) between the 40° and 60° latitudes in both the northern and southern hemispheres. These optimal wind zones will produce the highest concentrations of wave power — a low-frequency energy that can be converted to a 60-Hertz frequency. The waters off California and the United Kingdom are regarded as the best potential sites. "California's coastal waters are sufficient to produce between 7 and 17 MW per mile of coastline."[5]

Tapping clean, renewable wave energy has its own set of obstacles. The energy potential of deep ocean sites is 3 to 8 times the wave power at coastal sites, but constructing and mooring the site plus transmitting the electricity to shore is costly. Generating electricity by harnessing the power of waves is significantly less efficient than using wind turbines to produce the power. And once in place, the device can be a dangerous obstacle to ship crews that cannot see or detect it on radar.

Wave energy has received little attention in comparison to other renewable sources of energy. Although 12 broad types of wave-energy systems have been developed — combinations of fixed or moveable, floating or submerged, onshore or offshore — scientists have not fully investigated this technology. Many research and development goals remain to be accomplished. These include cost reduction, efficiency and reliability improvements, the identification of suitable sites in California, as well as interconnection with the utility grid. Environmentalists want engineers to better understand the impact of the technology on marine life and the shoreline. Also essential is a demonstration of the ability of the equipment to survive the salinity and pressure environments of the ocean as well as weather damage over the life of the facility. Even a successfully built and operated wave-power facility will not provide extra power for peak demand, nor would it be a reliable source of energy.

There are a handful of wave-energy demonstration plants operating worldwide, but none produces a significant amount of electricity. Projects have been discussed for various sites in California — San Francisco, Half Moon Bay, Fort Bragg, and Avila Beach — but no firm plans have been made.

> [The low specific work of thermodynamic cycle and large mechanical components] involves high plant installation costs and, consequently, a doubtful economical feasibility of the OTEC power plants.[B]
> — *Journal of Energy Resources Technology*

> The United States and Japan have each spent more than $100 million on OTEC research.[C]
> — Walter Youngquist

Though government agencies in Europe and Scandinavia are sponsoring research and development, wave-energy conversion is not yet commercially available in the United States. This developing technology is not expected to be available within the near future due to limited research and lack of federal funding.

Ocean Thermal Energy Conversion (OTEC) seems to be a promising source of renewable, non-polluting energy for the future. The oceans comprise over two-thirds of the Earth's surface, and they collect and store an enormous amount of solar energy. Some experts predict that if even 0.1% of this stored energy could be tapped, the output would be 20 times the current daily energy demands of the United States.[6]

Ocean thermal-energy conversion exploits the temperature gradient between the varying depths of the ocean, requiring at least a 36°F difference from top to bottom, as is found in tropical regions. This difference in temperature is the "heat engine" for a thermodynamic cycle. Cool water brought up from the ocean depths is used to condense the vapor, just as the cooling towers of other power plants do. There are three types of OTEC designs: open cycle, closed cycle, and hybrid cycle. In an open cycle, seawater is the working fluid. Warm seawater is evaporated in a partial vacuum, expanding through a turbine connected to an electrical generator. The steam then passes through a condenser that uses cold seawater from the depths of the ocean, and the result is desalinated water that can be used for other purposes. New seawater is used in the next cycle. In a closed cycle, a low boiling point liquid such as ammonia or refrigerant is used as the working fluid, vaporized by warm seawater. After expanding through a turbine connected to an electrical generator, cold seawater is used to condense the vapor back into a liquid to start the process again. A hybrid cycle combines the two processes in which flash-evaporated seawater creates

*The National Renewable Energy Laboratory predicts that OTEC may become competitive in select tropical markets within a decade.*

The date by which OTEC might be expected to become commercially viable is too far off to attract entrepreneurs.[D]

— *Issues in Science and Technology*

steam, which, in turn, vaporizes a working fluid in a closed cycle. The vapor from the working fluid powers the turbine while the steam is condensed for desalinated water, as in an open system. The hybrid system continues to process seawater and produce electricity.

OTEC taps energy in a consistent fashion, producing what "is probably *the* most environmentally friendly energy available on the planet today."[7] Unfortunately, the realization of this promising potential is largely experimental in nature at this time. In fact, the only ocean thermal energy conversion plant in the US was an experimental facility — the Natural Energy Laboratory of Hawaii (NELHA), which was closed at the end of a successful test in 1998. (There are no OTEC sites adjacent to the continental US because off-shore waters are too cool. OTEC potential can operate only in tropical or semi-tropical waters.)

> There must be a sustained effort to educate and inform the public to achieve acceptance for the necessary measures to develop renewable marine energy.[E]
> — *Environment News Service*

The technology still has a long way to go before becoming a viable alternative energy source. The facility on the Kona Coast of Hawaii, for instance, produced the highest amount of electricity to date with a 210 kW open-cycle OTEC experimental facility that operated from 1992 to 1998. When considering the capacity of conventional combustion turbines, ranging from a typical output of 25 MW to a maximum 220 MW, this technology is not cost competitive.[9] It is most applicable on small islands that rely heavily on imported fossil fuels. This system would render an island more self-sufficient, improving sanitation and nutrition standards and providing desalinated water that could be used to grow aquaculture products.

*OTEC technology still has a long way to go before becoming a viable alternative energy source.*

Overcoming the obstacles to developing this energy source will challenge engineers. It's been estimated that a 250-megawatt OTEC plant (about 20% of a conventional coal-fired plant) would have to use pipes nearly 100 feet in diameter and circulate an amount of water equivalent to the discharge of the Mississippi River. Other problems include salt-water corrosion, algae and barnacle growth, and the susceptibility of tropical regions to damaging hurricanes. Building OTEC facilities could supply electricity for small tropical islands, but environmentalists worry about the effects of pumping up large quantities of nutrient-rich cold water, which may stimulate algal growth.

Researchers believe it will be quite some time before OTEC technology is in a position to partially phase out the use of fossil fuels in limited regions of the world. The large amount of equipment, chronic maintenance problems, and very low efficiency rates suggest that OTEC power generation will not be replacing fossil fuels any time soon.

> To compensate for its low thermal efficiency, OTEC has to move a lot of water.[F]
> — *Popular Science*

## FOOTNOTES

1. <http://renewable.greenhouse.gov.au/technologies/ocean/tidal.html> The Australian Renewable Energy website offers descriptions and images regarding tidal power technology. The generation of electricity from tides is very similar to hydroelectric generation, except that water is able to flow in both directions and this must be taken into account in the development of generators.

2. <http://194.178.172.97/class/ixb13.htm> Tidal Energy information by the International Energy Administration. This page lists 1994 data from France, the UK, China, Canada, and Russia on general information, cost, environmental, and implementation characteristics and requirements of tidal energy plants and their outputs. It was last updated in April 1997.

3. <http://zebu.uoregon.edu/1998/ph162/l15.html> Fact sheet explaining what OTEC and tidal energy are, the best locations for extracting energy globally, links to and photos/maps of current and proposed sites. It is lecture #15, May 19, 1998, from a physics class on alternative and renewable energy sources taught by Professor Greg Bothun at the University of Oregon.

4. <http://bigbro.biophys.cornell.edu/~duesing/work/ba.html> Descriptions of tidal and wave energy focusing on the Bay of Fundy and the Severn Estuary as prime sites for tidal energy extraction. Peter Duesing, a postdoctoral student in an unrelated field at Princeton, is the author of this page on his personal website.

5. <www.energy.ca.gov/development/oceanenergy/index.html> Ocean Energy is a page off the California Energy Commission's website that describes the different types of wave-energy facilities and the drawbacks of this source of power, mainly due to lack of research and uncertainty of the equipment's potential harm. Includes several related links. Its most recent reference is 1995, but it was updated in August 1998.

6. <www.nrel.gov/otec/what.html> "What is Ocean Thermal Energy Conversion?" Article is located on the United States Department of Energy's National Renewable Energy Lab website. Explains the basics of OTEC with a map showing the temperature gradients of the world's oceans. Links at the bottom of the page contain information on achievements in OTEC technology, research needed to make this technology economically viable, information on plant design and location, the benefits of OTEC, and the markets for OTEC.

7. <www.nelha.org/otec.html> Ocean Thermal Energy Conversion(OTEC) is a process which utilizes the heat energy stored in the tropical ocean. The world's oceans serve as a huge collector of heat energy. The OTEC plants at NELHA utilize the difference in temperature between warm surface seawater and cold deep seawater to produce electricity.

## SIDEBAR FOOTNOTES

A  The National Renewable Energy Laboratory (NREL) is the United States' leading center for renewable energy research. They are developing new technologies to benefit both the environment and economy. To learn more about the NREL and Ocean Thermal Energy Conversion (OTEC) on the Internet, check out <www.nrel.gov/otec/what.html>

B  M. Gambini, "Improving the OTEC Power Plant Performance by Metal Hydride Energy Systems," *Journal of Energy Resources Technology* (June 1997), p. 145.

C  Walter Youngquist, GeoDestinies: *The Inevitable Control of Earth Resources over Nations and Individuals*," (Portland, Oregon: National Book Company, 1997), p. 257.

D  William H. Avery and Walter G. Berl, "Solar energy from the tropical oceans," *Issues in Science and Technology* (Winter 1997), p. 4.

E  Bill Eggertson, "UK Explores Offshore Wind, Wave & Tidal Energy," *Environment News Service* (June 16, 1999), p. 1.

F  Mariette DiChristina, "Sea Power," *Popular Science* (May 1995), p. 72.

# Section III
# The Energy Revolution: Technologies on the Horizon

IN 1953, SIR CHARLES GALTON DARWIN WROTE:

"The fifth revolution will come when we have spent the stores of coal and oil that have been accumulating in the Earth during hundreds of millions of years....It is to hoped that before then other sources of energy will have been developed...but without considering the detail [here] it is obvious that there will be a very great difference in ways of life....Whether a convenient substitute for the present fuels is found or not, there can be no doubt that there will have to be a great change in ways of life. This change may justly be called a revolution, but it differs from all preceding ones in that there is no likelihood of its leading to increases of population, but even perhaps to the reverse."[1]

Research scientists across the United States and around the world are pursuing exotic carbonless technologies that will help humanity survive the fast-approaching energy crisis. Sometime within the next two decades, global demand for cheap oil will exceed supply as the world's proven oil reserves reach their midpoint of depletion, to be followed by steadily declining petroleum production. World energy production per capita averaged 3.45% in annual growth from 1945 to 1973 but slowed to less than 1% annual growth between 1973 and 1979, at which point, global oil production per capita peaked. Oil production

> The transition to alternative fuels will not be simple nor as convenient as is the use of oil today, and it will involve much time and financial investment.[A]
> — Walter Youngquist

per capita has declined an average 0.33% a year since 1979 because producers can't keep up with the booming world population growth and our ever-increasing demand for energy.[2]

The international community is banding together to pressure wealthy industrialized nations to reduce their fossil fuel gas emissions (especially the United States, which contributes 25% of the world's total), for the health of the planet's life-sustaining atmosphere and ecosystems. In order to cut greenhouse gas emissions, which directly harm the human family and are contributing to global warming, the United States is switching to natural gas for power generation and as a transportation fuel. Although relatively cleaner than coal and oil, natural gas is also a limited nonrenewable resource that does contribute to air pollution. Coal reserves are plentiful in some regions of the world, but this fossil fuel is a leading cause of respiratory illness and should not be relied on for electricity generation in the 21st century. Once touted as an energy source "too cheap to meter," nuclear power is waning in the United States and Europe. Due to strong public pressure, the German government has banned new nuclear power plants within their borders. Spent fuel rods and other nuclear waste products remain hazardous for centuries; worldwide, there are about 200,000 tons of radioactive nuclear waste in temporary storage.

*California's troubled Rancho Seco 913-megawatt nuclear power plant was shut down by a voter referendum in 1989.*

There are promising cutting-edge technologies that could have profound effects on how and where our future energy is produced. Twenty-first century energy systems should be robust, pollution-free, renewable, decentralized, and have an acceptable scientific description of how the device works. The three sources that appear most favorable are a new type of chemical energy, energy from low-energy nuclear reactions (LENR), or the abundant energy in space (Zero Point Energy). Unfortunately, the only real new-energy development that the Department of Energy (DOE) has funded has been about $500 million a year for nuclear fusion research, which is estimated to be decades away from producing commercial electric power. The DOE's long-range plan for electricity generation still relies heavily on coal and nuclear fission.

The Alternative Energy Institute, Inc., wholeheartedly supports traditional renewable energy systems like wind, solar, hydroelectric, and geothermal, but they alone will not replace the petroleum gap in sustaining the quality and standard of living Americans have grown accustomed to. The following section focuses on a medley of new-energy and propulsion sources that may contribute to our collective energy supply, as well as improve the way we generate power for our cars, homes, and energy-intensive lifestyle. Developing these new energy systems will take time, money, and a coordinated effort, as well as professional and institutional support. These crucial ingredients for success have been sorely lacking in the field of frontier science.

If nothing else, the excitement over cold fusion and the subsequent media bashing of electrochemists Pons and Fleischmann by skeptics taught new-energy researchers the danger of releasing experimental results prematurely and piece-meal. The March 23, 1989, announcement of a successful cold fusion experiment by two electrochemists working at the University of Utah shocked the scientific world. The chemically assisted nuclear reaction (CANR) claimed by Stanley Pons and Martin Fleischmann apparently produced excess heat and by-products of nuclear fusion at room temperature. When other laboratories around the

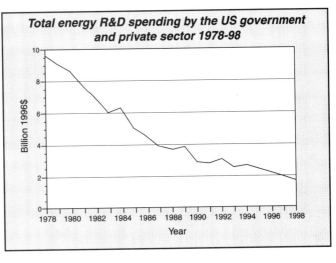

*The dimishing trend in R&D investment is global in scope.*

Science is a culture with a strong interest in protecting its borders, both from those who would invade from the outside and from those inside who deviate from condoned behavior and thought.[B]
— *Science, Technology, & Human Values*

world failed to duplicate their results immediately, Pons and Fleischmann were lambasted for sloppy science. Cold fusion researchers have been excluded from professional journals and rejected by the traditional scientific community, despite subsequent experiments that have made it clear that, although the CANR does not produce robust energy, it may yield much promise in the transmutation of dangerous radioactive elements into safe and inert products. Edmund Storms, a retired Los Alamos National Laboratory scientist, lists more than 50 published examples of excess heat production and nuclear reactions that can be cited, with more than ample peer-reviewed experimental verification.

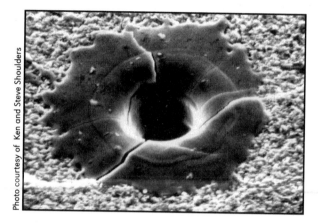

*Optical micrograph of a 10 micron diameter electron cluster borehole entrance into lead oxide glass with evidence of wave action and sloshing.*

Critics of "free" energy research cite the apparent violation of the principle of conservation of energy in which energy output cannot exceed energy input in a closed system. Supporters do not argue the concept of energy conservation but point out that if an energy system can tap into the sea of energy in the "vacuum" of space all around us, the system is no longer "closed," and virtually unlimited power may become available. Scientists call the energy produced by fluctuations of the electromagnetic and gravitational force fields in the vacuum Zero Point Energy (ZPE). All of space is filled with an extraordinary amount of this vacuum energy that can theoretically be accessed via many different technological methods, such as electromechanical, chemical, thermal, magnetic, etc. It is thought that the excess (overunity) electric or thermal energy repeatedly measured by many researchers is the result of tapping the larger system, which includes the vacuum energy. If even one of these techniques that is now in the research and development stage becomes commercialized, it will launch an energy revolution that will change the world.

Quantum electrodynamics and experimental results show that ZPE is the essence of everything in the Universe, but there is radical disagreement over just how much of it is available for possible extraction. Nobel Laureate Richard Feynman and one of Einstein's protégés, John Wheeler, calculated that there is more than enough energy in the volume of a coffee cup to evaporate all the world's oceans. Proponents of ZPE point out that, although the particles in the zero-point field are minute, there are so many possible modes of propagation (frequencies, directions) in open space, the total ZPE summed up over all possible modes may be enormous. In recent years, evidence suggests that the electromagnetic zero-point field is not merely an artifact of quantum mechanics but a real entity with major implications for gravity, astrophysics, and technology.

Among other tenets, the scientific method is based on several fundamental assumptions that hold that the speed of light is the upper limit for all velocities in the Cosmos; that the speed of light is everywhere constant; that gravitational force can only be described as a function of mass; and that there is a fundamental distinction between matter and energy. New experimental research indicates that these fundamental assumptions are incorrect, which means that physicists do not completely understand how light or gravity really work or the basic relationship between matter and energy. Today, there are several mathematical models circulating within the scientific community that describe the inner workings of gravity, as well as different interpretations for its potential in terrestrial and space travel. Scientists and engineers, some working for NASA, are experimenting with gravity modification and shielding. The concept of gravity modification is being investigated by NASA and others because shielding earthbound objects from the effects of gravity will change physics and may radically alter the economics of space travel and transportation in general.

The rocky road to new energy is fraught with financial and professional peril. Not only is it difficult to probe the elusive recesses of physics, but researchers often face an obstinate US Patent and Trademark Office that refuses to

> The scientific revolution now getting into gear is faster than the Copernican revolution and broader than the one initiated by Einstein.[c]
> — *The Whispering Pond*

protect their hard work with a patent. Walls of silence surround the academic and traditional spheres of physics and engineering, which offer pioneer researchers no professional or financial support, only cutting criticism. Editors at professional scientific journals will not publish research papers that mention cold fusion or tapping zero point energy, meaning little information about experimental results is shared among the isolated scientists searching for the answer to the coming energy crisis. The obstacles to success are serious and institutionalized, making them difficult to change, but these nascent technologies have the potential to shift fundamentally the way humans access and distribute energy.

The energy revolution will be a global event that will help humanity solve some of its most pressing problems and will usher in a more promising and equitable era for the billions of people who live in poverty and squalor. But the energy revolution will continue to struggle without the help and support of mainstream scientists and government. The US government should play a major role in assuring that these promising new technologies develop and successfully reach the commercial market. Ralph Waldo Emerson once said, "This time, like all times, is a very good one, if we but know what to do with it." If we are going to make the transition from polluting fossil fuels to clean, decentralized power sources without economic and environmental upheaval, the time to act is now!

## FOOTNOTES

[1] Charles G. Darwin, *The Next Million Years* (Garden City, New York: Doubleday, 1953), p. 52. Sir Charles Galton Darwin was a leading physicist, pioneer in nuclear studies, and grandson of English naturalist Charles Darwin who wrote *On the Origin of Species by means of Natural Selection*.

[2] Richard C. Duncan, "The Peak of World Oil Production and the Road to the Olduvai Gorge." Pardee Keynote Symposia, Geological Society of America Summit 2000, Reno, Nevada, November 13, 2000. Richard Duncan's Olduvai theory is defined by the ratio of world energy production (use) and world population. It states that the life expectancy of industrial civilization is less than or equal to 100 years: 1930-2030. Duncan believes that "the collapse will be strongly correlated with an 'epidemic' of permanent blackouts of high-voltage electric power networks — worldwide." Richard Duncan's oil forecasting models, the application to run them, and a User's Guide are all available free at www.halcyon.com/duncanrc/

## SIDEBAR FOOTNOTES

[A] Walter Youngquist, "Alternative Energy Sources: Water and Energy, the Basis of Human Society: Are They Globally Sustainable Through the 21$^{st}$ Century?" (L. C. Gerhard edition, 2000), pp. 45-57, *Kansas Geological Survey Open-File Report 2000-51*.

[B] Dale L. Sullivan, "Exclusionary Epideictic: *NOVA*'s Narrative Excommunication of Fleischmann and Pons," *Science, Technology, & Human Values* (Sage Publications, Inc., Summer 1994), Vol. 19 No. 3, pp. 283–306.

[C] Ervin Laszlo, *The Whispering Pond: A Personal Guide to the Emerging Vision of Science* (Boston, Massachusetts, Element, 1996), p. 125.

# Cold Fusion: Fact or Fiction?

## 13

Cold fusion has been called the Holy Grail of the new energy field. In 1989, two widely-respected electrochemists, Stanley Pons and Martin Fleischmann, shocked the world with claims of excess heat energy and evidence of nuclear fusion, both of which occurred in a small electrochemical cell at room temperature, thus the term "cold" fusion. The scientific community protested that their own experiments exhibited no excess heat and nuclear byproducts, and Pons and Fleischmann were ridiculed into obscurity. But despite vociferous objections that cold fusion is "pathological science," hundreds of qualified researchers around the world are working to prove that low-energy nuclear fusion is no hoax.[1]

THE MODERN FIELD of chemically assisted nuclear reactions (CANR), initially called cold fusion, was announced to the world at a press conference on March 23, 1989, by Stanley Pons and Martin Fleischmann, two electrochemists working at the University of Utah. Cold fusion is the nuclear fusion of deuterium, a heavy isotope of hydrogen, at or relatively near room temperature.[2] It is a reaction that occurs under certain conditions in supersaturated metal hydrides (metals with large amounts of hydrogen or heavy hydrogen dissolved in them). The process produces excess heat, helium, and a very low level of neutrons. Cold-fusion

> Like Jesus, Fleischmann and Pons were proclaimed from the deserts, followed by the crowds, inspected by the orthodox, forsaken by the unfaithful, rejected by authorities, and set outside the walls.[A]
>
> — *Science, Technology, & Human Values*

effects have occurred with palladium, titanium, nickel, and some superconducting ceramics.[3] The Pons and Fleischmann claims were based on a proposed fusion reaction occurring near room temperature after a large concentration of deuterium had dissolved in palladium metal. This phenomenon generates excess heat and has been proposed as a method to produce cheap, renewable, and pollution-free energy, as well as a way to reduce the radioactivity of nuclear waste through transmutation. However, before these very beneficial possibilities can be realized, conventional science will have to accept the reality of the claims.

Professional, scientific, and academic acceptance of room-temperature fusion will not be achieved until researchers develop convincing and repeatable experiments with definitive proof of excess heat and nuclear byproducts, which advocates of cold fusion insist have occurred. If the proof of low-energy nuclear reactions (LENR) is already confirmed, then why has the conventional scientific community continued to vilify research into cold fusion? The reasons for animosity toward the field are partly scientific and partly psychological. The scientific objections are based mainly on the conflict between the observations and a well-accepted theory of nuclear reaction that appears to forbid such behavior. The psychological basis stems from the way Pons and Fleischmann made their information public through the media and the subsequent exaggeration of expectations. To make matters worse, important information was withheld for legal reasons, which made early efforts to duplicate the claims very difficult.

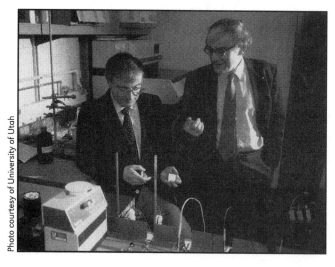

*Electrochemists Stanley Pons and Martin Fleischmann (on right) opened a Pandora's Box with their 1989 claim of a successful cold-fusion experiment.*

Pons and Fleischmann were not the first to experiment with the idea of low-energy atomic fusion, nor were they the last. The concept can be traced back at least 2,000 years to the age of alchemy, when earnest but futile attempts were made to transmute base metals into gold. The alchemists were ridiculed as charlatans, and today's cold-fusion advocates have fared little better. In the last 200 years, scientists have progressed in their understanding of the atom and its electrically charged components. Persistent effort by mathematicians and researchers to develop the related physics and technical tools has slowly unlocked the secrets of the atom. A cursory glimpse into the history of fusion and fission reveals that the field took a giant leap forward in 1803 when British chemist John Dalton theorized that each element is composed of identical atoms whose properties determine the chemical reactivity of that element. A century later, Albert Einstein furthered the effort with his theory of Special Relativity, in which mass and energy are shown to be equivalent.[4]

> If [cold fusion] remains the poor stepchild of science, starved into obscurity, we'll never have a chance to learn what we may be missing.[B]
> — Wired

Electrochemistry and nuclear physics advanced rapidly during the 20th century. The development of more sophisticated equipment helped scientists delve deeper into atomic manipulation. In 1915, American chemist William D. Harkins suggested that hydrogen atoms could be fused to produce helium, with a fraction of a percent of the original mass being converted to energy. Four years later, physicist Ernest Rutherford transmuted nitrogen to oxygen in his alpha particle bombardment experiments — considered the first confirmed successful transmutation of elements in scientific history. In 1932, chemist-physicist John Tandberg tried to create nuclear fusion by electrolyzing heavy water with high energy electrical discharges in a palladium wire. Two years later, Otto Hahn and Fritz Strassman in Germany bombarded uranium with neutrons and produced the first man-made fission of an element.

Nuclear research hit intellectual and financial pay dirt on December 6, 1941, when President Roosevelt launched the Manhattan Project. The best and brightest of America's scientific minds were marshaled together in a race against Nazi Germany to be the first to develop a nuclear weapon of mass destruction. With unlimited funding and ardent political support, the do-or-die effort to build an atomic bomb pushed our understanding of nuclear physics ahead at light speed. After World War II, physicists at Los Alamos theorized utilizing magnetic confinement for plasma to create controlled hot fusion, in effect, to hold it in a magnetic bottle. But it took another two decades before Soviet scientists developed the hot fusion reactor concept. Initially derided by Western physicists, the hot fusion reactor is known as a "tokamak" and has become the modern icon of hot fusion technology.[5]

*The decommissioning cost-estimates for California's defunct Rancho Seco nuclear reactor will exceed $500 million.*

The difference between hot and cold fusion is like night and day. Hot fusion occurs at temperatures in the millions of degrees, when the nuclei of hydrogen atoms can overcome their natural tendency to repel one another and join or fuse to form helium nuclei. It's the same nuclear reaction that powers the Sun and the stars. It releases enormous energy but also deadly radiation in the form of radiation-neutrons and gamma rays. On the other hand, cold fusion releases excess energy in the form of heat, not ionizing radiation. The process produces trillions of times fewer neutrons than plasma fusion or fission, which makes many scientists skeptical that it is really a nuclear reaction. But it can't be a chemical reaction because it consumes no chemical fuel, and it produces no chemical ash. Because the heat energy produced in a successful cold-fusion experiment is hundreds of thousands of times what an ordinary chemical reaction could possibly yield, it has led researchers in the field to believe that is a heretofore unknown form of benign nuclear reaction. The lack of dangerous radiation gives cold fusion a priceless advantage over hot fusion for commerce and the environment.

Cold-fusion research got its first real jolt of attention in the late 1940s when physicists Andrei Sakharov in the Soviet Union and F.C. Frank in England independently suggested the possibility of muon-catalyzed cold fusion. This process would use muons to confine hydrogen nuclei, so they can fuse to release energy, but the muons remain behind to induce further fusion cycles.[6] In 1956, Louis Alvarez and his colleagues at UC Berkeley accidentally stumbled upon muon-catalyzed cold fusion as well. In the late 1960s, electrochemist Martin Fleischmann and some colleagues embarked on research involving the separation of hydrogen and deuterium isotopes, work that eventually led to his controversial CANR experiments in the 1980s. But research into low-energy fusion lagged as chunks of funding were gobbled up in expensive hot-fusion research and development (R&D).

In the 1970s, American scientists were focused on the development of inertial confinement (laser) fusion. Much time and money has been spent on this process in which scientists use lasers to generate dense, high-temperature plasmas by bombarding tiny solid or gas-containing pellets of fusion fuel to create momentary little fusion blasts — thermonuclear microexplosions. Inertial confinement fusion really took off after Princeton University astronomer Professor Lyman Spitzer, Jr., proposed using magnetically confined plasmas to control high-temperature fusion reactions. Spitzer's vision generated new enthusiasm for hot fusion research and, despite its huge costs and technical difficulties, many countries have built these complex experimental tokamak reactors.

Today, despite the tremendous financial investment in hot fusion ($15 billion by the US alone), there is a real concern whether the US Department of Energy's tokamak approach will ever result in a commercially viable technology. In fact, even a demonstration-scale, power-producing hot-fusion reactor still appears decades away. The controlled hot-fusion program has made enormous strides, but most nuclear scientists believe that engineering a practical power plant will take another 25 to 30 years before a reactor could safely generate reliable power. Hot-fusion scientists in the US, who spend half a billion dollars annually on research projects, are now seeking increased funding to build a gigantic, complex reactor call ITER (International Thermonuclear Experimental Reactor). Recent successes by the Joint European Torus (JET) project, located near Oxford, England, have been encouraging to supporters of

> So far, however, the process of using [hot] fusion on a controlled commercial scale has eluded some of the world's best scientists and engineers.[c]
> — Walter Youngquist

hot fusion, but, once again, the earliest commercial power plant remains decades away.[7]

Meanwhile, few scientists noticed in 1978 when Japanese researchers Tadahiko Mizuno and Takayuki Kurachi reported observing an anomalous 200-ml evaporation event in a palladium and heavy-water electrolytic cell. Mizuno and his colleagues scratched their heads and dubbed the incident an "unsolvable mystery." At the same time, Stanley Pons' own research into isotopic separation in palladium electrodes was also yielding puzzling results. For many physicists, the anomalous data in these experiments did not confirm low-energy fusion. The research failed to impress the US government and was mostly ignored by a scientific community fixated on hot fusion as the technology of choice. It seemed that American research into cold fusion was dealt a mortal blow when Congress passed the US Magnetic Fusion Energy Engineering Act of 1980 and recommended doubling the magnetic hot fusion budget within seven years.

Nevertheless, in 1984, Pons and Fleischmann initiated a joint venture in an electrochemical experiment to search for clear and unequivocal evidence of fusion. In the late 1980s, the fervor for low-energy fusion gained momentum when researchers at Brigham Young University began their own experiments. In July 1989, *Scientific American* published an article titled "Cold Nuclear Fusion" in which the process of muon-catalyzed fusion reactions were described.

Cold-fusion research was getting press, but hot fusion was getting most of the money. In 1988, the $300 million Tokamak Fusion Test Reactor (TFTR) at Princeton University attained a plasma temperature of 300 million degrees K. Both the Princeton tokamak and another one in Britain were close

> The biggest stumbling block may be the term *cold fusion* itself, which seems to engender hysterical opposition on its own.[D]
>
> — Tadahiko Mizuno

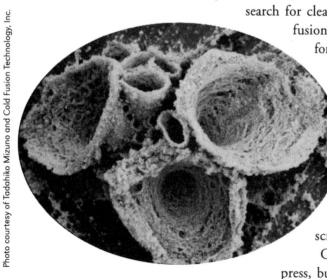

*Products found on the surface of a gold cathode after electrolysis.*

Photo courtesy of Tadahiko Mizuno and Cold Fusion Technology, Inc.

to break-even conditions, the point where the reactor supplies the same amount of energy that it consumes in the fusion process.

Research into low energy fusion reactions (LENRs) was beginning to heat up in February 1989 when Pons and Fleischmann set up their experimental apparatus in a laboratory at the University of Utah. Stephen E. Jones was trying to produce a fusion reaction using an electrolytic cell at Brigham Young University in Utah, and he mentioned to the visiting chemists that he intended to submit an article about his own research for publication. Jones also expected to deliver his paper to the upcoming American Physical Society Meeting on May 4, 1989. Pons and Fleischmann stated that they needed another 18 months of research before they would be ready to publicize their experiments, but their professional report was written and submitted in a rush to the *Journal of Electroanalytical Chemistry and Interfacial Electrochemistry* by March 11. The two electrochemists had several years of study invested and many apparently successful experiments under their belts, but this unfortunate series of events had forced them to reveal their results prematurely.[8]

> The whole generation of people who have actually seen something with their own eyes will have to die before cold fusion dies.[E]
> — Michael McKubre

On March 23, 1989, Pons and Fleischmann announced at a press conference at the University of Utah in Salt Lake City that a simple experiment they had conducted had resulted in sustained fusion at room temperature for the first time. The technique used by Pons and Fleischmann involved passing current through an electrolyte consisting of lithium dissolved in heavy water ($D_2O$) in which were suspended two metal electrodes, one of palladium and the other of platinum. Current was passed between these electrodes such that palladium was negative, and platinum was positive. Electrolytic action released deuterium ions from the solution at the palladium electrode, and these ions dissolved in the metal until a very high concentration was achieved. Pons and Fleischmann claimed that the electrical energy input into the system produced some heat, and the cold-fusion reaction had apparently generated additional or excess heat.

At a conference called by the University of Utah staff who were besieged by the media due to the premature distribution of Pons and Fleischmann's review paper, the press demanded more information regarding the experimental results. Lacking a peer-reviewed paper published in a refereed journal from which the experiment's details and results could be studied carefully and objectively, scientists around the world grabbed what limited data they could and hunkered down in their labs in an all-out attempt to confirm or reject the electrochemists' claim. In the US, millions of dollars were tentatively approved for aggressive research projects, and major efforts were initiated in Japan and India as well. The state of Utah established the National Cold Fusion Research Institute (NCFI) in 1989 with an endowment of $5 million. Staffed by an enthusiastic group of top-notch electrochemists, NCFI proved that cold fusion produced tritium, a nuclear product. Plausible theories and claims of success in detecting neutrons from fusion as well as excess heat flooded the scientific and popular media. Despite the hoopla, then-President George Bush said that the Utah claims were not fusion but should be investigated anyway.

> Science is about knowing. It's not about believing.[F]
> — Time

One month after the Pons and Fleischmann announcement, Secretary of Energy Admiral James Watkins established a review panel to evaluate the cold-fusion claims. Experiments at highly regarded institutions such as Yale University and the Massachusetts Institute of Technology (MIT) appeared to produce results in conflict with the Pons and Fleischmann data. Skeptical researchers at MIT's Plasma Fusion Center (PFC) called the chemists' data incorrect and then held a "Wake for Cold Fusion" theme party. On May 1, a group of scientists from Caltech led by Nate Lewis presented papers at a session of the American Physical Society in Baltimore, Maryland, that convinced their peers that cold fusion was an illusion and deserved no further study.[9] After five weeks of repeating every experiment, looking for procedural or data errors, the Caltech posse rejected the possibility of cold fusion. Although a later examination of the Caltech heat measurements revealed several errors that hid the existence of some excess energy, the credibility of cold fusion was crushed at that conference and others.

In fact, more than any other experiments in 1989, the negative results from three prestigious laboratories — Caltech, Harwell Laboratory (UK), and the MIT Plasma Fusion Center — convinced the scientific community that cold fusion was a bust. But the calorimetry work of these labs has since been criticized and determined to be technically flawed. In addition, failure to produce the anomalous effects in many of the early experiments was likely caused by a failure to achieve a sufficiently high deuterium concentration, a limitation unfortunately suffered by most palladium samples and which is now known to be required to produce a larger effect. The variables that influence this critical composition are only now becoming understood.

By July 12, 1989, the US Department of Energy had found no solid evidence for cold fusion, and it rejected plans for funding a formal research center for low-energy fusion. As more and more physicists, journals, and newspaper articles lambasted Pons and Fleischmann for sloppy and errant science, optimism for cold fusion quickly clouded over. Respected scientists lined up one-by-one to bury the claims by Pons and Fleischmann. In 1991, MIT Professor Ronald Ballinger, a faculty member of the departments of Nuclear Engineering and Materials Science and Engineering, wrote in the March/April issue of *Gordon Institute News*:

> "It would not matter to me if a thousand other investigations were to subsequently perform experiments that see excess heat. These results may all be correct, but it would be an insult to these investigators to connect them with Pons and Fleischmann."

In defense of his colleagues' professional lynching of Pons and Fleischmann, David Goodstein, a friend of Nate Lewis and Vice Provost and Professor of Physics and Applied Physics at Caltech, later suggested that because of the way the cold-fusion announcement was made, many people had an opportunity to make up their minds before what was taking place could be understood well enough

> If you had asked me in the first year or two of the controversy, I could not have imagined that a field as officially discredited and excoriated as cold fusion could hang on for so long.[G]
> — *Chemical & Engineering News*

> It would take at least 30 to 40 years to scale [Cold fusion] up, build plants and make this energy source a significant factor in the energy supply of the United States.[H]
>
> —Walter Youngquist

to be counted on to produce results reliably. It killed off funding and made cold fusion a pariah field, with many American scientists dismissing it out of hand merely because of theoretical objections.

Despite being rejected by mainstream science, the elusive search for "fusion in a jar" continued in laboratories worldwide throughout the 1990s. Researchers at the University of Hawaii reported multiwatt excess power output from electrochemical cells incorporating molten salts and deuterium. Scientists at the Los Alamos National Laboratory found electrolytic tritium production (a nuclear byproduct) in their cold-fusion experiments. Fifty-three electrolytic cells of various configurations and electrode compositions were examined for tritium production. Significant tritium was found in 11 cells at levels between 1.5 and 80 times the starting concentration after enrichment corrections were made. An October 1990 conference on "Anomalous Nuclear Effects in Deuterium/Solid Systems" at Brigham Young University also presented confirming evidence of energetic tritons in cold-fusion experiments. The conference produced indications that cold fusion can occur in deuterated systems without being catalyzed by muons.

> If the history of science reveals anything, it is that science does not grow and encompass new knowledge by *expectation*![I]
>
> —Eugene F. Mallove

Experiments conducted by the University of Texas and the US Naval Air Warfare Center Weapons Division "showed that eight electrolysis gas samples collected during episodes of excess power production in two identical cells contained measurable amounts of $^4$He whereas six control samples gave no evidence for helium. This places the $^4$He production rate at $10^{11}$ to $10^{12}$ atoms per watt of excess power, which is the correct magnitude for typical fusion reactions that yield helium as a product."[10] Over the years, scientists in the US, England, France, Italy, Russia, Japan, and India have also reported anomalous heat and low level radiation in their research. Several thousand technical papers now exist in the areas of cold fusion, LENR, and CANR; many of them published by reputable universities and laboratories from around the world.[11]

The emotionally charged debate has inspired books and articles either supporting or ridiculing the claims of

CANR researchers. In 1991, Frank Close authored a book *Too Hot to Handle*, which, among other things, accused Pons and Fleischmann of ethical violations. Several months later, MIT's chief science writer, Eugene Mallove, released a book, *Fire from Ice: Searching for the Truth Behind the Cold Fusion Furor*.[12] Mallove called the evidence for low-energy fusion "overwhelmingly compelling" and claimed that MIT's PFC staff had doctored their results to hide measurements of excess heat. The battle over cold fusion grew so acrimonious that Mallove angrily resigned his position at the MIT News Office, citing ethical violations by the MIT researchers who had attacked cold fusion. On August 18, 1991, Eugene Mallove filed a formal request for an investigation into MIT's mishandling of data in their 1989 PFC cold-fusion calorimetry experiment.

In 1992, Japanese researcher Akito Takahashi of Osaka University reported excess heat and low-level neutrons, and he gave a seminar at MIT on his heat and neutron-producing cold-fusion experiments. Shortly after, physical chemist Edmund Storms of Los Alamos National Laboratory announced that he had replicated Takahashi's results of excess heat using palladium from the same batch. In October 1992, scientists attending the Nagoya Third International Conference on Cold Fusion in Japan confirmed the reproducibility of excess heat and nuclear phenomena. In 1994, E-Quest Sciences claimed large amounts of helium production in tests of their ultrasonic cold-fusion reactor at the Los Alamos lab. The ongoing controversy as to the legitimacy of CANR gave birth to countless magazine and newspaper articles, television programs, and documentaries, as well as two Hollywood movies (*Breaking Symmetry* and *The Saint*).

Despite repeated claims of experimental success, prominent physicists continue to call cold fusion the scientific fiasco of the century. In contrast, futurist Arthur C. Clarke said that ignoring cold fusion "is one of the greatest scandals in the history of science." The PFC staff at MIT had led the initial charge against Pons and Fleischmann in 1989, but, in August 1992, Mitchell R. Swartz, an MIT graduate, published his "Reexamination of a Key Cold Fusion

> This field is subject to hype and disappointment, yet I'm seeing evidence now that hints that we may be on the verge of an energy breakthrough.[j]
> — Arthur C. Clarke

> The phenomenon claimed by Pons and Fleischmann is real, but it is only a small part of a much larger picture. [K]
>
> — Edmund Storms

Experiment."[13] In his careful review of the calorimetric testing done by the staff at the MIT Plasma Fusion Center, Swartz found numerous faults in their original analysis, as well as indications that data were inappropriately handled.[14] Ironically, in May 1995, MIT received a significant US patent (#5,411,654) relating to cold fusion: "Method of Maximizing Anharmonic Oscillations in Deuterated Alloys."

Funding agencies of the United States Government have ignored the cold-fusion field, but research has continued nonetheless. Early on, the Electric Power Research Institute (EPRI) had set aside some money for R&D, just in case, spending more than $6,000,000 in research at SRI International.[15] Additional funding was made available overseas. Pons and Fleischmann were asked to resign their university positions but quickly found work in a laboratory set up for them in France by the Toyota Company of Japan. This laboratory recently shut down because the Japanese have redirected their approach to the field.[16] At the Sixth International Cold Fusion Conference (ICCF-6) held in Japan in 1996, Pons presented the results of seven experiments, reporting 250% excess-heat in one and 150% in another. At the very same conference, the French Atomic Energy Agency reported a successful duplication using the same boil-off technique: i.e., a cold-fusion cell operating at near boiling temperatures.

*James Patterson and Jim Redding of Clean Energy Technologies, Inc., demonstrated the Patterson Fuel Cell on the television show* Good Morning America.

For more than a decade now, CANR researchers around the world have held many conferences and symposiums in order to advance their understanding of these unexplained nuclear effects. Because they lack strong political, financial, and professional support, there are few traditional scientific organizations where researchers in this field can share their results without being censured or criticized. Virtually all peer-reviewed professional journals refused to publish results from cold-fusion experiments. But, over the years, cold-fusion advocates have stubbornly resisted efforts by detractors to pull the plug on CANR research, and interesting results have kept rolling in.

In 1995, a company called Clean Energy Technologies, Inc. (CETI), was founded by inventor James Patterson to develop a novel cathode design, which he claims makes excess energy from ordinary water. In December of that year, the company showcased its Patterson Power Cell at Power Gen '95 in Anaheim, California.[17] The Patterson cell apparently generated 1,300 watts with an input of only 1.4 watts. Earlier in the year, a CETI demonstration of the power cell at the 16th Biannual Symposium on Fusion Energy produced a power ratio of 80:1. In 1997, the Office of Energy Intelligence at the DOE wrote CETI praising them for "the development of what looks to be a most promising innovation in energy production." Shortly after, the US Patent and Trademark Office (USPTO) issued patent #5,672,259 to James Patterson for his electrochemical device: "System with Electrolytic Cell and Method for Producing Heat and Reducing Radioactivity of a Radioactive Material by Electrolysis."[18]

The company is unique in that it now has at least seven granted patents describing methods that are derived from the original Pons and Fleischmann electrolytic approach. However, the excess energy does not result from "cold fusion," according to Patterson, but from another yet-to-be-determined reaction involving normal hydrogen. Unfortunately, the results have been found hard to duplicate. Despite that difficulty, the Patterson Process has gained attention as a way to neutralize radioactive waste. According to CETI, low-level nuclear transmutations take place on the cathode surface, which may be used to render dangerous radioactive waste relatively benign by turning them into more stable elements. This aspect of the technology has piqued interest at the Hanford Nuclear Plant

*Paul Brown is developing a technology that can stabilize dangerous radioactive waste.*

in Washington State, where millions of Curies of radioactive waste have been stored over the past 50 years. In June 1997, Norm Olsen, of the Hanford site, joined James Patterson and CETI executive Jim Redding in a demonstration of the Patterson Fuel Cell on the television show *Good Morning America*. Olsen, referring to the Patterson device, said, "If this technology works out as advertised, it means that we could significantly reduce the radioactivity of nuclear waste in the US and the world."

Mainstream scientists are unsure of how to deal with the growing claims that excess energy can be generated in these unusual chemical compounds. Clearly, the so-called "cold fusion" reaction is not like the well-known "hot fusion." This is both a curse and a benefit. It is a curse because it is hard for conventional science to account for the lack of radiation, but it's also a benefit because the expected dangerous radiation and radioactive waste are absent. It appears that scientists can study and use the effect without these major handicaps. Besides fusion, several other kinds of nuclear reactions apparently are initiated, some of which also produce excess energy. Researchers are faced with a growing realization that many nuclear transformations are possible and that some of these might be used to eliminate the radioactive waste created by nuclear fission and weapons production. In that vein, in 1999, nuclear engineer Paul M. Brown presented his paper "Effective Radioactive Waste Remediation" at the First International Conference on Future Energy held in Bethesda, Maryland.[19] Brown is a leader in cutting-edge nuclear technologies; his Colorado-based company utilizes photo-remediation to neutralize hazardous radioactive waste materials.[20]

Some scientists find it ironic that their experimental research is rejected because the results do not exhibit the expected nasty products. They suggest that with more de-

*Radioactive waste is nasty stuff. The US government estimates cleaning up the nation's nuclear weapons program will take at least 50 years and cost between $168 billion and $212 billion.*

velopment, low energy nuclear reactions may eventually be used to undo the damage caused by conventional nuclear processes. The current methodology for the transmutation of radioactive waste, a process that converts long-lived radioisotopes to short-lived isotopes, uses neutrons generated by a particle accelerator (ATW). In late 1999, the Department of Energy released a report that concluded that the implementation of ATW "will require additional research and will require a significant investment in research and development funding. In addition, complex institutional and public acceptance issues regarding acceptance of the technology would have to be addressed."[21] The US has adopted a policy of placing spent radioactive waste in a repository storage site, raising safety concerns for the long-term future. However, if by transmutation, the waste toxicity were reduced by a factor of about 1,000, then deep underground geologic storage would become unnecessary. There would no longer be a need to identify one very expensive, best site, and almost every area in each waste-producing country could identify several sites to handle that region's nuclear waste.[22]

A great deal of controversy currently surrounds CANR research because claims of heat excess and nuclear byproducts have been hard to duplicate and because they violate current understanding of nuclear reactions. These initial limitations are slowly being eliminated by many successful demonstrations and by a new understanding of the relationship between a nucleus and the surrounding chemical structure. Nevertheless, a very powerful institution-based skepticism remains, one that continues to view the studies as examples of poor science. The reasons +for this reluctance to view the accumulating evidence with an open mind are complex and not uncommon in science. In the case of cold fusion, the negative reaction is more extreme and of longer duration than is normal in face of such overwhelming evidence and such a great need for the possible benefits of cheap, renewable, and pollution-free energy.

There are strict protocols for confirming experimental data in the scientific community. For a new claim to be

> The best laboratories, like Mitsubishi, reported that seven out of seven experiments over the past two years produced excess heat and gamma rays.[L]
> — *Skeptic*

It's time for the US program to abandon its dead-end focus and to explore alternative paths to practical fusion power.<sup>M</sup>

— *Issues in Science and Technology*

believed, scientists require replication by other independent professionals, and they require that the same patterns of behavior can eventually be explained. Such duplication is sometimes difficult initially because the variables and processes causing the phenomenon are not understood. Only gradually, as more effort is applied, are the necessary conditions produced and the claimed results seen more often. During this exploratory phase, scientists need support from their peers and an ability to communicate for a better understanding of the results. Without institutional and financial support, cold-fusion research has progressed in fits and starts. As of year 2000, the requirements of reproducibility have been met. The anomalous results consisting of excess energy production, radiation from nuclear reactions, and the formation of elements that were not previously present have all been duplicated by hundreds of researchers worldwide using nine different techniques. The same patterns of behavior have been reported when similar measurements were made using a variety of detection devices. In addition, the theoretical understanding of cold fusion is now well underway.

Edmund Storms, a retired Los Alamos National Laboratory scientist, lists more than 50 published examples of excess heat production and nuclear reactions that can be cited, with more than ample peer-reviewed experimental verification. Many have passed the peer review process required of an acceptable replication.[23] This problem of acceptability demonstrates the conflict that any new-energy investigator will experience — early full disclosure may allow the scientific community to duplicate the claims, but this same disclosure will compromise the inventor's later attempts to obtain a patent and profit from the idea. Because the scientific community was not convinced the effect was real, the patent application filed by Pons and Fleischmann was totally rejected after much effort and money was invested.

*Ed Storms is concerned that palladium is a poor choice for cold-fusion research, and that platinum, gold, titanium, and nickel may work more reliably with fewer mechanical problems.*

Photo: Alternative Energy Institute, Inc.

This same fate has befallen hundreds of submitted patents in the US but not in many other countries.

Although not yet at a level to justify industrial development, several methods of generating excess energy are very reproducible in the hands of experts and are able to achieve a large energy amplification. With more research, this clean energy source has possibilities that are compelling. No reasons have been found that would prevent construction of a personal energy generator that could supply all energy required in a home for years. Hot water and electricity would be generated within a box perhaps the size of a refrigerator without any other sources of energy being used or needed. Meanwhile, pollution would diminish or be removed by the same process. The social consequences of this energy source are difficult to predict, but clearly the phenomenon has the potential to improve the lives of many people. We need only the will to explore this possibility.

Besides the electrolytic method, eight other techniques have been found to produce anomalous results.[24] In addition, most of these methods work when either heavy hydrogen (deuterium) or light hydrogen is used, although the nuclear products are different. It is hard to imagine how different anomalous results would be found using two very similar isotopes if the results were caused by error or delusion as the skeptics suggest. Claiming anomalous energy production is not the only issue. To further claim that this energy results from a nuclear reaction, a nuclear product must be identified. Production of normal helium has been found to result when anomalous energy is produced, either by using the original electrolytic method or by using the newer technique of gas loading finely divided palladium powder. In addition, the relationship between energy production and helium generation is consistent with the known energy of the fusion reaction. Although helium is not the only element found after a CANR reaction, it is the only one for which a quantitative relationship has been demonstrated.

> The suggestion that cold fusion could produce useful power moved it from the province of scientific curiosity into political and economic realms beyond the narrow scope of scientists.[N]
> — *Contemporary Physics*

> We have to evade being deceived by the junk science first, but the real trick is in knowing the difference.[O]
> — *R & D*

> An unusual network of what might be called "cold fusioneers" has sprung up around the country.[p]
> — *Popular Science*

Despite abundant and convincing evidence that cold-fusion effects are real, reproducibility has been difficult, and challenging political and professional obstacles still exist. Stiff opposition by the scientific community has cast researchers of CANR into the frustrating and unprecedented position of having to identify all reaction mechanisms and to provide totally reproducible evidence upfront before they can get a fair review of their research. The inability to reproduce results consistently is common with new discoveries as scientists work out reliable techniques and the physics behind the phenomenon. Many researchers in the field now believe that there is no simple cure for this problem: that is, no single experiment or commercial demonstration will convince a majority of scientists or even the skeptical American public.

Some proponents of cold fusion have not helped themselves or anyone else in the field by making wildly optimistic predictions that the energy can be produced simply. One researcher wrote, "Credibility in the field has been hurt by a large number of poor, uncritically done experiments and by a large body of poorly considered theoretical approaches."[25] Unfortunately, sloppy research or insupportable claims of viability and potential application reflect negatively and generally on all research associated with the cold-fusion effect. It is time for more rigorous and higher standards with quality experiments and sufficient funding. Today's leading cold-fusion scientists are among the world's most renowned electrochemists, including Martin Fleischmann, Fritz Will, and John O'M Bockris. Technical disagreements have split the field into competing rivalries, and improperly conducted experiments by unprofessional researchers have muddied the waters considerably for those reviewing the disparate results from various experiments.

Cold fusion remains difficult to reproduce, and many attempts fail, even today. But it is hoped that ongoing research will create a technique that will someday produce a predictable level of heat without the risk of harmful radiation or greenhouse gas emissions. Institutional skepticism and rogue researchers have hurt the credibility of professional and experienced scientists who have dared to spend their time, money, and expertise pursuing cold-fusion effects. Although cold-fusion scientists have developed a siege mentality and have been exiled by the status quo, they must learn that the public will listen if they take their message to them through lectures and popular media. They must open up and allow reporters and investors into their laboratories for a closer look. A working cold-fusion cell would instantly attract millions of dollars toward additional funding if presented properly. Technical progress and good experiments alone will not convince the public, and public support for cold fusion is essential to its success.

## FOOTNOTES

1. <www.altenergy.org/4/iccf8/iccf8.html> — Alternative Energy Institute's review of the 8th International Cold Fusion Conference held from May 21 to 26, 2000, in Lerici, Italy. AEI reports on the latest research being conducted in laboratories around the world in the field of cold fusion. Includes photographs, background profiles, and contact information.

2. <www.encyclopedia.com> — Definition of cold fusion from Electric Library Presents Encyclopedia.com. Brief description of fusion as a possible energy source and comments regarding the Pons and Fleischmann controversy. Contains links to other definitions including hydrogen, nuclear energy, and nuclear reactor.

3. <www.mv.com/ipusers/zeropoint/IEHTML/FAQ/faqbody.html> — This frequently Asked Questions (FAQs) webpage includes questions and answers pertaining to the subjects of Cold Fusion and New Energy Technologies. If you are new to this field, you will find the information interesting, and, if you have a question, ask them, and they'll try to answer it. This page is hosted by *Infinite Energy*.

4. Editor, "A Chronology of Cold Fusion," *Infinite Energy: Cold Fusion and New Energy Technology* (Issue 24, 1999), pp. 58-62. Detailed chronology of cold fusion from the Age of Alchemy — 500 BCE (Before Common Era) to 1800 ACE (After Common Era). Includes references to numerous conferences, publications, and professional debate both pro and con for chemically assisted nuclear reactions.

5. <wwwofe.er.doe.gov/> — US Department of Energy Sciences Program website offers the latest information on hot fusion, documents, education, program news and fusion, and the environment.

6. <www.triumf.ca/muh/pres/cam_ovhd/cam_ovhd.html> — Abstract and list of collaborators for Muon Catalyzed Fusion with a Muonic Tritium Beam, an October 1994 TRIUMF Muonic Hydrogen Collaboration project. Preliminary results are reported for a time-of-flight technique was used to measure the energy dependence for the rate for muon molecular formation in d-t fusion. Explains theory and motivation, as well as the formation and emission of muonic atoms in this process.

7. <www.jet.efda.org/> — Cooperative effort between the European Fusion Development Agreement (EFDA) and the Joint European Torus (JET) program. This website offers the latest information on the JET hot-fusion project and its history and role as a future energy provider. Includes scientific publications and worldwide links. Sign up for the EFDA-JET bulletin.

8   <www.infinite-energy.com> — Eugene F. Mallove, *Fire From Ice: Searching for the Truth Behind the Cold Fusion Furor* (John Wiley & Sons 1991/Reprinted 1999 by Infinite Energy Press), pp. 34–75.

9   <www.its.caltech.edu/~dg/fusion.html> — "Whatever Happened to Cold Fusion?" by David Goodstein. In-depth background to the Caltech investigation and the Pons and Fleischmann 1989 announcement.

10  Melvin H. Miles, Benjamin F. Bush, and Joseph J. Lagowski, "Anomalous Effects Involving Excess Power, Radiation, and Helium Production during $D_2O$ Electrolysis using Palladium Cathodes," *Fusion Technology* (July 1994), pp. 478-486. These experiments with precise helium measurements reported show simultaneous evidence for excess power, helium production, and anomalous radiation.

11  Eugene F. Mallove, "Key Experiments That Substantiate Cold Fusion Phenomena," *Infinite Energy* (Issue 24, 1999), pp. 29-34. Editor Mallove has selected 34 publications of experimental results with the salient conclusions of each study extracted. The presentation is designed to give a feel for the kinds of results and the variety of institutions reporting them. Mallove defies any previously uninvolved MIT student or graduate to examine the 34 references and conclude that the information is not strong enough to warrant further investigation and action.

12  Eugene F. Mallove, *Fire From Ice: Searching for the Truth Behind the Cold Fusion Furor* (John Wiley & Sons 1991/Reprinted 1999 by Infinite Energy Press).

13  <world.std.com/~mica/cft.html> — Cold Fusion Times website contains the "Journal of the Scientific Aspects of Loading Isotopic Fuels into Materials." This well-organized and comprehensive website produced by Mitchell Swartz and Gayle Verner covers New Hydrogen Energy and Low Energy Nuclear Reaction [LENR/FC] developments worldwide. Site also offers books for sale, reports, quality links to worldwide news services, and abundant information on the current status of cold-fusion research.

14  Mitchell Swartz, "Re-Examination of a Key Cold Fusion Experiment: Phase-II Calorimetry by the MIT Plasma Fusion Center," *Fusion Facts* (August 1992), pp. 27-40. Swartz is now the publisher and editor of the quarterly *Cold Fusion Times: The Journal of the Scientific Aspects of Loading Isotopic Fuels Into Materials*. For more information, contact Swartz at P.O. Box 81135, Wellesley Hills, MA 02181, USA.

15  <www.epri.com/> — The Electric Power Research Institute (EPRI) is recognized as a world leader in creating science and technology solutions for the energy industry and for the benefit of the public. EPRI serves more than 1000 energy organizations

worldwide. EPRI is the result of America's private and public utilities banding together to develop an industry-organized alternative to the US Senate's 1971 proposal to create a federal agency to conduct electricity-related research and development.

[16] <http://fomcane.nucl.eng.osaka-u.ac.jp/jcf/infoe.html> — The Japan CF-Research Society is focused on condensed-matter (solid state) fusion, coherently induced fusion, and cold fusion. (All terms refer to a nuclear reaction inside a solid-state body.) The Society's main goal is to investigate the nuclear reactions that occur in the solid-state and, ultimately, to develop techniques to extract useable energy from these reactions. The Society believes cold-fusion research requires an interdisciplinary/multidisciplinary approach involving scientists from many different fields.

[17] <http://members.nbci.com/hartiberlin/ceti.htm> — Latest information via *Infinite Energy* about Clean Energy Technologies, Inc. (CETI), and the Patterson Power Cell (PPC). Professor Miley from the University of Illinois and Professor Clinton Bowles of the University of Missouri have both independently verified the results of excess energy from the PPC cell and from their own rebuilt cells. This website contains photographs of the PPC cell and the transcript from ABC television's February 7, 1996, *Good Morning America* program, which profiled CETI and the PPC cell.

[18] <www.altenergy.org/3/new_energy/cold_fusion/patterson/patterson.html> — Profile on Clean Energy Technologies, Inc., (CETI) and the patented Patterson Power Cell developed by James Patterson. The cell is modeled on the work of Pons and Fleischmann, but there are noticeable differences. Both techniques involve the absorption of an isotope of hydrogen into a metal lattice, but the Patterson fuel cell uses a pile of metal-coated plastic beads as the cathode instead of a solid piece of palladium metal.

[19] <www.altenergy.org/4/cofe/cofe.html> — Paul Brown presented his paper "Effective Radioactive Waste Remediation" at the First International Conference on Future Energy, held in Bethesda, Maryland, in 1999. Brown proposed using a linear accelerator of the monochromatic type to "burn-up" spent fuel from fission reactors. Photo-fission reduces the half-life and increases the stability of radioactive waste fuels such as Cs-137 and Sr-90. Chemical separation of the spent fuel isotopes is not necessary.

[20] <www.globalatomics.com/> — Global Atomics, Inc., develops and implements advanced solutions for nuclear applications. Their cutting-edge technologies are intended to eliminate the threat of nuclear waste production and containment, as well as to facilitate the complete neutralization of radioactive waste.

[21] <www.pnl.gov/atw/> — This report is a synthesis of information gathered from worldwide experts, national laboratory staff, and individual consultants on developing accelerator transmutation of waste (ATW) technology and is available on the Energy Department's homepage.

[22] Charles D. Bowman, "Accelerator-Driven Systems for Nuclear Waste Transmutation," *Annual Review of Nuclear and Particle Science* (Volume 48, 1998), pp. 505-507.

[23] <www.padrak.com/ine/NEN_5_10_4.html> — Fusion Report Summary: "Cold Fusion: An Outcast of Science," *21st Century Science & Technology*, Volume 10, No. 4, Winter 1997-98, pp. 19-26 by Edmund Storms. Additional descriptions of these efforts can be found in "Cold Fusion Revisited," *Infinite Energy*, Vol. 4 No. 21 (1998), p. 16, and in "Review of the Cold Fusion Effect," *J. Scientific Exploration*, Vol. 10, No 2 (1996), p. 185.

[24] <http://home.netcom.com/~storms2/> Edmund Storms' website offers the latest review of current research in cold-fusion technology. See in particular: "Ways to Initiate a Nuclear Reaction in Solid Environments."

[25] Jed Rothwell, "Comments on the Tenth Anniversary Contributions," *Infinite Energy* (Issue 24, 1999), p. 23. Quote by Oriani.

## SIDEBAR FOOTNOTES

[A] Dale L. Sullivan, "Exclusionary Epideictic: NOVA's Narrative Excommunication of Fleischmann and Pons," *Science, Technology, & Human Values*, Vol. 19, No. 3 (Sage Publications, Summer 1994), p. 300.

[B] Charles Platt, "What If Cold Fusion Is Real?" *Wired* (November 1998), p. 230.

[C] Walter Youngquist, *GeoDestinies: The Inevitable Control of Earth Resources Over Nations and Individuals* (National Book Company 1997), p. 227.

[D] Tadahiko Mizuno, *Nuclear Transmutation: The Reality of Cold Fusion* (Concord, New Hampshire: Reprinted by Infinite Energy Press, 1998), p. 104. (Translated by Jed Rothwell.)

[E] David Kestenbaum, "Cold fusion — science or religion?" *R & D* (April 1997), p. 54.

[F] Dick Thompson, "Putting the Heat on Cold Fusion," *Time* (May 15, 1989), p. 63.

[G] Ron Dagani, "Cold fusion lives — sort of," *Chemical & Engineering News* (April 29, 1996), p. 69.

H  Walter Youngquist, *GeoDestinies: The Inevitable Control of Earth Resources over Nations and Individuals* (Portland, Oregon: National Book Company, 1997), p. 228.

I  Eugene F. Mallove, *Fire From Ice: Searching for the Truth Behind the Cold Fusion Furor* (Reprinted by Infinite Energy Press, 1999), p. 211.

J  Interview with author and futurist Arthur C. Clarke, "God, science, and delusion," *Free Inquiry* (Spring 1999), p. 36.

K  Edmund Storms, "My Life With Cold Fusion As A Reluctant Mistress," *Infinite Energy* (Issue 24, 1999), p. 47. (Talk given at the Cold Fusion and New Energy Symposium, October 11, 1998, Manchester, New Hampshire.)

L  Jed Rothwell, "Cold Fusion Still Hot," *Skeptic* (Spring 1999), p. 22.

M  Robert L. Hirsch, Gerald Kulcinski, and Ramy Shanny, "Fusion research with a future," *Issues in Science and Technology* (Summer 1997), p. 60.

N  P. T. Greenland, A book review of *Cold Fusion: The Scientific Fiasco of the Century*, (Volume 35, No. 3. 1994), p.210. Book written by J. R. Huizenga, (Oxford University Press, 1993).

O  Fred Jueneman, "Fusion and Diffusion," *R&D* (August 1999), p. 11.

P  Jerry E. Bishop, "It ain't over till it's over…Cold Fusion," *Popular Science* (August 1993), pp. 47–51.

# Zero-Point Energy: Sailing the Sea of Energy

## 14

Scientists working at the frontier of human knowledge are tantalized by the lure of a potential carbonless energy source known as Zero-Point Energy (ZPE). ZPE fills the fabric of space with unlimited quantities of energy — the trick is how to tap it. Henry T. Moray and Nikola Tesla described the nature of zero-point energy and designed and built equipment in an attempt to engineer its properties. If scientists can effectively cohere this enigmatic but omnipresent energy source, the consumption of fossil fuels will plummet, and ZPE will help provide an unlimited supply of inexpensive, nonpolluting electrical power extracted from the vacuum of space.[1]

CLASSICAL PHYSICS DICTATES that the vacuum is devoid not only of matter but also of energy.[2] The word usually suggests uninteresting empty space, but to modern quantum physicists, the vacuum has turned out to be rich with complex and unexpected behavior. They describe it as a state of minimum energy where quantum fluctuations, consistent with the uncertainty principle of physicist Werner Heisenberg, can lead to the temporary formation of particle-antiparticle pairs.[3] Before the advent of quantum theory, classical physics taught that any simple, real-world oscillator, such as a pendulum, when excited, would eventually come to rest if not continuously energized by some outside force, such as a spring. Friction kills momentum. Then quantum theory came along declaring that an oscillator does not come to total rest but that it actually continues to "jiggle" randomly about its resting point with a very small amount of energy always present. Scientists call the energy produced by fluctuations of the electromagnetic and gravitational force fields in the vacuum *zero-point energy* (ZPE).

In quantum-field theory, the vacuum is considered to be the state having the least energy density. Scientists use quantum-field theory to conduct modern particle physics investigations, and the concept of vacuum energy appears in certain computations of this complex theory. In the real world, particles interact with one another through a variety of forces, but researchers use the simplified, so-called "free-field theory" in which particles do not interact at all to develop mathematical models.

Conventional physicists use the free-field theory and its associated ZPE values as a mathematical tool to help them understand the accepted interacting theory. The concept of vacuum energy shows up in many of these computations in quantum field theory, the paradigm used to conduct modern particle physics research. Because the vacuum state of the interacting theory is the state of least positive energy in reality, mainstream scientists insist that there is no practical way to extract and tap the vacuum energy as an energy source. Tapping ZPE is a worthwhile goal, but no one has yet discovered how to exploit this cosmic jitter with a genuine device that can deliver sizable amounts of energy.[4] However, Frank Mead, who works for the Advanced Propulsion Air Force Laboratory, received a patent on a zero-point energy antenna (#5,590,031) in 1996 for a "system for converting electromagnetic radiation energy to electrical energy." Though it is theoretical, the invention nicely capitalizes on the fact that the captured ZPE becomes more powerful as the receiver gets smaller, causing the resonant frequency to increase.

There is no doubt that vacuum fluctuations affect the behavior of microscopic particles and the world around us. In 1968, Russian physicist Andrei Sakharov speculated that

> A special class of aether theories describes space as a sea of fluctuating energy. They are significant because quantum physics predicts that vacuum fluctuations exist and gives them the name zero-point energy.[A]
> — Moray B. King

gravitation might not be a fundamental interaction but rather a secondary effect associated with other nongravitational fields. Sakharov theorized that gravity might be an induced effect brought about by changes in the ZPE of the vacuum due to the presence of matter.[5] His pioneering insight was based on the assumption that the electric component of the zero-point field causes charged particles to oscillate, and this vibration gives rise to a secondary electromagnetic field. The secondary electromagnetic field then reflects back onto the primary field. This subatomic cosmic dance results in an attractive force between the particles, thus causing gravitation.

Greek philosophers considered the vacuum to be devoid of matter and energy, but, late in the 19th century, it became apparent that the region still contained thermal radiation. The "zero" in zero-point refers to the fact that if an evacuated container (with no air left) were chilled to absolute zero (-273 Celsius), the lowest possible energy state, there still exists in the container a nonthermal radiation zero-point radiation. Defined another way, ZPE results from a finite, minimum amount of motion (kinetic energy) in all matter, even at absolute zero. Therefore, an "empty vacuum" is actually a seething cauldron of energy with particles flashing into existence for microseconds before being absorbed back into an invisible sea of energy. For example, the jittering of an electron, called *zitterbewegung*, occurs at frequencies higher than $10^{21}$ Hz. Proponents of ZPE point out that, although the particles in the zero-point field are minute, there are so many possible modes of expression (frequencies, directions) in open space, the total ZPE summed up over all possible modes may be enormous.[6] But many mainstream physicists wonder how we'll ever obtain it.

Scientists radically disagree over just how much ZPE is available for possible extraction. Jordan Maclay, a former professor of electrical engineering at the University of Illinois in Chicago, who has secured funding from NASA to study the energy of the vacuum, has calculated that a region of vacuum the size of a proton could contain as much energy as all the matter in the entire Universe.[7] Nobel

> All the fundamental characteristics we normally associate with matter are vacuum interaction products: inertia, mass, as well as gravity.[8]
> — *The Whispering Pond: A Personal Guide to the Emerging Vision of Science*

prize-winning physicist Steven Weinberg, presently at the University of Texas in Austin, has an opposing view. Weinberg agrees that the total amount of ZPE in the Universe is abundant, but he believes that all of the zero-point energy available in a space the size of Earth is equal to the energy contained in one gallon of gasoline.[8] Nobel Laureate Richard Feynman and one of Einstein's protégés, John Wheeler, calculated that there is more than enough energy in the volume of a coffee cup to evaporate all the world's oceans. Energy density in the zero-point field is just one of the hotly contested questions bouncing around the physics' community regarding this topic.

The physical effect of zero-point energy in vacuum fields has been detected, but just barely. In theory, vacuum energy takes the form of particles that are constantly forming and disappearing on a constant but subatomic scale; these particles are difficult if not impossible to observe. The energy randomly pops out of empty space in the form of particle-antiparticle pairs that destroy each other nearly instantly. In flat space-time, destruction follows closely upon creation; the pairs are said to be "virtual" because there is no apparent source of energy to give the pair permanent existence. These particles include lighter electrons and photons, as well as heavier particles such as protons, neutrons, mesons, and baryons. Theorists conjecture that the gravity field generates massless graviton-antigraviton pairs. It is an elusive exotic stew, but there is no doubt as to the existence of ZPE. This energy field produces verifiable effects; it excites the atoms in fluorescent lamps, generates low-level noise in microwave receivers, plays a role in the surface tension of liquids, and takes part in other physical phenomena. To study ZPE, scientists have used the Casimir force of attraction between two metal plates in a vacuum and the Lamb shift in the wavelength of spectral lines of atoms.

Modern zero-point energy research has languished with relatively little conventional interest or much funding, but scientific curiosity actually predates the formalism of quantum mechanics in 1925, which confirmed the existence of ZPE. Quantum mechanics asserts that just considering the fluctuation of the electromagnetic force, any given volume

> The Casimir effect is important evidence for the physical reality of vacuum fluctuations and vacuum zero-point energy, but the Casimir force can be derived from points of view that do not employ the concept of vacuums.[c]
> — *Studies in History and Philosophy of Science*

of empty space could contain an infinite number of vacuum-energy frequencies — and, therefore, an infinite supply of energy. Proponents believe that ZPE is energy from the vacuum continuum and is responsible for gravity and inertia as well as the Lamb shift and Casimir force. Researchers have called ZPE "zero-point electromagnetic radiation energy" or referred to it as "a flux of virtual particles," but, most important, nearly all agree that the quantum mechanical zero-point oscillations are real.

Few contemporary physicists deny the validity of the zero-point energy field theory, but most consider vacuum-energy calculations unrealistic and professionally naïve. They also wonder how anyone will trap and harness the energy from the flickering particles that are not only subatomic but sub-elementary in size before they disappear. ZPE has inspired passionate arguments throughout its history. Historically, scientists believed that space was comprised of a material aether capable of supporting the propagation of light waves. Paradoxically, this aether would have to be as stiff as a dense solid to manifest the high velocity of light yet be tenuous enough to allow matter to travel through it.

In the late 1800s, physicists A. A. Michelson and E.W. Morley used a sensitive optical device in an attempt to detect the velocity of the Earth with respect to the hypothetical luminiferous aether. They reasoned that if the speed of light were constant with the proposed aether through which the Earth was moving, that motion could be detected by comparing the speed of light in the direction of the Earth's motion and the speed of light at right angles to the Earth's motion. No difference was found. The negative results from the Michelson-Morley experiment discredited contemporary aether theories, an outcome that may have come as a relief to a scientific community attracted to the postulates of relativity. For more than a century, most scientists have considered empty space to be a void, a view typically supported in today's textbooks. The idea of an energetic aether was discarded quickly, which ultimately led to Albert Einstein's 1905 theory of relativ-

> The "zero" in zero-point refers to the fact that if you were to cool the Universe to absolute zero, its lowest possible energy state, some energy would remain... rather a lot of energy.[D]
> — *New Scientist*

ity in which the speed of light is assumed to be a universal constant.

However, in 1987, Ernest Silvertooth published the results of an experiment that clearly showed that opposed laser interference fringes were minimum when the beams were aligned along a certain direction in the sky (toward the constellation Leo). When the path of the opposed laser beams was rotated away from that heading, the fringes spread apart to greater distances. He concluded that his "New Michelson-Morley Experiment" marked the direction of the Earth's motion through the aether. Furthermore, by measuring the minimum fringe spacing, Silvertooth determined that the Earth moves through the aether toward Leo at a speed of about 378 (±19) kilometers per second.[9]

Physicist Maxwell Planck's second theory of blackbody radiation, published in 1912, marked the birth of the concept of zero-point energy. It counted Albert Einstein and German physicist Werner Heisenberg among its early adherents. Although the Einstein-Stern theory of specific heats invoked ZPE in a way that later turned out to be incorrect, the experimental implications of ZPE were recognized in vibrational spectroscopy and x-ray diffraction.

Max Planck's equation expressing ZPE is correct according to modern theory, although his route to it was not. Planck's zero-point electromagnetic energy in the equation is also correct from the standpoint of modern quantum theory, but just two years after he published his second theory, Planck became convinced that ZPE would be of no experimental value. Despite Planck's dismissal of ZPE, the concept attracted much attention and soon came to play a major role in the work of Einstein and Stern, who showed that it could be used to derive the Planck spectrum from largely classical considerations.[10] Given the equivalence of mass and energy expressed by Einstein's $E = mc^2$, vacuum energy must be able to create particles. Ironically, it was the

*German physicist Maxwell Planck (1858 – 1947) won a Nobel Prize for proposing that electromagnetic radiation can be emitted and absorbed only in definite units, which he called quanta. His explanation laid the foundations for quantum theory.*

great simplicity in Einstein's 1917 equations that ended speculations about the role of ZPE in blackbody radiation emission, although it is now believed that the zero-point energy of the electromagnetic field affects the coefficient for spontaneous emission. Furthermore, this ZPE appears to be important in connection with the Einstein fluctuation formula, historically the first indicator of wave-particle dualism and Bose-Einstein statistics.[11]

Interest in ZPE declined after Albert Einstein's 1917 paper, but it was not completely abandoned. Subsequent research by R. Mulliken provided direct spectroscopic evidence for the reality of ZPE in 1924, just before it appeared in the quantum formalism established during 1925 and 1926. During the 1930s, the concept of ZPE was applied to a variety of problems in physical chemistry, but most physicists paid little attention to this vague energy field. Scientific curiosity increased in the late 1940s when American physicist Willis Lamb measured a slight frequency alteration or shift (Lamb shift) in the spectral line of an excited hydrogen atom. In 1947, through sophisticated measurements, Lamb showed that the two possible energy states of hydrogen differed by a very small amount, contradicting predictions by English physicist Paul Dirac, who had posited that these energy states were equal. The discovery of the Lamb shift required physicists to revise their understanding of the interaction of the electron with the ZPE electromagnetic radiation.[12] Since the early days of quantum mechanics, Dirac had theorized that the vacuum was actually filled with particles in negative energy states, a physical vacuum that was not empty at all. This prediction of Dirac led to the discovery of the positron, the antiparticle of the electron.

The ground state of the electromagnetic field is not zero, but rather the field undergoes vacuum fluctuations that interact with the electron. In 1948, researchers at the Philips Laboratory in the Netherlands studying the van der Waals force — a weak attraction between neutral atoms — discovered that at long distances, the van der Waals force weakened unexpectedly. Philips scientists Hendrick Casimir and Dik Polder figured that this weakening re-

> For decades after Planck and Heisenberg described the zero-point energy, physicists preferred to ignore it.[E]
> — *Science*

sulted from correlated zero-point fluctuations in the electric field, which would propagate from atom to atom at the speed of light (Casimir-Polder effect). Because of the lag, the likelihood that the atoms would be influenced by fluctuation would diminish with distance.

Casimir and Polder proposed testing for the presence of ZPE by placing two thin uncharged metal plates very closely together (as a capacitor). Casimir argued that the vacuum is filled with particles of varying wavelengths, and, if some of the virtual photons were excluded from the tiny gap between the plates, the photons outside the plates would generate a minute pressure to push the plates together. The closer the plates were pushed together, the stronger the attraction would be. This "Casimir effect" was confirmed experimentally by M. J. Sparnaay in 1958 with nonconductive plates.

More recently, a widely noted experiment by physicist Steven K. Lamoreaux of the Los Alamos National Laboratory gave a very precise and unambiguous confirmation of the existence of the Casimir force.[13] To measure the Casimir effect, Lamoreaux positioned two gold-coated quartz surfaces less than a micrometer apart, one of them attached to a torsion pendulum while the other was fixed. The narrow gap between the two surfaces allowed only small, high-frequency electromagnetic particles of the vacuum energy to squeeze in between, while, outside that tiny space, the full complement of virtual particles popped in and out. In much the same way that an airplane wing creates low pressure on one side, the zero-point energy on the outside outweighed the ZPE inside, forcing the surfaces together.[14]

Lamoreaux's results agreed to within 5% of Hendrick Casimir's prediction for that particular plate separation and geometry. Critics pointed out that the ZPE influence on the plates generated only about 100 microdynes of force, which Lamoreaux wrote, "corresponds to the weight of a blood cell in the Earth's gravitational field."[15] Lamoreaux's textbook experiment extracted a paltry amount of energy from the ZPE, but his measurement of an energetic vacuum confirmed a very basic prediction of quantum electrodynamics (QED), that there is an all-pervading vacuum that con-

> In the last 50 years, the study of Casimir effects has evolved from an esoteric effort dealing with one of the least intuitive consequences of quantum electrodynamics to a very active and multifaceted research area.[F]
>
> — The American Physical Society: *Physical Review B* 1999

tinuously spawns particles and waves that spontaneously pop in and out of existence. This simple experiment should help physicists accept that the subatomic world is as strange as quantum mechanics predicts and that ZPE deserves more research.[16]

Physicist, author, and President of Integrity Research Institute Thomas Valone has written, "Physical theories predict that on an infinitesimally small scale, far smaller than the diameter of an atomic nucleus, quantum fluctuations produce a foam of erupting and collapsing virtual particles, visualized as a topological distortion of the fabric of space-time."[17,18,19] The churning quantum foam extends throughout the Universe, even filling the empty space within atoms. Valone posits the experimental evidence for the existence of ZPE is: 1) the Casimir effect, 2) the Lamb shift, 3) van der Waals forces, 4) diamagnetism, 5) spontaneous emission, 6) microdegree liquid Helium, 7) quantum noise, and, most recently, 8) cosmological antigravity. Awaiting experimental verification is that inertia and gravity are also proof of zero-point energy.[20] Valone also emphasizes that a recent discovery by Eberlein that sonoluminescence (blue light from cavitation) occurs too fast to be caused by the usual electron transitions and that the light emission spectrum matches only a ZPE spectrum.

Demonstrating the existence of zero-point energy in a laboratory is one thing; extracting useful amounts of energy from it is a different challenge altogether. Because it exists in a vacuum, ZPE is homogeneous (uniform) and isotropic (identical in all directions) as well as ubiquitous (exists everywhere). Skeptics of ZPE theory argue that if the entire Universe is filled with zero-point energy, its gravitational field should affect the structure and evolution of the Universe. In relative theory, all energy is associated with mass and, therefore, with gravity.

The conventional view is that the energy in the vacuum is limited because, if it were infinite, the structure of the Universe would be completely different. The fabric of space and time, though slightly curved near objects, is essentially flat overall. Because energy is equivalent to matter, and matter exerts a gravitational force, scientists expect

> In the Casmir effect, the negative energy density between the plates can persist indefinitely, but large negative energy densities require a very small plate separation.[G]
>
> — Scientific American

that an energy-rich vacuum would create a strong gravitational field and distort space and time. Consequently, if vacuum energy were indeed abundant, it would take more than the so-called "cosmological constant" first developed by Albert Einstein to counteract the gravitational force generated by ZPE.[21] In 1998, astronomers announced that, indeed, ZPE was the only candidate for the antigravity force they detected in distant galaxies that were found to be accelerating *away* from each other. Apparently, ZPE has affected the evolution of the Universe, and the cosmological constant of Einstein's may be right after all.[22]

The quantum and classical pillars of modern physics are not architecturally compatible in that the electromagnetic zero-point field (ZPF) does not appear to produce the expected classical gravitational effects. In response to this cosmological conundrum, new theories have been proposed. Hal Puthoff, who is president and CEO of EarthTech International, Inc., and director of the Institute for Advanced Studies in Austin, Texas, has proposed that ZPE is not something set by the initial conditions of the Universe — like the cosmic microwave background radiation associated with Big Bang.[23,24] (Some cosmologists have speculated that at the beginning of the Universe, when conditions were similar to the inside of a black hole, vacuum energy was very high, but after the Big Bang, energy levels dropped off.) Puthoff suggests that the motion of charged particles generate ZPE and that charged particles, in turn, get their vibrational energy from the zero-point field.[25] Hal Puthoff's novel and controversial ideas are predicated on his as-yet-unproved hypothesis that ZPE is what keeps electrons in an atom orbiting the nucleus.[26] Classical physics dictates that circulating charges, like an orbiting electron, lose energy through radiation, but Puthoff believes that it is ZPE the electron continuously absorbs in order for it to keep zipping around the nucleus.[27]

Puthoff explains it this way:

> [It] is shown that the electron can be seen as continually radiating away its energy as predicted by classical theory, but simultaneously absorbing a

> In *Scientific American's* PBS broadcast, [Harold Puthoff] predicted that just as the twentieth century was known as the nuclear age, so will the next millennium be known as the zero-point energy age.[H]
>
> — *Skeptical Inquirer*

compensating amount of energy from the ever-present sea of zero-point energy in which the atom is immersed, and an assumed equilibrium between these two processes leads to the correct values for the parameters known to define the ground-state orbit. Thus the ground-state orbit is set by a dynamic equilibrium in which collapse of the state is prevented by the presence of ZPE. The significance of this observation is that the very stability of matter itself appears to depend on the presence of the underlying sea of electromagnetic zero-point energy.[28]

Puthoff exhibits an indomitable enthusiasm for harnessing the power of ZPE. Over the last decade, the Institute for Advanced Studies at Austin has inspected about ten devices for tapping the energy of space, but, upon close examination, all of them have failed to provide overunity energy.

Puthoff believes that the connection between the cosmological constant and ZPE is much more complex than previously realized and notes that the predictions of quantum mechanics have repeatedly proven that there are still missing pieces to the puzzle. Measuring and tracking the behavior of subatomic particles is problematic. In 1927, Werner Heisenberg determined that it is impossible to learn both the position and the momentum of a particle to a high degree of accuracy; if the position is known perfectly, then the momentum is completely unknown, and vice versa. ZPE emerges from Heisenberg's uncertainty principle, which limits the accuracy of these measurements. The virtual particles flash briefly into existence and then disappear within an interval dictated by the uncertainty principle.

*Hal Puthoff specializes in fundamental electrodynamics and has led the development of a theoretical framework within which zero-point energy can be understood.*

Time and energy, like position and momentum, are also subject to the uncertainty principle. If the time during which energy is measured is known exactly, the amount

> As a self-organization system with energy and matter exchanging externally, excess energy can be generated due to torsion coherent with zero-point energy in the vortex state on the tips of electrodes.[1]
>
> — *Journal of New Energy*

of energy becomes uncertain. The shorter the time interval, the greater the uncertainty. If an electron is trapped in an increasingly confined space and its position becomes more accurately known, the uncertainty relation ensures that its momentum will increase as it strikes the containment walls with greater force and frequency. This pressure of electrons within every atom preserves the atom in what is called its "ground state." The electron cannot be totally motionless because, then, its position would be precisely known. Heisenberg's famous uncertainty forbids it to become motionless.

University of Waterloo, Ontario, Canada, physics professor Paul Wesson agrees with some of Puthoff's ideas, but he insists that the resulting energy density of the vacuum must, in fact, decrease as the Universe expands, pushing back the ultimate source of the energy to the origin of the Universe itself. Wesson has shown that the gravitational influence of this energy cannot be ignored, so either the zero-point field (ZPF) does not gravitate (in violation of relative theory), or somehow the virtual photons of the ZPF do not have time during their brief lives to interact gravitationally with ordinary matter. Professor Wesson states that:

> "Zero-point fields are predicted by interactions that are well-described by quantum mechanics, but the energies involved do not manifest themselves in terms of the curvature of space-time as formalized in general relativity or in the energy density of the cosmological vacuum as measured by the cosmological constant."

He considers this peculiar situation arguably unique in modern physics.[29]

Zero-point fields predicted by particle physics are many orders more intense than evidence gleaned by astrophysical observation. Wesson points out that there are several possible resolutions for this contradiction, but the consensus is that the ZPFs associated with the interactions of particles must in some way cancel, perhaps due to the operation of a physical principle such as super-symmetry. Paul Wesson indicates that research over the last decade has shown that problems exist about how to reconcile the zero-point fields that follow from quantum mechanics with the energy conditions built into classical gravitational theories such as general relativity. Wesson's independent research, which is funded by NASA and the Natural Sciences and Engineering Research Council, has identified 11 topics wherein past research has run into difficulties. Wesson believes these problems are, in principle, surmountable, and he recommends six topics that are important for future pursuit. These suggestions should lead to a resolution of the difficulties, which exist in the amalgamation of quantum mechanics and classical field theory. Professor Paul Wesson's conclusion is that "research into the zero-point field is justified because it is of fundamental academic importance and of potential importance to technology."[30]

Thomas Bearden, Director of the Association of Distinguished American Scientists, has ignited controversy with his proposal that generators and batteries do not furnish any of their internal energy to their external circuit but only dissipate it internally to perform work on their own internal charges to form a source dipole. Thomas Bearden has spent many years attempting to develop an operational overunity electrical engine, as well as other alternative uses for electromagnetic energy. In a recently-written position paper, the former nuclear engineer with degrees in mathematics and aerospace engineering, states that:

> "...once [it's] formed, the dipole's broken symmetry extracts observable energy from the virtual particle exchange between dipole charges and active vacuum. The tiny *Poynting* fraction intercepted by the surface charges enters the circuit to power it, while the huge non-intercepted *Heaviside* fraction misses the circuit and is wasted. [In this way] electrical loads are powered by energy extracted from the vacuum, not by chemical energy in the battery or shaft energy input to the generator."[31]

Bearden insists that self-powering systems that extract electrical energy from the vacuum to power themselves

> [The zero-point field theory] is one emerging area of physics that has the potential for a breakthrough in our basic understanding of science with implications for novel energy and transportation applications.[J]
> — *Breakthrough Energy Physics Research Program Plan*

and their loads can be developed whenever the scientific community decides to fund them.

Thomas Bearden, a retired Army Lieutenant Colonel, has more than 30-years experience in air defense systems, technical intelligence, and Soviet electromagnetic weaponry.[32] But this former military man breaks rank with mainstream physicists when he asserts:

> "Electrical energy can be extracted copiously, easily, and cheaply from the vacuum at will — anywhere, anytime. The reason this has escaped notice for so long is that two changes to electrodynamics were made by Lorentz, and those changes — still applied to limit things to a subset of electrodynamics — have hidden it for a century."[33]

In another paper, Bearden argues:

> "What has remained hidden (since Lorentz arbitrarily discarded it from electrodynamics theory, stating it was "physically insignificant") is that every generator and battery already extracts and pours forth enormous EM (electromagnetic) energy — directly from the vacuum. [Only] about one ten-trillionth of the energy poured forth and filling all space surrounding a circuit actually strikes the circuit and gets diverged into it to power it up."[34]

Brian K. Schimmoller, *Power Engineering* managing editor, describes a future world powered by zero-carbon technologies, space-based solar power, and nuclear fusion in "Magicians Wanted," an August 2000 editorial opinion.[35] Both of these expensive, potential technologies have questionable technical viability, as well as commercialization schedules measured in decades, not years. In a written response to Schimmoller's article, Thomas Bearden states that extracting usable EM energy from the vacuum is "pure magic in a sense, but it is also a very rigorous kind of physics."[36] In his 15-page letter to *Power Engineering*, Bearden cites 19 references from refereed publications, plus nearly two dozen serious scientific papers on extracting EM energy from the vacuum. He points out that electrodynamicists have largely ignored and arbitrarily discarded EM from the energy flow theory for a century and

> Demonstrating the existence of zero-point energy is one thing; extracting useful amounts is another.[K]
> — *Scientific American*

have ignored the broken symmetry of a dipole — proven in particle physics for nearly 50 years — in its vacuum flux exchange.

Working with Myron Evans and Mendel Sachs at the Alpha Foundation's Institute for Advanced Study (AIAS), Bearden believes that they have deciphered the vacuum source of the enormous nondiverged energy flow component and a mechanism that provides it from the active vacuum.[37] They have shown that in the proper model, it is very clear that electrodynamic energy can readily be taken from the vacuum in copious quantities. Bearden insists that the way to solve the energy problem is two-fold. Scientists must first update the century-old notions in electrodynamic theory of how an electrical circuit is powered and, second, rid the classical electrodynamics model of serious foundational flaws. To that end, Thomas Bearden and his colleagues have prepared a series of scientific papers. Three scheduled for publication in Russia deal with the exact mechanisms, including the experiments that prove the excess energy is there and can be intercepted and utilized.[38]

Bearden's controversial concepts occasionally elicit stern rebuttal among the traditionally inclined, but he is not the only one pushing the boundaries of contemporary physics. Research in high-density charge cluster (HDCC) technology has yielded evidence of excess energy and low-energy nuclear transmutation when certain materials are bombarded by bursts of electron clusters. High-density charge clusters can be formed in a near vacuum by a short pulse of negative potential applied to a specially designed cathode. A typical charge cluster will impact a witness plate (a thin, metal foil placed near the anode) and leave various-sized holes or blisters in the metal foil. Single clusters, as produced in the lab, may vary in size from less than a micron to several microns. Higher energies can create a necklace of clusters.[39]

*Thomas Bearden believes that tremendous amounts of electromagnetic energy are wasted in contemporary electrodynamics.*

Charge clusters are formed by many types of electrical discharges and are evident when a spark impacts a metal surface. HDCCs are potent micron-sized plasmoids that contain a net charge of about 100 billion electron volts. They are created in the laboratory by the application of a short pulse of a few hundred to tens of thousands volts to a cathode positioned adjacent to a dielectric. The electron clusters (EVs) maintain a stable configuration even though they consist primarily of electrons. There is some unknown binding force at work that holds the electrons together in a cluster. One analysis suggests that its anomalous stability is due to a thin, helical vortex ring filament that possesses an extraordinary rotational velocity. (A spherical electron cluster is unstable and would tend to form into a toroid by a force balance relationship.) The charge cluster has been described as a toroidal electron vortex that can exist at various combinations of electron densities, electron velocities, and cluster sizes.[40] The cluster's toroidal form indicates a high dynamic in that the electrodynamic forces are stronger than the repulsive forces of the electrons themselves. Many ideas regarding ZPE considered esoteric and speculative in the West have been taken seriously, theoretically developed, and experimentally investigated by Russian scientists. The Russian literature on torsion fields describes solutions along the torsion filament exceeding light speed.[41]

> The new physics of like-charge clustering in bundles under low power conditions opens a wide range of applications including spacecraft maneuvering microthrusters.[L]
>
> — Thomas Valone

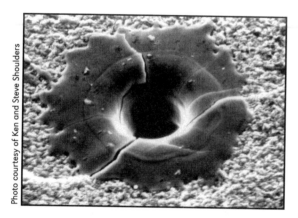

*Photo courtesy of Ken and Steve Shoulders*

*When interacted with solid material, charge clusters cause a low-energy phase transformation type of atomic disruption that liquefies the lattice and propels the material to a high-velocity without apparent signs of conventional heating.*

The size and number of EVs are determined by formation parameters, especially by the magnitude and shape of the electrical pulse used to create the cluster. When created and launched in the presence of a strong electric field, a charge cluster is subject to the same accelerating potential as a single electron placed in the same electric field. The cluster may be carrying a large number of positive ions, but the ratio of electrons to positive ions is so large that the positive ions embedded in or attached to the cluster have very little effect on the velocity imparted to the EV by the electric field.⁴² Some researchers believe that the excess energy in these experiments is derived from the zero-point energy field.

Scientist Ken Shoulders states that it is not kinetic energy alone that provides the powerful energy content observed in EV impacts. It is like the EV was a truckload of TNT that does not have to be moving very fast to do a lot of damage upon collision. When interacted with solid material, these highly organized, micron-sized clusters of electrons perform a low-energy phase transformation and atomic disruption that liquefies the lattice and propels the material to a high velocity without apparent signs of conventional heating. Using an ordinary thermal interpretation, Shoulders reports that "a thermal gradient for bulk material greater than 26,000 degrees C per micrometer would be required to achieve the observed effects."⁴³ There is no relevant theoretical model to predict the behavior of EV phenomena, which means progress can be gained only by slow, careful laboratory research. The fact that only empirical methods can be used to solve the mystery of these reactions is one indication that a new class of phenomena is being investigated.

Ken Shoulders is credited as the inventor of EV technology and the leading pioneer in HDCC research and development. In 1980, he discovered the physical mechanism whereby electrons were seen to cluster into a very dense structure resembling miniature ball lightning. The electrons were packed as tightly as atoms in a solid, and yet they didn't repel each other as expected under space-charge repulsion laws. Ken and his son, Steve Shoulders, have spent two decades working on the electronic aspects of what they call EVs, a Latin acronym for "electrum validum" or strong electron. During their experiments, they have witnessed many intense effects that seem out of place in an electronic technology. One example of an unusually energetic characteristic is that these electron clusters have been found to bore their way through very

*Ken Shoulders has been researching high density charge clusters (HDCC) since 1980 and has published several papers describing his results. Four independent scientists in three countries have also discovered HDCC.*

refractory materials, such as silicon carbide that has a high melting point of 4,700 degrees Fahrenheit. They produce a precise hole in the carbide, having a diameter of about ¼ that of a hair, and propel the fluidized material to a high velocity, all with no apparent thermal heat.

In 1987, Ken Shoulders published a book that described the discovery process and electronic devices he used in researching HDCC.[44] On May 21, 1991, the US Patent Office elated zero-point energy research advocates when it issued Patent 5,018,180 to Ken R. Shoulders for "Energy Conversion Using High Charge Density." The patent states:

"An EV passing along a traveling wave device, for example, may be both absorbing and emitting electrons. In this way, the EV may be considered as being continually formed as it propagates. In any event, energy is provided to the traveling wave output conductor, and the ultimate source of this energy appears to be *the zero-point radiation of the vacuum continuum*"[45]

*The President of Emerging Energy Marketing Firm, Hal Fox (right center), uses a table-top demonstration unit to explain charge cluster phenomenon.*

During 1991 and 1992, the US Patent Office issued four more patents to Shoulders: one for his methods and apparatus for producing and manipulating HDCC; one for energy conversion techniques using HDCC; and two for circuits responsive to and controlling HDCC. The excess energy phenomena generated by charge clusters may not be a direct extraction of useable energy from the physical vacuum, but research in this field is well beyond the tinkering stage and is already leading to practical applications. This new technology has been developed solely by private funds, and parts of it have been discovered or invented by five independent scientists in three different countries (US, Belarus, and Russia).[46] Properly produced and controlled, HDCC can provide more than ten times the amount of energy output as required to produce the charge clusters.[47]

In order to record and study the HDCC phenomena, Shoulders' team makes photographic images of EV strikes. The effects produced can be very dramatic if the conditions are right. Scanning electron micrographs show EV borehole production, dual EV existence, and an electrically driven, sloshing type of material reflection. Shoulders reports that if the source of material propelled by EV action is viewed sideways, small shock cones can be seen coming out of a micro-nozzle. The spacing of these cones is typically 70 micrometers at atmospheric pressure. That represents a very high velocity and high specific impulse. The electrical energy input used to push a 20-micrometer diameter by 100-micrometer long slug of material to this velocity is only 20 micro-Joules.[48]

Some of the micrographs show evidence of a low-energy nuclear reaction that has produced nuclear transmutations by using a nuclear cluster reaction process. Shoulders has written that this abnormal behavior introduces the notion of energy gain produced through a low-energy atomic and molecular phase change coupled with high recombination energy release. Data from one experiment shows an EV strike on a foil of palladium that had been previously "loaded" with deuterium. When the foil was examined in detail with an x-ray energy dispersive analyzer, evidence indicated clean palladium in all places except those bombarded with EVs. Many of the bombarded areas, but not all, showed that nuclear conversion had taken place. The new materials were mostly silicon, calcium, and magnesium. Despite these interesting results, Ken Shoulders writes:

> "Theories and ideas have almost no value in the real world, and laboratory demonstrations are worth very slightly more. These notions stem from trying to sell paper inventions and finding it is a process that

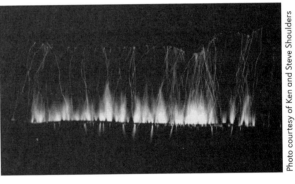

*Time exposure of a side view of a 0.001-inch thick aluminum foil, coated with silicon carbide, being sparked by an induction coil from a moving electrode located at the top of the photo. The small jets seen coming from under the foil have penetrated through it.*

stops creative work. To my way of thinking, it is only a properly engineered device that has real value."[49]

One of the efforts to harness excess energy of HDCC technology involved firing electron clusters into water to produce steam. Although many materials can be converted to a vapor state by electron clusters, water has a unique advantage in that it has a moderately high initial density and dependably returns to its original state. The electrical properties of water are also favorable for EV interaction. The generation process can be a violent one, so one of the principle goals of this project was to confine the violence to the water and not allow erosion of the electrodes responsible for generating and terminating the electron clusters. This project addressed a set of properties related to strikingly powerful EV energetics and used a different set of properties. The goal of building a steam generator that used an electric spark to make steam in a unique way suffered the same limitations of many earlier ideas, namely the electrodes eroded at such a high rate as to render the device impractical. The process appears to have significant potential as an energy source, but there are serious engineering obstacles to be overcome.

> The technical problem for me, and always has been, is how to keep from progressively destroying the highly ordered machine built to produce such good results.[M]
>
> — Ken Shoulders

According to electrical and systems engineer Moray B. King, the primary hypothesis for tapping the ZPE is that stimulating a plasma into a self-organized coherent form induces a similar coherence in the virtual plasma of the quantum foam.[50] He suggests that glow plasma be subjected to one or more stimulations to induce a coherent nonlinear self-organization, such as an abrupt EM pulse, bucking EM fields, or counter-rotating EM fields. King also points out that several new energy researchers utilize the abrupt discharge in their inventions to manifest a plasmoid form as well as excess energy. Paulo Correa, Henry Moray, and J. Papp apply the discharge directly to a glow plasma. Shoulders pulses a liquid metal electrode, whereas Graneau pulses a small cylinder of water.[51]

Shoulders, Correa, and Graneau have observed plasmoid formations via photographic methods.[52] Shoulders, Correa, and Moray have tapped the excess energy via rectifying an output pulse; Papp and Graneau have focused on tapping the large mechanical reaction force via a piston. Echoing Thomas Bearden, Moray King asserts that if such experiments repeatedly produced large energetic effects, then it would be easy for the scientific community to replicate a self-running device that directly taps the zero-point energy.

Moray King postulates that "in order to theoretically allow the coherence of the zero-point energy, two fields of modern physics must combine; theories of the zero-point energy interacting with matter, and theories of system self-organization."[53] King cites 63 references to support his point that the dynamics of the ZPE can fulfill the conditions for self-organization. It is easy to forgive his idealism when he writes:

> "It is hoped that many inventors, by working with these ideas, would freely donate a frugal, 'generic' experiment that could be readily repeated. For it is the repetition of an experiment that will spread across our planetary consciousness like a wave and usher in the forthcoming golden age."[54]

Tom Valone suggests that the Yater patented process, also explained by a series of *Physical Review* articles, will work to rectify ZPE in the same way it was designed for rectifying thermal noise. The scientists from the Nanotechnology Lab at IBM reported to the 1999 AAAS meeting that they are pursuing a similar course of research to develop a "nano-rectifier" to power nano-motors. Therefore, small ZPE-powered engines may be on the horizon.

Most modern physicists understand that the vacuum is not a tranquil void but a quantum state with fluctuations having observable consequences. In recent years, evidence suggests that the electromagnetic zero-point field is not merely an artifact of quantum mechanics but a real entity with major implications for gravity, astrophysics, and technology. ZPE may represent a fundamental link about the relationship between relativity theory and quantum physics. Top-ranked physicists insist that research into zero-point energy is justified and should continue to be supported.[55]

Peter Milonni and M. L. Shih note that:

> I am enough of an artist to draw freely upon my imagination. Imagination is more important than knowledge. Knowledge is limited. Imagination encircles the world.
> — Albert Einstein
> (1879 – 1955)

"The idea of the vacuum as a quantum state with zero-point energy and fluctuations of physical consequence is now commonplace, with implications ranging from optical communications to quantum chromodynamics to inflationary models of the Universe. From relatively humble origins in the second decade of this century and after long periods of neglect, the concept of ZPE has become firmly ingrained in the worldview of contemporary physicists."[56]

## FOOTNOTES

[1] <www.padrak.com/ine/BGH97_1.html> — "Survey and Critical Review of Recent Innovative Energy Conversion Technologies," written by Patrick Bailey, Toby Grotz, and James Hurtak and posted on the Institute for New Energy website. This 1997 summary reviews the experiments, motors, generators, devices, and demonstrations that have been reported in the past several years to produce near-unity or over-unity operation. The paper concludes that "it remains to be seen if new forms of potential energy can be demonstrated and successfully utilized within the near future for the ultimate benefit of the human race."

[2] <www.physlink.com/index.cfm> — This website is an online education and reference tool for physics and astronomy. The reference section is loaded with information: commonly used constants in physics and engineering, online unit conversion, nuclear and particle data, as well as other resources for teachers and students.

[3] <www.britannica.com> — Encyclopedia Britannica website. Excellent source of reliable information on Werner Heisenberg, quantum mechanics, particle physics, and his famous "uncertainty principle." Use the search tool to locate the topic of interest.

[4] <www.sciam.com/askexpert/physics/physics32.html> — *Scientific American*'s "Ask the Expert" in physics. Question: "What is the 'zero-point energy' in quantum physics? Is it really possible that we could harness this energy?" Answers given by John Obienin, a materials science researcher at the University of Nebraska at Omaha; John Baez, a member of the mathematics faculty at the University of California at Riverside; and Paul A. Deck, assistant professor of chemistry at Virginia Polytechnic Institute and State University. Answers posted July 28, 1997.

[5] A. D. Sakharov, "Vacuum Quantum Fluctuations in Curved Space and the Theory of Gravitation," *Soviet Physics Journal*, 1968. Puthoff also suggests the discussion in C. W. Misner, K. S. Thorne, and J. A. Wheeler, *Gravitation* (Freeman, San Francisco, 1973), p. 426.

[6] <www.padrak.com/ine/QUANTFLUX.html> — "Quantum Vacuum Fluctuations: A New Rosetta Stone of Physics?" Article by Harold E. Puthoff, Institute for Advanced Studies, January 1989. Referring to his 1987 paper, "Ground State of Hydrogen as a Zero-Point Fluctuation-Determined State," *Physics Review*, D 35, p. 3266, Puthoff writes, "It is shown that the electron can be seen as continually radiating away its energy as predicted by classical theory, but simultaneously absorbing a compensating amount of energy form the ever-present sea of zero-point energy in which the atom is immersed, and an assumed equilibrium between these two processes leads to the correct values for the parameters known to define the ground-state orbit."

7 <www.newscientist.com> — Henry Bortman, "Energy unlimited," *New Scientist* (January 22, 2000), pp. 32–34. "Empty space is seething with huge quantities of energy — if only we could tap it." Jordan Maclay has formed a company called Quantum Fields in Richland Center, Wisconsin, to develop his innovative idea that by carefully balancing the vacuum pressure of a cavity and the elasticity of a spring, he can build a tiny oscillator driven by Casimir forces. Funded by NASA, Maclay and associate Rod Clark intend to make an array of hundreds of topless cavities on a substrate, and then create a single lid that fits over the entire array. Cosmic jitter will run the tiny oscillator.

8 Martin Gardner, "Zero-Point Energy and Harold Puthoff," *Skeptical Inquirer* (May/June 1998), pp. 13–15. In "Notes of a Fringe-Watcher," Gardner shares details about Harold Puthoff's background, including his previous work in ESP, psychokinesis, and precognition. Puthoff has also been an active Scientologist. Gardner criticizes Puthoff's logic published in his Rosetta Stone article. The article also covers Puthoff's theory on ZPE and how it revolves around a search to obtain unlimited free energy. Gardner surmises that Puthoff's theory is destined to become "as pseudoscientific as his earlier work on remote viewing of the atom interior or the efficiency of the E-meters." Martin Gardner was a columnist for *Scientific American*; he retired from the magazine in 1981.

9 In 1992, his experiment was repeated and published in *Physics Essays* (Vol. 5, 1992), pp. 82-89.

10 Peter W. Milonni and M. L. Shih, "Zero-point energy in early quantum theory," *American Journal of Physics* (Vol. 59, No. 8, August 1991), pp. 684–698. In modern physics, the vacuum is not a tranquil void but a quantum state with fluctuations having observable consequences. This paper discusses the roots of QED in the blackbody research of Planck and Einstein-Stern theory of specific heats, and the experimental implications of zero-point energy recognized by Mulliken and Debye in vibrational spectroscopy and x-ray diffraction. This article is based on the introductory chapter of *The Quantum Vacuum*, written by Milonni and Shih, published in 1992.

11 <http://hyperphysics.phy-astr.gsu.edu/hbase/indx.html> — Excellent index to hyperphysics terms and definitions. The Bose-Einstein Distribution describes the statistical behavior of integer (spin) particles (bosons). At low temperatures, bosons can behave very differently than fermions because an unlimited number of them can collect into the same energy state, a phenomenon called condensation.

12 <http://eta.pha.jhu.edu/~rt19/hydro/node8.html> — This webpage offers the equations behind the spectral shift and Feynman loop diagrams, showing some effects

that contribute to the Lamb shift and how much each of these contributes to energy splitting. Willis Eugene Lamb, Jr., an American physicist and Nobel Prize winner, made groundbreaking discoveries about the energy states of the hydrogen atom. Lamb and German-born American physicist Polyharp Kusch shared the 1955 Nobel Prize in physics for their independent research. According to the Dirac and Shrodinger theory, the energy levels of the hydrogen electron should depend only on the principal quantum number $n$. In 1951, Willis Lamb discovered that this was not so after noting a slight shift of the corresponding spectral line (the Lamb shift). The effect is explained by the theory of quantum electrodynamics.

[13] *Physical Review Letters*, Vol. 78, No. 1, pp. 5-8, January 6, 1997.

[14] <http://xxx.lanl.gov/abs/quant-ph/9907076> — "Experimental Verifications of the Casimir Attractive Force between Solid Bodies" by Steve K. Lamoreaux. A brief review of the recent experimental verifications of the Casimir force between extended bodies is presented. Becasuse of modern techniques, it now appears feasible to test the force law with 1% precision. Lamoreaux addresses the issues relating to the interpretation of experiments at this level of accuracy. Although Lamoreaux's experiment tapped only a very minute and impractical level of energy using Casimir plates, it was an important proof of principle. Thermodynamic analysis has shown that it is apparently possible, in principle, to extract energy from the quantum vacuum.

[15] Philip Yam, "Exploiting Zero-Point Energy," *Scientific American* (December 1997), pp. 82–85. Staff writer Yam explores the likelihood and possibility of tapping ZPE. Yam mentions NASA's *Breakthrough Propulsion Physics Workshop* program and their consideration of zero-point energy as a possible propulsion force. Yam writes that "some conventional scientists have decried the channeling of NASA funds to a meeting where real science was lacking."

[16] Charles Seife, "The Subtle Pull of Emptiness," *Science* (January 10, 1997), p. 158. Review of the Steven Lamoreaux Casimir force experiment.

[17] Thomas Valone, "Inside Zero-point Energy," *Infinite Energy* (Issue 26, 1999), p. 53. Physicist, author, and carbonless energy proponent, Thomas Valone writes, "For the first time in history, a lot of media attention is being paid to the sea of energy that pervades all of space. It just happens to be the biggest sea of energy that is known to exist and we're floating inside of it." (Valone credits *The Sea of Energy* by T. Henry Moray for the idea.) This article gives a review of the latest developments in ZPE as well as an introduction to the topic for those who are nonspecialists.

[18] <www.altenergy.org/4/cofe/cofe.html> — Thomas Valone was instrumental in organizing the controversial First International Conference on Future Energy (COFE), held in Bethesda, Maryland, in 1999. COFE was cosponsored by *Infinite Energy*

and Alternative Energy Institute, Inc. Valone is a physicist, author, retired college teacher, and licensed professional engineer. He is also president of the Integrity Research Institute.

[19] <www.integrity-research.org> — Integrity Research Institute (IRI) is a nonprofit corporation "dedicated to the research and development of energy sources that are carbonless future energy technologies, not contributing to global warming, stabilizing the Earth's climate for our descendents." This website offers a selection of new energy books, videos, and other information for sale.

[20] Valone, *op. cit.*, "Inside Zero-point Energy," p. 55.

[21] <http://map.gsfc.nasa.gov/html/lambda.html> — The Microwave Anisotropy Probe (MAP) is part of the Explorers Program selected by NASA in 1996 to probe conditions in the early Universe. MAP measures temperature differences "anisotropy" in the cosmic microwave background radiation. MAP helps to answer three of the most fundamental questions in cosmology. Einstein's cosmological constant theory predicted that the Universe must either expand or contract. Einstein thought the Universe was static, so he added this term to stop the expansion. Later in life, Einstein viewed the cosmological constant term as his "greatest mistake."

[22] As a result, this discovery was labeled the "Breakthrough of the Year" in *Science* (Dec. 18, 1998).

[23] <www.earthtech.org/mission.htm> — EarthTech International is dedicated to the exploration of new frontiers in physics. Their activities are primarily centered around investigations into various aspects of the zero-point field. EarthTech performs evaluations of reported "overunity" energy devices and specializes in performing accurate power-balance measurements using calorimetry.

[24] Harold E. Puthoff, "Gravity as a Zero-Point-Fluctuation Force," *Physical Review* Vol. 39 (1993), p. 2333, and "On the Source of Vacuum Electromagnetic Zero-Point Energy," *Physical Review* Vol. 40 (1989), p. 4857. Puthoff's professional background spans more than three decades of research at General Electric, Sperry, the National Security Agency, Stanford University, and SRI International. He has published numerous papers on electron-beam devices, lasers, and quantum zero-point energy effects.

[25] John Gribbin, "Energy of 'nothing' upsets cosmology," *New Scientist* (August 31, 1991), p. 18.

[26] Gardner, *op. cit.*, "Zero-Point Energy and Harold Puthoff," p. 13.

[27] <www.altenergy.org/3/new_energy/zero_point_and_other/puthoff/puthoff.html> — Alternative Energy Institute profile of Harold E. Puthoff. "There is perhaps no other

figure in the field of new energy today who has done more to provide a theoretical framework within which ZPE can be understood than Harold Puthoff."

[28] Puthoff, *op. cit.*, "Quantum Vacuum Fluctuations: A New Rosetta Stone of Physics?" p. 2.

[29] <www.calphysics.org/> — "Zero-Point Fields, Gravitation and New Physics," a 1999 report by Professor Paul S. Wesson for the California Institute for Physics and Astrophysics (CIPA), pp. 1 - 21. CIPA is an academically oriented research organization comprised of scientists studying topics in physics and astrophysics with emphasis on electrodynamics, relativity, gravitation, inertia, and the zero-point field of the quantum vacuum. Abstract: Research over the last decade has shown that problems exist about how to reconcile the zero-point fields that follow from quantum mechanics with the energy conditions built into classical gravitational theories such as general relativity. (This is at odds with the majority of astronomers, who are in agreement about the recent discovery of "Cosmological Antigravity," the title of a *Scientific American* article published in January 1999.)

[30] *Ibid.*, p. 16. Wesson has six recommendations for research: (1) the status of the cosmological constant needs to be addressed; (2) the existence of a real electromagnetic ZPF is conceptually distinct from the questionable existence of a quasi-Newtonian law of gravity derived from ZPF perturbations, and future work should focus on ZPF physics; (3) a large body of data exists that supports standard gravitational theory of general relativity, and it is essential that effort be made to match ZPF theory and Einstein theory; (4) the discussion of the Davis-Unruh effect should be dropped; (5) theoretical studies should be made of the Casimir effect for various configurations as a precursor to practical measurements of force that might be of technological value; (6) following experimental verification of the Casimir effect, work should be continued to identify technological applications of a ZPF.

[31] Thomas E. Bearden, "On Extracting Electromagnetic Energy from the Vacuum," a well-referenced paper published by the Alpha Foundation's Institute for Advanced Study (1999). Poynting and Heaviside independently discovered the flow of energy through space. Bearden says that Poynting erroneously considered the energy flow component entering the circuit, and completely missed the huge additional component shown by Heaviside that missed the circuit. "Hence," writes Bearden, "Poynting got the direction of the energy flow in error by essentially 90 degrees, and was later corrected by Heaviside. In this paper, we call the neglected nondiverged component of energy flow the Heaviside component."

[32] <www.altenergy.org/3/new_energy/zero_point_and_other/bearden/bearden.html> — Alternative Energy Institute profile of Thomas Bearden. The Association of Dis-

tinguished American Scientists' (ADAS) latest position paper, dated June 24, 2000, "The Unnecessary Energy Crisis: How to Solve it Quickly" can be read on this webpage.

[33] Email to Alternative Energy Institute and John Shahabian, Executive Director for the Clean Power Campaign from Thomas Bearden congratulating both organizations for making "a significant contribution to help alleviate the oil and electricity crunch, and clean up the biosphere by switching to clean energy." Bearden believes that tremendous amounts of electromagnetic energy are wasted in contemporary electrodynamics. To further understand this potentially enormous energy flow, Bearden suggests reading John Kraus' *Electromagnetics*, 4$^{th}$ Edition, McGraw-Hill, New York, 1992.

[34] <www.nobel.se/physics/laureates/1902/> — Background biographies on Professors Hendrik Antoon Lorentz and Pieter Zeeman and why they won the Nobel Prize in Physics in 1902 for their pioneering work on the connection between optical and electromagnetic phenomena. The Lorentz transformations are a set of equations in relativity physics that relate the space and time of coordinates of two systems moving at a constant velocity relative to each other. The Lorentz equations formally express the relativity concepts that space and time are not absolute; that length, time, and mass depend on the relative motion of the observer; and that the speed of light in a vacuum is constant and independent of the motion of the observer or the source.

[35] <http://pe.pennwellnet.com/home/aboutthesite.cfm?Section=ABOUT> — *Power Engineering* magazine was established in 1896 and is the comprehensive voice of the power-generation industry. It provides readers with the critical information they need to remain efficient and competitive in today's market.

[36] Thomas Bearden, letter dated August 29, 2000, to Brian K. Schimmoller, managing editor of *Power Engineering* magazine in response to the article "Magicians Wanted." Bearden presents ten questions and their extended answers in proposing a solution to the long-vexing critical problem of the association of the fields and potentials with their source charges.

[37] <www.ott.doe.gov/electromagnetic/> — The Alpha Foundation's Institute for Advanced Study (AIAS) does not have a website of its own. Instead, quite a few of its papers (about 90) are carried on a controlled Department of Energy website for Advanced Electrodynamics. Access to the site and to the papers is controlled by David Hamilton of DOE. Legitimate scientists can obtain limited access by contacting Hamilton at <David.Hamilton@ee.doe.gov> Many of the papers (about two-thirds) have been published in leading journals such as *Foundation of Physics, Foundations of Physics Letters, Physica Scripta, Optik,* and others. Sixty of the papers were published by *Journal of New Energy* as a special issue in late 1999.

[38] A very important milestone paper in the alternative energy effort is the Alpha Foundation's Institute for Advanced Study group paper written by M. W. Evans, P. K. Anastasovski, T. E. Bearden, et al., "Classical Electrodynamics without the Lorentz Condition: Extracting Energy from the Vacuum," *Physica Scripta*, Vol. 61(2000), pp. 513–517. In this paper, Bearden and his colleagues managed to put the notion of practical extraction of electromagnetic energy from the vacuum on the "acceptable science" map for the first time. They give 15 or so different legitimate approaches, including the Bohren experiment, which produces 18 times as much energy output as is input by conventional calculations (which do not include the Heaviside energy-flow component).

[39] <www.padrak.com/ine/FB97_1.html> — "High-Density Charge Clusters and Energy Conversion Results," an updated and revised version of a paper by Patrick G. Bailey and Hal Fox from the Institute of New Energy, originally submitted in May 1997 to the 1997 Intersociety Energy Conversion Engineering Conference, held July 27 – August 1, 1997, in Honolulu, Hawaii. Paper covers several recent developments of devices that produce low-energy nuclear reactions. Presents evidence of the application of the control and use of high-density charge clusters for Plasma-Injected Low-Energy Nuclear Reactions in the production of low-cost, nonpolluting, abundant thermal energy.

[40] Shang-Xian Jin and Hal Fox, "High-Density Charge Cluster Collective Ion Accelerator," *Journal of New Energy*, Vol. 4, No. 2 (Fall 1999), p. 97.

[41] A. E. Akimov, "Heuristic Discussion of the Problem of Finding Long Range Interactions, EGS-Concepts," *Journal of New Energy*, Vol. 2 (1998), pp. 3-4, 55-80; A. E. Akimov & G. I Shipov, "Torsion Fields and their Experimental Manifestations," Ibid., Vol. 2, pp. 67–84. Overview of torsion field research (predominantly in Russia), includes 177 references. Akimov models the vacuum as a lattice of "phytons," counter-rotating, charged entities sized at the Planck length ($10^{-33}$ cm). Each phyton can polarize in three different ways to manifest 1) electric fields via charge polarization, 2) gravitational fields via oscillating, longitudinal spin polarization, and 3) torsion fields via transverse spin polarization. (See *Infinite Energy* — Issue 28, 1999, p. 68.)

[42] Bailey, Fox, *op. cit.*, "High-Density Charge Clusters and Energy Conversion Results," p. 2.

[43] Ken Shoulders and Steve Shoulders, "Charge Clusters in Action." Paper presented by father and son research team, Ken and Steve Shoulders, for the First International Conference on Future Energy (COFE) — Document available for download from AEI at <www.altenergy.org/4/cofe/cofe.html>. This well-illustrated presentation in-

cludes several interrelated discoveries that, taken together, may greatly improve the ability to produce and manipulate energy and material. Charge clusters perform a low-energy atomic disruption that liquefies the lattice and propels material to a high velocity with no apparent signs of conventional heating.

[44] Ken R. Shoulders, "*EV — A Tale of Discovery*," (Austin, Texas: Jupiter Technologies, 1987). A historical sketch of early EV work. Well-illustrated with photos and drawings, the book is available from the author at P.O. Box 243, Bodega, California, 94922.

[45] US Patent #5,018,180, issued May 21, 1991, to Kenneth R. Shoulders for "Energy Conversion Using High Charge Density." Patent excerpt reprinted in *Journal of New Energy*, Vol. 4, Fall 1999, p. 103. *Proceedings of the Institute for New Energy's 1999 Symposium on Low Energy Nuclear Reaction and New Energy Research.*

[46] G. A. Mesyats, "Ecton Processes at the Cathode in a Vacuum Discharge," Proceedings for the 17th International Symposium on Discharges and Electrical Insulation in Vacuum, (1996), pp. 720–731. Russian research is presented, analyzing the discovery of charge clusters, called "ectons." "Ectons often arise from micro explosions on the surface of the cathode, where surface imperfections such as micro protrusions, adsorbed gases, dielectric films, and inclusions play an important role. The simplest way to initiate ectons is to cause an explosion of cathode micro protrusions under the action of field emission current. Experiments confirm that micro protrusions jets can form from liquid or melting metal. The breakdown of thick dielectric films in their charging with ions also plays an important role in the initiation of ectons. A commonly used way to initiate an ecton is to induce a vacuum discharge over a dielectric in contact with a pointed, metal cathode. An ecton can readily be excited at a contaminated cathode with a low-density plasma, but a clean cathode requires a high plasma density." — Moray B. King, "Charge Clusters: The Basis of Zero-Point Energy Inventions," *Infinite Energy* (March-June 1997), p. 101.

[47] Hal Fox, President, Emerging Energy Marketing Firm, Inc. (EEMF), April 5, 2000. Press Release: "Firm, with headquarters in Utah, named as exclusive agent for marketing and licensing six new-energy patents." Fox is editor of *Journal of New Energy*, published quarterly by EEMF, Inc. Fox points out that charge cluster research is "also known as the Shoulders-Puthoff-Gleeson-Ilyanok-Mesyats technology." Additional note: *New Energy News*, a monthly email newsletter edited by Fox's colleague, Patrick Bailey, covers new-energy developments and is published by the Institute for New Energy.

[48] K. Shoulders and S. Shoulders, *op. cit.*, "Charge Clusters in Action," p. 8.

[49] *Ibid.*, p. 12.

⁵⁰ <www.altenergy.org/4/ine-99/king/king.html> — Alternative Energy Institute, Inc., profile of Moray B. King and his lecture presentation at the Institute for New Energy's 1999 Symposium in Salt Lake City, Utah.

⁵¹ King, *op. cit.*, "Vortex Filaments, Torsion Fields, and the Zero-Point Energy," p. 66.

⁵² Alternative Energy Institute, Inc., *op. cit.*, First International Conference on Future Energy.

⁵³ Moray B. King, *Tapping The Zero-Point Energy*, (Provo, Utah: Paraclete Publishing, 1989). Moray B. King supports, in conjunction with the standard physics literature, the speculation that a zero-point energy coherence can be induced by technological means.

⁵⁴ King, Ibid., p. 88.

⁵⁵ Wesson, *op. cit.*, "Zero-Point Fields, Gravitation and New Physics

⁵⁶ Milonni and Shih, op. cit., "Zero-point energy in Early Quantum Theory," p. 696.

## SIDEBAR FOOTNOTES

A  Moray B. King, *Tapping the Zero-Point Energy*, (Provo, Utah: Paraclete Publishing, 1989), p. 3.

B  Ervin Laszlo, *The Whispering Pond: A Personal Guide to the Emerging Vision of Science* (Boston, Massachusetts: Element Books, Inc., 1996), p. 190. (Awarded "Outstanding Academic Book" *Choice*, 1997, by the American Library Association.)

C  H. Zinkernagel, S.E. Rugh, and T.Y.Cao, "The Casimir effect and the Interpretation of the Vacuum," *Studies in History and Philosophy of Science* (March 1999), p. 111.

D  Henry Bortman, "Energy Unlimited," *New Scientist* (January 22, 2000, v.165 No. 2222), p. 32.

E  Charles Seife, "The Subtle Pull of Emptiness," *Science* (January 1997, v. 275), p. 158.

F  F. Pinto, "Engine Cycle of an Optically Controlled Vacuum Energy Transducer," *Physical Review B* (The American Physical Society, December 1999), Vol. 60 No 21, pp. 740–755.

G  Lawrence H. Ford and Thomas A. Roman, "Negative Energy, Wormholes and Warp Drive," *Scientific American* (January 2000, v. 282 No. 1), p. 52.

H  Martin Gardner, "Zero-point energy and Harold Puthoff," *Skeptical Inquirer* (May-June 1998, v.22 n3), p. 60.

I  Li-jun Han, Jin-shi-lei, and Xing-liu Jiang, "Torsion Field and Tapping the Zero-Point Energy in an Electrochemical System," *Journal of New Energy* (Fall 1999), p. 93. Proceedings — Institute for New Energy 1999 Symposium on Low Energy Nuclear Reaction and New Energy Research.

J  Phil Carpenter, David Hamilton, & Dave Goodwin, et. al., *Breakthrough Energy Physics Research (BEPR) Program Plan* (Office of Energy Efficiency & Renewable Energy, Office of Science, and the Office of Nuclear Energy, Science and Technology — Joint Exploratory Research (October 2000 – Draft for Agency Comment Only), p. 20.

K  Philip Yam, "Exploiting Zero-Point Energy," *Scientific American* (December 1997), p. 83.

L  Thomas Valone, "Future Energy Technologies," (Integrity Research Institute - 2000).

M  Ken Shoulders, "Open Letter," mailed December 2, 2000, p. 5.

# Universal Forces: Blackholes, Electrogravitics, & Deep Space Propulsion

## 15

Gravitation is one of the four known forces responsible for maintaining the structure of the Universe. The others include the strong nuclear force, which binds the particles in atomic nuclei; the weak nuclear force, responsible for some radioactive decay; and electromagnetism. Gravitation is the weakest known force in nature and plays no role in determining the internal properties of everyday matter; but gravity shapes the structure, evolution, and trajectories of the planets, stars, and galaxies. Twentieth-century engineers reached the limits of conventional rocket technology, so scientists are studying methods of gravitational modification and electromagnetic propulsion to conquer the incredible distance associated with space travel.

According to the Big Bang theory, a single force existed at the beginning of the Universe, and, as the Universe expanded and cooled, this force separated into gravity, electromagnetism, and the strong and weak nuclear forces. Physicists are searching for the elusive Unified Field Theory that will merge quantum mechanics and Einstein's law of gravity but have not found one yet. In fact, the mathematical structures of General Relativity and quantum mechanics, the two great theoretical achievements in 20th century physics, seem utterly incompatible. Gravitation is the least understood of all the fundamen-

> Despite more than 200 years of experiments, the best laboratory determinations of the force of gravity have yielded no better than a ballpark figure.[A]
> — *Science News*

tal forces, but mass and space, which are governed by gravity, are the building blocks and fabric of the Universe.

As the most ubiquitous force in the Universe, gravity affects all matter at some level.[1] Whether it is the gentle pull of Earth that keeps our feet planted firmly on the ground or the beautifully balanced force that maintains the planets orbiting around the Sun, gravity is the vital cosmic glue that holds our world together. Scholars and philosophers have grappled with a changing concept of gravity since before the days of Aristotle in the 4th century BC. The development of gravitational theory has come in fits and starts and even today is hampered by energy values that are difficult to nail down precisely. The Gravitational Constant, called "big G" by physicists, is still an elusive quantity since scientists have not been able to measure precisely the minuscule gravitational attraction exerted by objects that are small enough to experiment with.[2] Some physicists believe that the standard gravitational model is built on equations that contain too many constants of nature and on values that seem arbitrary.[3]

One of the most compelling cosmological enigmas confronting scientists today is "Why hasn't gravity slowed or stabilized the expansion of the Universe?" Recent measurements by an international scientific team found evidence that expansion has actually speeded up, not slowed down. Researchers are now suggesting that space is permeated by an unknown repellent force that counteracts the pull of gravity on a cosmic scale, and some have proposed an antigravity hypothesis to support these findings.[4] There is another mystery that puzzles physicists. For ten years, NASA has observed an unknown force that influences the trajectories of their spacecraft. Research scientists at the Jet Propulsion Laboratory in Pasadena, California, have noticed that space probes using the gravitational field of Earth to slingshot outward into the solar system gain an extra boost in velocity by some unidentified force. The satellites Cassini, NEAR, and Galileo were all similarly affected, but NASA has no idea what source or mechanism is generating the bump in speed. Researchers hope to confirm definitively the phenomena in 2001 when the spacecraft Stardust whips around our planet on its seven-year journey to intercept the comet Wild-2 and return material samples from it to Earth in 2006.⁵ Gravitation remains one of the great mysteries of science, and there is still much to be learned about this force before physicists can resolve the apparent discrepancies between Einstein's theories and quantum mechanics.

Although gravity plays a vital role in the physical world around us, it has only reluctantly revealed its secrets. Humans have grappled with the concept of gravity since the earliest days of civilization. Classical Greek philosophers had no idea that celestial bodies were influenced by gravity. Aristotle believed that these objects were unaffected by any external force and that heavenly objects tended toward a natural motion in a perfect circle with a velocity proportional to its weight. He also postulated that a body moving at constant speed requires a continuous force acting on it and that the force must be made by actual contact rather than by an invisible interaction over space and distance.⁶

Aristotle's erroneous concept of natural motion prevailed for hundreds of years until the 17th century when Galileo supplied the first experimental evidence that constant motion does not require any force or acceleration. (At the Earth's surface, the acceleration of gravity is about 32 feet per second; for every second an object is in free fall, its speed increases at that rate.) Galileo, an Italian mathematician, astronomer, and physicist credited with inventing the telescope, contributed greatly to the understanding of natural motion. By experimenting with inclined planes and pendulums, Galileo realized that force is necessary to change motion but not to maintain constant motion — a concept called inertia. Inertia is the property of matter that causes it to resist any change of its motion in either direction or speed; furthermore, weight measures how local gravity affects an object, and an object's mass is independent of weight.⁷ Legend has it that Galileo simultaneously dropped objects of different weights from the leaning Tower of Pisa and observed them hitting the ground at the same time. He concluded that gravity affects all objects the same way, regardless of mass, leading

*Italian physicist Galileo Galilei (1564 – 1642) was charged with heresy for challenging his contemporaries' erroneous scientific paradigm that the Sun and planets revolved around Earth. Galileo spent the last 8 years of his life under house arrest for his belief in Copernican theory and his commitment to the freedom of scientific inquiry.*

> Weight measures how local gravity affects an object, and an object's mass is independent of weight.[8]
>
> — *The Physics Teacher*

him to postulate his law of free fall and the impetus theory that without resistance, a violent motion will persist with constant velocity. But, like the Greek philosophers before him, Galileo saw no connection between the force behind planetary motion and gravitation on Earth.

During the same time period Galileo was experimenting with gravity, German mathematician and astronomer Johannes Kepler was studying data on planetary movements based on measurements made by the Danish astronomer Tycho Brahe. He used the new information to expound on the Copernican perspective of a heliocentric model of the Universe.[8] Kepler formulated three important laws of planetary motion that described the planetary orbits with simple geometric and arithmetic relationships, one of which postulated that planets follow elliptical orbits not circular.

Galileo was one of the first scientists to agree publicly with Copernican theory and its radical notion that the planets revolved around the Sun, not the Earth. This contrasted sharply with the long-accepted belief that Earth was fixed as the center around which the planets and stars orbited. Before the advent of modern science, this cosmological viewpoint had become embraced as an incontrovertible religious teaching in Europe. As punishment for Galileo's outspoken challenge to his contemporaries' religious/scientific paradigm, he was charged with heresy and found guilty by trial. He spent the last eight years of his life under house arrest for standing by his faith in scientific reason. More than 400 years later, Galileo's historic commitment to truth remains an inspiration to the innovative researcher battling with authority for the freedom of scientific inquiry.

Over the centuries, many gifted scientists and natural philosophers have contributed to our understanding of gravitation, velocity, and mass. Isaac Newton's and Albert Einstein's ideas dominate the development of gravitational theory and form the foundation of the contemporary interpretation of this enigmatic force. In 1687, Sir Isaac Newton, an English mathematician and physicist, published *Philosophiae Naturalis Principia Mathematica*, a magnificent synthesis of natural philosophy. To develop his theory of gravitation, Newton first had to devise the science of forces and motion called *mechanics*, as well as a new branch of mathematics known as calculus. His law of inertia proposed that the natural motion of an object is motion at a constant speed in a straight line and that it takes a force to slow down, speed up, or change the path of a moving object. Newton studied the orbit of the Moon and discovered the relationship between the motion of the Moon and the motion of any falling body on Earth. His gravitational theory helped explain Johanne Kepler's laws and established the modern quantitative science of gravitation.[9]

In his *Principia*, Newton posited that the force between two bodies is proportional to the product of their masses and the inverse square of their separation and that the force depends on nothing else. In other words, Newton believed that every particle in the Universe attracts every other particle in the Universe with a force that depends on the product of the two particles' masses divided by the square of the distance between them. The gravitational force between the two objects is expressed by the equation $F = GMm/d^2$ in which $F$ is the gravitational force, $G$ is a constant known as the universal constant of gravitation, $M$ and $m$ are the masses of each object, and $d$ is the distance between them.[10]

Newton tested his assumptions with pendulum experiments that confirmed the principle of equivalence (*forces produced by gravity are equivalent to forces produced by acceleration*). With his third dynamic law, Newton postulated that every force implies an equal and opposite reaction force. The force of gravity on an object is considered the object's weight, and this force depends on the mass or quantity of matter within the object. Therefore, the weight of an object is equal to its mass multiplied by the acceleration due to gravity. If the mass of one or both particles increases, the force of attraction increases. If the distance between the particles increases, then the attraction decreases, based on the square of the distance between their centers of mass. Doubling the distance between two particles will make the force of attraction one quarter as great as it was. Until Newton's findings, it was not realized that the movement of celestial bodies and the free fall of objects on Earth are both determined by the same force.[11]

> [Isaac Newton] realized that the motion of all bodies, from a falling apple to an orbiting planet, could be explained via a law of universal gravitation.[c]
> — *The Sciences*

Early in the 20th century, Albert Einstein developed several scientific theories that revolutionized the world of contemporary physics. Conceived by just one man and, at the time, based on pure speculation on the nature of space and time without experimental confirmation, Einstein's General Relativity is still the most fundamental theorem of science about the nature of gravity. Despite his groundbreaking work on relativity, Einstein received his only Nobel Prize in Physics for work on light in which he showed that light can be considered as consisting of particles, and the energy carried by any particle (photon) is proportional to the frequency of radiation. His illuminating proposal that light energy is transferred in individual units (quanta) overturned a 100-year theory that considered it a continuous process. He concluded that an observer peering into the distant reaches of the Universe is seeing things from the past because light particles take time to travel to Earth.

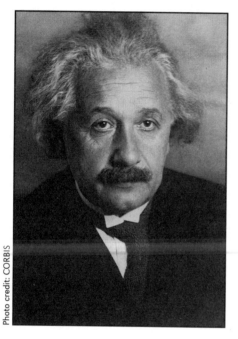

*The modern description of the Universe is based on physicist Albert Einstein's theory of gravity, which provides the equations that describe the evolution of the expanding Universe and indicate that it was born at a definite moment in time known as the Big Bang.*

Einstein's theory of General Relativity assumes that we exist in a four-dimensional Universe called "space-time." The force of gravity results from the curvature of space-time by matter, and all objects — heavy, light, or even massless — are similarly affected as they move through the same curved space-time. Einstein's revolutionary concept meant that the orbits of the planets are not caused by gravitational attraction to the Sun; they are simply following straight lines in curved space.[12]

Unlike his contemporaries, instead of seeing the path of a lightwave being bent by the force of gravitation, Einstein imagined photons flying in straight lines and merely following the shortest route in a background space-time that is curved. Einstein's model of the Universe theorized that particles and light rays travel along the geodesics (shortest paths) of a four-dimensional geometric world, which countered the Euclidean postulate that parallel lines never cross. This approach infers that all bodies accelerate the same under the force of gravitation and follow their natural trajectories. Einstein believed that the total volume of space was finite, yet it had no boundary or edge. He posited that the gravitation of the mass contained in cosmologically large regions might warp perceptions of space and time. In Einstein's view, curved space-time tells mass-energy how to move, whereas mass-energy tells space-time how to curve.[13]

The foundation of Einstein's theory of gravitation rests on two empirical findings that he elevated to the status of basic postulates. In the relativity principle, local physics is governed by the theory of special relativity, which relies on the constancy of the speed of light measured in a vacuum (186,000 miles per second). His equivalence principle states that there is no way for an observer to distinguish locally between gravity and acceleration. In other words, the measure of a body's response to a gravitational field is exactly equal to the measure of its resistance to motion. Constant acceleration gives one the same sensation as does standing on a gravitating body of appropriate mass; inertia and gravity are different aspects of the same reality. When viewed through relativity theory, inertial mass is a manifestation of all forms of energy in a body according to $E = mc^2$, with $E$ representing the total content of a body, $m$ the inertial mass of the body, and $c$ the speed of light.[14]

Gravitation is considered the weakest universal force; indeed, if it were as strong as the electric force that pulls oppositely charged particles together, humans on Earth would be crushed flat by their own weight. Despite gravity's relatively weak level of force, when the gravitational field is generated by an object with great mass, its power can be tremendous. The idea that light waves could be attracted gravitationally was thought implausible until 1783, when British astronomer and geologist John Michell presented his seminal paper to the Royal Society of London. In his abstract, Michell posited that an object seeking to escape the gravitational clutches of the Sun would need a speed only about 1/500 that of lightspeed. Based on Newtonian gravity, he determined that the escape velocity would be much greater for a body with more density than the Sun. Michell theorized that the gravitational pull from a very dense mass might be so great, light could never escape. Even on Earth, light traveling upward from the surface loses energy as gravity tries to pull it back.[15]

If gravity were as strong as the electric force that pulls oppositely-charged things together, a bathroom scale would read out your weight in a number some 40 digits long.[D]

— *Science*

Albert Einstein's theories allowed for unimaginably strong gravitational fields in collapsed bodies of mass. Scientists gained a better idea of how gravity behaved around a spherical mass in 1916 when German astronomer Karl Schwarzschild applied Einstein's equations to calculate the deflection of light by a powerful gravitational mass. To achieve such high attraction, the mass must be gigantic, or extremely dense. Extremely dense celestial bodies result from the demise of stars. Stars are like nuclear furnaces powered by fusion, and, depending on their size, they collapse into white dwarfs, neutron stars, or Black Holes when their nuclear fuel is exhausted. (Schwarzschild's theory about the existence of such gravitational masses predated the recognition of neutron stars by half a century.)

Over a life span that lasts billions of years, stars are slowly evolving, ever changing in size, density, and luminosity. They are actively forming, living, and dying.[16] In the terminal phase of its solar life, when a star about as heavy as the Sun runs out of fuel and collapses from its own gravity, it turns into a "white dwarf," a hot glowing remnant about the same size as Earth. For an aging solar mass 1.5 to 3 times heavier than our Sun, the greater internal gravitational pull will transform it into a dense neutron star with a diameter less than 20 miles, surrounded by a substantial gravity field. Ultimately, a star with at least 3 times the mass of our Sun will collapse into itself with such force that it metamorphs into a Black Hole — infinitely small but possessing a gravitational pull of incredible magnitude.[17] Russian theorists Yakov Zeldovich and I.D. Novikov, who studied time distortion in the vicinity of collapsed objects, called these bodies of dense mass "frozen stars," but the term *Black Hole*, coined by American physicist John A. Wheeler in 1968, was the name that stuck. The term *Black Hole* is appropriate because nothing in the Universe, not even light, can with-

*Stars, clouds of gas, and dust are held together by gravity to form galaxies that orbit around a center similar to how the planets in our solar system orbit around the Sun.*

stand the deadly attraction from these incredibly dense, collapsed-star remnants.

Gravity may be the weakest universal force, but cosmologists gawk in awe when they consider the overwhelming gravitational field evidenced by the all-consuming Black Hole, a formidable level of attraction which nothing can escape. Celestial bodies with the greatest mass and smallest diameters exert the most gravity, with Black Holes exerting the greatest gravitational force known to science. The intense gravitation generated by these cosmic conundrums is awesome and graphically illustrates the energy potential associated with gravitational fields. Any matter that comes within a certain distance of a Black Hole, the point-of-no-return called an "event horizon," cannot escape. Planets and objects outside the event horizon are safe and may orbit the Black Hole in the same way planets orbit a star, but anything that passes beyond it is destined to be crushed inside the gravitational maw, including visible light, X-rays, or any other form of electromagnetic radiation.[18] And, according to Einstein, the deeper one travels into a gravitational field, the slower time progresses, and time itself stops at the center of a Black Hole.

> The weakness of gravity is an illusion. Not all of its strength is exercised in the perceptible Universe.[E]
> — *The Economist*

Gas and dust particles hurtling at high velocity into the Black Hole heat by friction and compression to temperatures near one hundred million degrees Celsius. Anything drawn in only adds to a Black Hole's mass and increases its gravitational attraction. In fact, some cosmologists theorize that super-massive Black Holes that have consumed many millions of stars lie at the center of most large galaxies, where X-ray observations indicate a large amount of energy is being produced. The biggest may have diameters as large as our whole solar system and manifest themselves as quasars or as intense sources of cosmic radio emission. In 1974, University of Cambridge physicist Stephen W. Hawking postulated that Black Holes may actually evaporate by emitting radiation at a rate inversely proportional to the square of its mass. Hawking provided an important and crucial link between the concept of negative energy, Black Holes, and the laws of thermodynamics. In order to satisfy the conservation of energy principle, the production of positive energy, which astrophysicists observe as radiation, must be accompanied by a flow of negative energy into the Black Hole. The negative energy is generated by the extreme space-time curvature near the hole, which disturbs vacuum fluctuations. Hawking's theory on radiation unifies Black Hole physics with modern thermodynamics and puts them into a thermal equilibrium with their environment.[19]

Because Black Holes are invisible to the human eye, astronomers identify intense, high-energy X-ray generation as the cosmic signature of these stellar ghosts. Just before gas and particles disappear forever beyond the event horizon, they heat up and emit gamma rays and X-rays. First discovered in 1895 by German physicist Wilhelm Roentgen, X-ray photons have energy hundreds to thousands of times as high as the optical photons to which human eyes are sensitive. Indeed, light exists in many wavelengths, including radio waves, microwaves, infrared radiation, visible light, ultraviolet light, X-rays, and gamma radiation. Very low temperatures (far below zero Celsius) produce mostly low-energy radio and microwave photons, whereas the human body produces mainly infrared radiation. X-rays are a highly energetic form of light that find their source in the hottest gases, the strongest gravity, and the most energetic explosions that occur in the Universe. Only objects at extremely hot temperatures (millions of degrees Celsius) emit their energy as X-rays. Since a Black Hole cannot be directly observed from Earth (no visible light escapes, and our atmosphere absorbs nearly all of the X-rays), astronomers have turned to examining X-ray binary systems, such as Cygnus X-1. In this binary system, a visible star is in close orbit around an invisible super-compact celestial body that is sucking matter away from the visible star and producing vast fluxes of X-rays. In fact, the most distant object yet measured by astrophysicists at the Lawrence Livermore National Laboratory is a quasar, a probable active Black Hole 12.4 billion light years from Earth, identified only by the powerful radio waves that it emits.[20]

The National Aeronautic and Space Administration (NASA) is delving into the fascinating world of Black Holes and gravitational energy fields as they aggressively search for better ways to launch and propel satellites through space. Part of NASA's responsibility, as outlined by the National

> Cosmic enigmas no more, Black Holes are places where gravity has grown so strong that not even light can escape.[F]
> —*Astronomy*

Science and Technology Council, is to enhance the economic competitiveness and scientific and technical capabilities of the United States.[21] Knowledge is power in today's competitive world, and NASA is always pushing ahead in the field of space technologies. In July 1999, NASA used the Space Shuttle Columbia to deploy the Chandra X-ray Observatory, the most sophisticated X-ray observatory ever sent into orbit.[22]

This X-ray observatory, first proposed to NASA in 1976, took teams of top scientists, engineers, and technicians at numerous government centers, universities, and corporations, more than 20 years to design and build. The $1.3 billion spacecraft will spend 85% of its elliptical orbit above the belts of charged particles that surround the Earth and provide researchers with images 25 times sharper than previously launched X-ray telescopes. The spacecraft's four incredibly smooth cylindrical mirrors provide a high-resolution picture, powerful enough to identify two dimes side-by-side from two miles away or read a stop sign 12 miles distant. Imagine reflecting mirrors polished so smooth that the largest bump allowed was three angstroms, equal to the height of three hydrogen atoms piled on top of each other. One engineer who worked on the project pointed out that if the United States were scaled as smooth as these mirrors, no mountain would be taller than six feet high. On-board instruments are 50 to 100 times as sensitive as any other space-based X-ray telescope and are sharp enough to record X-rays from particles up until the last second before they are swallowed up by a Black Hole.

Chandra's on-board high-resolution camera is composed of one 4-inch-square cluster of 69 million tiny lead-oxide glass tubes. The tubes, which are one-twentieth of an inch long and one-eighth the thickness of a human hair, release electrons when X-rays strike them and enable Earth-bound scientists to measure precisely the source of the X-ray.

*NASA's Chandra X-ray Observatory was launched and deployed by Space Shuttle Columbia in July 1999. The space-based telescope has begun an exploration of the high-energy regions of the Universe and is capable of producing images 25 times sharper than previous X-ray telescopes.*

Chandra's improved instrumentation will allow scientists to study temperature variations across X-ray sources like vast clouds of hot gas drifting in intergalactic space and to make more detailed studies of Black Holes, supernovae, and the enigmatic dark matter. Dark matter gravitationally binds large concentrations of extremely hot X-ray-emitting gas in certain galaxies and most, if not all, clusters of galaxies.[23] Study of the intense magnetic fields around Black Holes and other cosmic phenomena has the potential to rewrite the physics books while increasing our understanding of the origin, evolution, and destiny of the Universe.

Data compiled by the Chandra X-ray telescope may give NASA's scientists new insight into the gravitational and magnetic fields that modern physics often struggles to explain. Our current perception of gravity is the space-time distortion produced by normal, positive energy or mass, but physicist Hawking's reference to the exotic force of negative energy opens the door to the possibilities of phenomena like wormholes and warp drives. Some researchers wonder if negative energy can distort the geometric fabric of space-time enough to enable humans to travel light-years worth of distance in an instant as in the plot of a science-fiction thriller.[24]

Although most mainstream physicists consider such exotic phenomena very unlikely, there are always scientists who keep pushing the frontier of contemporary thought. Albert Einstein's mathematical expressions of his ideas and theories are considered some of the most incredible artifacts of pure thought ever produced, but not all scientists completely agree with Einstein's relativity theories or that light speed represents the maximum velocity achievable. In fact, Einstein's theory has been only partially verified and is one of the least tested of all physical theories. In collaboration with scientists at Stanford University and the Marshall Space Flight Center, NASA is conducting gravitational physics research and has scheduled to launch a sophisticated gravity probe satellite on a Relativity Mission to confirm more of Einstein's theory.[25]

Today, there are several mathematical models circulating within the scientific community that describe the inner workings of gravity as well as different interpretations for its potential in terrestrial and space travel. The two most widely

> Other physical constants have been measured thousands of times more accurately [than gravity], and their values are known to within a few millionths of a percentage point.[G]
> — *The Economist*

accepted paradigms are the Newtonian and General Relativity models, which are the current mainstays of the conventional physicist. Under Newtonian law, the product of two masses divided by the distance squared times the gravitational constant equals the gravitational force. Relativity theorizes that the presence of mass curves the geometry of space-time, which causes masses to alter their paths through space, giving the appearance of the force of gravity. It also proposes that the speed of light is the maximum velocity at which mass can travel.

> It is a myth that Space Shuttle astronauts are weightless because they have traveled beyond the reach of Earth's gravity.[H]
> — *Discover*

But other intriguing and possibly groundbreaking theories regarding gravitation are turning heads in the new energy field. The Graviton Model of gravitation posits that masses emit and absorb large amounts of particles called gravitons. Gravitons are usually described as a particle of mass without charge or energy, forming a gravity wave. These gravitons cause the force to be transmitted between all masses. Since a huge amount of mass is required to produce any meaningful gravitational field, gravitons are assumed to be a very low-energy particle. Although the graviton model appears mathematically correct and able to predict certain effects, graviton particles have yet to detected.[26] Thomas Van Flandern, a professional research astronomer and former Chief of the Celestial Mechanics Branch of the US Naval Observatory, supports the idea that gravity is a rain of tiny particles (gravitons) that pass easily through most matter and that gravity exerts a pushing force, not a pulling force. He claims that the special relativity theory does not apply to binary stars with highly unequal masses.[27]

*Tom Van Flandern's experiments indicate that the speed of gravity is too great to measure, substantially faster than lightspeed.*

Among the new energy possibilities embedded in Van Flandern's perspective of a sea of gravitons, are space propulsion, gravitational shielding and sailing, and heat energy. Thomas Van Flandern concludes that space travel faster than light may be possible because experiments show that the force of gravity itself propagates orders of magnitudes faster than lightspeed.[28] He also believes that far from upsetting much of current physics, the main changes induced by this new perspective are beneficial to areas where physics has been struggling, such as explaining experimental evidence for non-locality in quantum physics, the dark matter issue in cosmology, and the possible unification of forces.[29]

A new gravitational theory called the Omicron Vector Field Model has been developed by James Tracy and Jurgen Schulz at Omicron Research in Sacramento, California.[30] Tracy and Schulz believe that mass causes particle space (aether) to move and accelerate inward. This movement of space causes resistance to be generated on other masses within this flow. Given the fact that all matter consists of mainly empty space, a large percentage of the flow passes unobstructed through the object. Only a tiny portion of the flow (or vector field) is actually used to accelerate the mass. Omicron theory posits that particle space moves through an object causing it to accelerate, but only a small amount of this flow is intercepted by any given atom. Tracy and Schulz view gravity as an extremely strong force but one that is intercepted by very little matter. According to Omicron's particle space theory, the phenomena of inertia is the resistance of mass against the attempt to accelerate it; therefore, only inertia is responsible for force. The theoretical possibilities imply that if scientists were able to intercept greater amounts of this gravitational flow by packing neutrons so tight that there is virtually no space between them, its downward acceleration would be tremendous.

The idea of gravity modification or propulsion from electrical fields has inspired innovative but controversial efforts by some researchers in the field. Magnetism was once regarded as separate from electricity until the electromagnetic effect was discovered. But is electricity a form of magnetism, or is magnetism a form of electricity? Electricity is a phenomenon associated with positively or negatively charged matter, either at rest or in motion. A charge at rest is static electricity; charged particles in motion are known as current. Magnetism is the force by which substances, such as iron, attract other similar substances by the energy in their surrounding magnetic field. In the early 19$^{th}$ century, scientists discovered important connections between electricity and magnetism. An electric current flowing through a wire pro-

> Scientists suggest that space is permeated by a mysterious repellent force that counteracts the pull of gravity on a cosmic scale.[1]
> — US News & World Report

duces a magnetic field, as does a magnet moving inside a loop of wire. Since electricity in motion produces magnetism, and magnetism in motion produces electricity, the phenomena is called electromagnetism. In 1865, Scottish physicist James Clerk Maxwell mathematically proved that since electromagnetic waves travel through space at the speed of light, light itself was an electromagnetic wave.[31]

Ever since Maxwell's profound discovery about electromagnetic waves, researchers have sought to understand better their potential in order to harness their energy. Hal Puthoff has suggested that incorporating the zero-point electromagnetic field as related to the quantum vacuum could unify gravity and electromagnetism and redefine mass. Puthoff says that charged particles oscillating due to the ZPE field emit electromagnetic waves whereas other charged particles are drawn towards secondary fields, causing gravitation.[32] Transportation engineers are developing practical applications that utilize electromagnetism to produce thrust and lift for high-speed trains and other propulsion systems. Japanese engineers have invented a novel lift and propulsion system that uses alternating-current electromagnets to produce not only a strong levitation force but also the propulsion force.[33] The Italians are also using electromagnetic lifting to develop high-speed transport systems in which devices for propulsion, levitation, and a contact-free, on-board electric power transfer are combined into a single electromagnetic component.[34] A bolder step towards the futuristic application of these concepts was taken in 1997 by two scientists at the University of Tokyo who are studying the possibility of levitating humans, using a specialized magnetic coil they developed.[35]

The search to tap electromagnetism has been going on for decades, and researchers have approached it both theoretically and in actual experimental application.

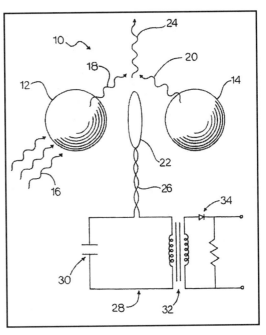

*Franklin Mead, Jr., and Jack Nachamkin were awarded US Patent #5,590,031 in 1996, for their "System for Converting Electromagnetic Radiation Energy to Electrical Energy."*

Electrogravitics is described as a synthesis of electrostatic energy and gravitics, which sets up a local gravitational force independent of the Earth's.[36] The essence of electrogravitic thrust is the use of a very strong positive charge on one side of a vehicle and a negative on the other, producing a driving force from a region of high-flux density to low-flux density. Scientists first began considering this propulsion concept in the 1920s, but the field literally took off after World War II when American physicist Thomas Townsend Brown constructed a prototype aircraft propelled by electrogravitic force. Brown designed and built a flying aircraft whose thrust was derived by an electrogravitic charge combination, a feat he repeated many times with demonstrations witnessed by scientists and the military.

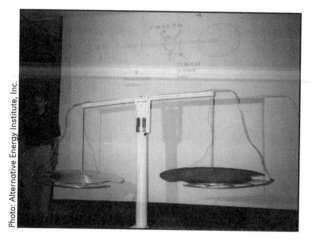

*This large demonstration unit based on T. T. Brown's Electrokinetic Apparatus is used to verify the anomalous forces of the Biefeld-Brown effect and the phenomenon of electrogravitics.*

In 1965, the US Patent and Trademark Office awarded T. T. Brown one of his many patents, this one being No. 3,187,206 for his Electrokinetic Apparatus, an "invention [that] relates to an electrical device for producing a thrust by the direct operation of electrical fields." Brown claimed that "mechanical forces are created which move the device continuously in one direction while the masses making up the environment move in the opposite direction."[37] Brown apparently used specially configured electrodes to shape an electrical field to produce a force upon the device that shaped the field in a way propulsion was possible.

Brown acquired about 30 patents worldwide, most of them based on the field of electrogravitics.[38] Despite more than 30 years of research and many demonstrations of his flying disks, both in the US and abroad, Brown retired in obscurity, his provocative research ignored. Interest in antigravity propulsion has not lessened, however, and the first to succeed in this field is guaranteed at least $10,000,000 by the X Prize Foundation.[39]

As a form of energy propulsion, gravity has advantages over its sister forces like electromagnetism or the strong nuclear force. Gravity is always attractive, as opposed to electromagnetism — which can be attractive or repulsive — and tends to cancel itself out. Although the strong nuclear force can't even reach from one atom to another, gravity has a power of attraction that can span vast distances of space. NASA already uses gravitational fields to slingshot satellites into space, where spacecraft travel for years following indirect trajectories, looping around several planets in order to gain velocity from gravity assists. But it would take NASA hundreds of years to send a satellite to another solar system using this method. Current rocket technology is limited and expensive because boosters are needed to carry enough fuel just to escape Earth's gravity field, let alone all the extra juice needed for deep space travel.[40]

It takes consistent financial funding, as well as decades of dedicated research and development, to make breakthrough advances in new transportation or energy technologies. Unfortunately, international competition, military interests, and the fear of falling behind technologically encourage tight-lipped secrecy and a minimal sharing of information among the many commercial and government laboratories around the world that are individually working towards energy solutions. The Chandra X-ray telescope project is a prime example of how long-term commitment and cooperation between various institutions and countries can push the technological frontier forward for the common good.

As we begin the 21st century, scientists worldwide are experimenting in the laboratory with antigravity research (professionals in this field prefer the term *gravity modification* or *gravity shielding*). During 1998 and 1999, NASA awarded $670,000 in research grants to Superconductive Components, Inc. (SCI), a private company that manufactures ceramic superconductor plates.[41] Selected after an extensive peer-review process, the company supplies NASA with specialized superconductive plates used for experiments in gravitational shielding.[42] Much of the technology driving the design of these gravity modification discs is based on controversial re-

> The historical difficulty in incorporating dramatically new approaches to science and technology programs is due to various factors, not the least of which has been continual budget pressures and the focus on near-term science and technology requirements.[j]
> — *Breakthrough Energy Physics Research Program Plan*

search developed by Russian émigré scientist Eugene Podkletnov.⁴³

Eugene Podkletnov received his master's degree from the University of Chemical Technology at Moscow's Mendeleyev Institute and then spent more than a decade working at the Russian Academy of Sciences. In 1988, at age 33, he was invited to Finland's Tampere University Institute of Technology to earn a Ph.D. in the manufacture of superconductors. By 1992, Podkletnov found that certain superconducting materials lost all resistance to electricity when chilled with liquid nitrogen and exposed to high frequency magnetic fields.⁴⁴ Podkletnov conducted an experiment in which he placed ceramic disks a few inches wide in a cold chamber and passed a magnetic field through them while spinning them rapidly. He was surprised to see that objects experienced a variable but measurable loss of weight (between <0.5% to around 2%) when placed over the rotating superconductor ceramic disks. The experimental values were small but, if accurate, would change the world forever.⁴⁵

In 1996, Podkletnov submitted a research paper detailing his experimental results on weak gravitation shielding properties that was accepted by the respected British *Journal of Physics-D*. Podkletnov, who holds a doctorate in materials science, never referred to his experimental results as "antigravity," but the press did. Unfortunately, Podlketnov's elation at finally getting his results published did not last long because his research paper was secretly leaked to the *British Sunday Telegraph* by a member of the editorial staff of the *Journal of Physics-D*. After reading the research paper, the *Telegraph*'s science correspondent Robert Matthews wrote an article about Podkletnov's experiment on September 1, 1996, calling it the world's first antigravity device. The skeptics had a field day.

Despite the fact that Podkletnov was working on *gravity shielding*, an entirely different notion than antigravity, the negative publicity forced the faculty at the Institute of Materials Science at the University of Tampere to dismiss the innovative researcher who had dared to challenge the status quo. Although Podkletnov insisted that he and his fellow researchers were very careful in measuring the weight change

---

Gravity is both the strongest and the weakest force in the Universe, depending on the range and scale you consider. It is a major force in the lives of universes, people, and stars; but virtually unfelt by atoms or even small spiders.ᴷ
— Sympathetic Vibrations: Reflections on Physics as a Way of Life

— they used metal shielding, nonmagnetic targets, and enclosed the target in a vacuum — he was abandoned by his friends and professional associates and was ridiculed by the scientific establishment. Mainstream physicists suggested that Podkletnov must have made a mistake, measuring magnetic fields or air currents instead of legitimate weight reduction. Critics insisted that huge amounts of energy would be required to shield gravity, but in a 1998 interview with Charles Platt of *Wired* magazine, Podkletnov argued:

> "We do not need a lot of energy. We don't absorb the energy of the gravitational field. We may be controlling it, as a transistor controls the flow of electricity. No law of physics is broken. I am not one crazy guy in a lab; we had a team of six or seven, all good scientists."[46]

The idea of controlling gravity has been considered so far out that Podkletnov's experiment was later mentioned in an episode of the popular science-fiction TV show *The X-Files*. Podkletnov insisted that his ideas were valid but pointed out that it was an entirely new field of knowledge that included physics, chemistry, ceramic technology, electrical engineering, and other applications. Podkletnov later withdrew his paper from peer-review after a fellow research scientist denied collaborating on the gravity shielding experiments with him.[47]

Podkletnov seriously compromised his reputation and the validity of his research because he had not shared his findings or the details of his gravity shielding apparatus with the University during the experiments. Many scientists dismissed Podkletnov's results as bogus because, according to accepted principles of physics, gravity cannot be shielded or modified. Although scientists can create shields for other fields, such as electromagnetic fields, conventional wisdom dictates that no combination of electromagnetic fields of any conceivable strength should modify gravity in any discernable way. One technical difficulty in determining minute gravitation changes in the laboratory is that there are so many known ways in which strong electric and magnetic fields can exert force on nearby objects, that these types of effects tend to dominate and skew weight measurements.

> We measured the weight in every way. We used metal shielding, we used nonmagnetic targets, we enclosed the target in a vacuum — we were very thorough.[L]
> — Eugene Podkletnov

Despite the skepticism surrounding Podkletnov's work, officials at NASA's Marshall Space Flight Center (MSFC) view it as a promising development in new physics propulsion research. MSFC has established a small laboratory and built an improved version of the Podkletnov apparatus. NASA's gravity modification program links researchers at Superconductor Components, Inc. (SCI), and the MFSC with scientists at the Oak Ridge National Laboratory. They are working together to try to duplicate Podkletnov's experiment in which he claimed to have shielded an object from the Earth's gravitational pull using a superconductor spinning at several thousand rpm in a magnetic field.[48] Researchers have developed a helium cryostat that holds the disk, levitation system, and rotation system in a cold gaseous helium environment for testing. Podkletnov's research results suggest superconductive toroids with a specific microstructure and specialized features may be able, under laboratory conditions, to shield gravity waves.

NASA established its Breakthrough Propulsion Physics Program (BPP) in 1997 to investigate exotic technologies with applications for space travel. To push the frontier further, SCI is collaborating with Podkletnov and the Electromechanics and Superconductivity Applications Section of the Energy Technology Division of Argonne National Laboratory.[49] The concept of gravity modification is being investigated by NASA and others because shielding earthbound objects from the effects of gravity will change physics and may radically alter the economics of space travel and transportation in general. The ability to shield an object from gravity has obvious commercial impact for space launches because the effect, reported to be up to 2% of the total gravitational pull on an object, is not trivial and may be cumulative for a stack of superconductive discs.

Chinese physicist Ning Li is a former NASA collaborator at the University of Alabama Huntsville's Center for Space Plasma and Aeronomic Research. A pioneer in the field of gravity modification for more than a decade, Li developed a theoretical formulation that established a connection for rapidly rotating superconductors within magnetic and gravitational fields. In collaboration with several colleagues, includ-

> NASA is also considering using electrodynamic tethers for upward propulsion in a system using solar panels to supply a flow of electricity through the tether to push against Earth's magnetic field.[M]
>
> — *Scientific American*

ing Douglas Torr, a senior scientist at the University of Alabama at the time, Ning Li published several peer-reviewed papers between 1990 and 1993, one of which appeared in the *Physical Review – D.* Her gravity-modification experiments were based on the interaction between a rotating superconductor in an electromagnetic field and the local gravity field. In 1997, Li, Ron Koczor, Associate Director of the Space Science Laboratory at Marshall Spaceflight Center, and others published another technical paper in which they collaborated on an experiment conducting static tests for a gravitational force to prove that gravity shielding is really possible.[50]

Ning Li, who has received $150,000 in funding from NASA, believes that Podkletnov's early results are consistent with her theories. She is dedicated to developing antigravity devices made by altering properties of superconducting materials. Although she no longer publicly shares her experimental results for fear competing scientists may take the lead in this exciting new field, Li believes that the results of her work will one day transform the world. In her laboratory, Li uses liquid nitrogen to reduce the temperature in a chamber down to 390 degrees below zero. Inside the chilled chamber spins a phonograph-sized disk made from an exotic ceramic material. The disk itself is levitated by powerful magnets and actually floats in midair. While NASA is focused on validating the basic experiments, Li is concentrating on applications. A practical antigravity device could mean rockets without propellant or power plants that run without fuel. Li hopes to start with an antigravity car, an accomplishment that this can-do scientist thinks is possible within the next decade.[51]

James F. Woodward, an adjunct Professor of Physics and Professor of History at California State University – Fullerton, has come up with an experimental device that may allow scientists to tap into the universal source of inertia and

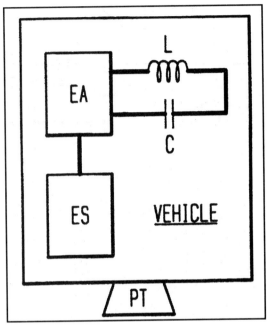

*US Patent #5,280,064 was issued to James Woodward in 1994 for his "Method for Transiently Altering the Mass of Objects to Facilitate Their Transport or Change Their Stationary Apparent Weights."*

> Both the impulse engine and negative mass effect terms hold out the promise of realizable propellantless propulsion.[N]
>
> — James F. Woodward

use it for deep space travel.[52] Woodward uses a bank of capacitors on top of a piezoelectric motion device to reduce the net weight of the apparatus by apparently reducing its mass. (Weight depends on the force of gravity; mass is an innate property of matter.) The simple experiment achieves an average mass reduction in the capacitors, which varies related to their energy level. In a series of published papers, Woodward has explained his system in great detail and even acquired a US patent for his device (No. 5,280,864 — issued January 25, 1994).[53] Although his reported measurements of weight reduction have been small, the results have attracted some serious attention. One of the possible implications is addressed in his paper "Mach's Principle and Impulse Engines: Toward a Viable Physics of Star Trek?" presented at NASA's Breakthrough Propulsion Physics Workshop in 1997. Woodward proposed that impulse engines could be explored experimentally with present technology and at reasonable cost.[54] Unlike current aerospace propulsion technology that relies on rockets expelling propellant mass to gain velocity, impulse engines can produce accelerations without ejecting any material exhaust. Woodward insists that the concepts involved in his theory are in line with established laws of physics and supported by experimental results already in hand.

*Schematic diagram of generic apparatus designed to measure transient mass and stationary forces described by Mach's Principle. Experiments by James Woodward at California State University Fullerton indicate that "impulse engines" theoretically could produce accelerations without ejecting any reaction mass.*

Woodward's theoretical and experimental research into propellantless propulsion relies on "Mach's Principle," which is one of the leading explanations for the origin of inertia. The term Mach's Principle was coined by Albert Einstein in 1918 while referring to Ernest Mach's thesis that contended all motion is relative to some material body and that acceleration with respect to the total matter distribution of the Universe produces inertial forces and effects.[55] Mach believed that inertial and gravitational mass must be the same because inertia is a gravitational effect. Albert Einstein accepted Mach's Principle and used its implications to formulate his Equivalence Principle, which asserts that gravitational and inertial mass are indistinguishable. Einstein's Equivalence Principle, a cornerstone of General Relativity, is generally accepted in physics, but its basis in Mach's Principle was never tested until Woodward's experiments.[56]

Mach's Principle explains inertial resistance as the combined gravitational attraction on the object from all other objects in the Universe. Physicists call this combined pull the gravitational potential, a huge force up to the square of the speed of light in magnitude, which Woodward has reduced to the equation $F = MA$ — force equals mass times acceleration. Although gravity is considered a much weaker force than electromagnetism, the gravitational potential is loaded. This large potential occurs because the Universe contains nearly equal quantities of positive and negative electric charges, which reduce the electric potential to near zero. Positive masses, however, have no apparent negative counterpart, making the gravitational potential enormous.

The dream of building an antigravity device has been around for a long time, and Thomas Townsend Brown was not the only individual who apparently devised a working "antigravity" device. John Roy Robert Searl first began developing his "Searl Effect" technology after World War II. The Searl Effect Generator (SEG) is an electric generator that he claims creates its own gravity field. Searl's company, Direct International Science Consortium, Inc. (DISC), is developing a large SEG with the intent of marketing it and working on a prototype flying craft called an Inverse-G-Vehicle (IGV) as well as other related technologies.[57] The IGV uses a Searl generator-induced gravity field to repel itself away from the Earth's own gravity field. In theory, the SEG causes an ionic breakdown of the air a few feet from the surface of the IGV, enveloping the craft in a vacuum. John A. Thomas, Jr., is a partner and Chief Executive Officer at DISC. In his 1993 book describing the Searl Effect, Thomas writes:

> "In ordinary high voltage generators the maximum potential is limited by the ionized breakdown of the air. Flashover occurs and accumulated energy is lost.

> Mach's Principle explains inertia — the tendency of an object to resist acceleration — by the sum of the gravitational attractions of all objects in the Universe.[o]
> — *Dictionary of the History of Science*

> Crazy as antigravity sounds, the idea was originally suggested by Einstein as a kind of add-on to his General Theory of Relativity.[P]
>
> — Time

The geometry and the arrangement of the field coils in the Searl generator is such that flashover is eliminated until the thing is in a vacuum and flashover is impossible."[58]

An outside source of energy must be supplied to the SEG until it reaches its electrical potential, at which point the Searl Effect apparently produces its own energy and levitation capabilities.

John Searl's claims are extraordinary, and he is no stranger to controversy and setbacks. He once spent a short time in prison for allegedly stealing electricity from his local utility while he publicly claimed he was powering his house using an SEG.[59] In 1999, two colleagues, working with a DISC team building the latest SEG prototype to provide a model for a manufacturer, allegedly walked off with large magnetized coils as well as confidential proprietary information. Although the two men signed non-disclosure contracts with DISC, the theft of the magnetized coils seriously set back the company's production timetable. One of the men has allegedly submitted a patent application that utilizes a method of magnetizing via a combination of D.C. voltage and A.C. voltage, technology that he apparently learned while working for DISC, Inc.[60]

Despite Albert Einstein's brilliance at weaving together space, time, and gravitation, profound theoretical problems remain at odds with modern physics and remain unsolved as we begin the 21st century.[61] Einstein himself remarked that the left-hand side of his field equation (describing the curvature of space-time) was solid as granite but admitted that the right-hand side (connecting space-time to matter) was weak like sand. The greatest scientific mind of the 20th century spent years trying to reconcile General Relativity with the rest of physics, but he was never satisfied. It will be up to the next generation of researchers to try to fit the assorted pieces of the cosmological puzzle together. Some question why the government spends tax money to fund such scientifically and technologically challenging research, but the ongoing studies build a critical base of knowledge about the physical world that surrounds us. There is no doubt that in order to make interstellar space travel realistic for humans, physicists will have to learn much more about gravitation and accept that movement faster than the speed of light is possible.[62]

Einstein's theory of General Relativity beautifully describes gravity and the structure of the Universe on a grand scale. However, Einstein never mentions the quantum effects found in the theory of quantum mechanics, which scientists now use to explain the physical world, and, in certain areas, quantum mechanics contradicts General Relativity. Although relativity theory accurately portrays gravitation in ordinary circumstances and distances, serious mathematical inconsistencies occur when the equations are applied to the distances between the smallest subatomic particles or, as Tom Van Flandern asserts, in binary stars with highly unequal masses. Physicists know that both theories can't be totally correct if they conflict with each other; therefore, at some point, Einstein's brilliant theory will have to be modified. The fact that the two fundamental theories of nature — quantum mechanics and General Relativity — are at odds with one another is attracting an increasing number of scientists who want to solve the physics of "quantum gravity" in order to reconcile the conflict. Recognition of a faster-than-lightspeed propagation of gravity, as indicated by existing experimental evidence, may be the key to taking conventional physics to the next level.

In the 21st century, physicists will strive to combine General Relativity and quantum mechanics into one grand unified theory that describes the structure of the Universe on all scales. This entails unifying gravity with the other three fundamental interactions (electromagnetism and the strong and weak nuclear forces). Some scientists speculate that all four of these interactions may be different aspects of a single force that is responsible for all the interactions between matter. Science is an inherently logical process, and problems and paradoxes are not so much in the science but in the way humans formulate it.[63] As Gregory Benford, a professor of physics at UC Irvine has said, "There's nothing impossible about gravity shielding. It just requires a field theory that we don't have yet. Anyone who says it's inconceivable is suffering from a lack of imagination."[64]

> A courageous scientific imagination was needed to realize fully that not the behavior of bodies, but the behavior of something between them, that is, the [electromagnetic] field, may be essential for ordering and understanding events.[Q]
>
> — Albert Einstein & Leopold Infeld

## FOOTNOTES

1. <www.curtin.edu.au/curtin/dept/phys-sci/gravity/index2.htm> — The Exploring Gravity website offers a set of tutorials suitable for advance high school and college students, covering a variety of topics relating to gravity.

2. Levin Santos, "Weighing the Earth," *The Sciences* (July 2000), Vol. 40, p. 11. Brief article covering the 200-year search for the value of "G" and the difficulty scientists have had ever since English physicist Henry Cavendish made the first credible attempt in 1798. (When Sir Isaac Newton first described the law of gravity, he postulated a fundamental constant, G, which relates force to mass and distance.) The Committee on Data for Science and Technology of the International Council for Science, which establishes the internationally accepted standard scientific values, has determined that measurements of the gravitational constant are more uncertain now than they were in 1987.

3. Steven Weinberg, "Will We Have a Final Theory of Everything?" *Time* (April 10, 2000), Visions 21/Science & Space. Steven Weinberg is a Nobel laureate in physics at the University of Texas. The article, which covers the lack of scientific principles addressing the force of gravity, is available at <www.britannica.com>.

4. Laura Tangley, "Found in Space, Antigravity," *US News & World Report* (March 9, 1998), Vol. 124, p. 8. Brief article on the recent results of a respected international scientific team that studied supernovae to measure changes in cosmic expansion over time.

5. <www.space.com/scienceastronomy/astronomy/mystery_force_0001126.html> — Leonard David, *May the Force Be With You? Mysterious Effect May Influence Spacecraft Trajectories*. Science writer David recently interviewed several Jet Propulsion Laboratory scientists who speculate that the effect may reveal something new in physics, including the possibility that the spacecraft becomes charged as it flies through the Earth's magnetic field.

6. <www.britannica.com> — Type "gravitation" in the Britannica search engine window to explore articles on the subject.

7. Richard C. Morrison, "Weight and Gravity — the need for consistent definitions," *The Physics Teacher* (January 1999), p. 51. Article explaining the differing, sometimes contradictory concepts of weight, gravity, and mass. Bodies of mass have two distinct aspects, gravitational mass and inertial mass. Gravitational mass generates and responds to gravitational fields. Inertia is the tendency of matter to resist acceleration, but, once velocity is gained, it requires an equally large counter force to slow it down again.

8. W.F. Bynum, E.J. Browne, and Roy Porter, *Dictionary of the History of Science* (Princeton University Press, Princeton, New Jersey, 1984), pp. 170–171.

9. *Ibid.*, pp. 170–171.

10. Rosemary Sullivant, "When the apple falls: Isaac Newton brought order and understanding to a Universe of apparent complexity," *Astronomy* (April 1998), Vol. 26, pp. 54–60. Article abstract: "The physical Universe was a virtual mystery until the emergence of Isaac Newton whose work connected previously unrelated phenomena under one coherent principle. Newton united the work of previous scientists and defined the laws that govern most of what is observed in the physical world."

11. <www.physlink.com/reference_edu.cfm> — Excellent online education and reference resource for physics and general science topics.

12. <http://library.thinkquest.org/27585/what/what7.html> — Article "General Relativity: The Art of Bending Nothingness," *From Apples to Orbits: The Gravity Story*. Short tutorial that explains what Einstein meant by curved space.

13. <www.britannica.com> — *op. cit.*

14. Bynum, Browne, and Porter, *op. cit.*, p. 171.

15. Martin Reese, "To the Edge of Space and Time," *Astronomy* (July 1998), Vol. 26, pp. 48–54. Article abstract: "Celestial bodies with the greatest mass and smallest diameters exert the most gravity and Black Holes exert the most overwhelming gravitational forces. Researchers have theorized on the behavior of Black Holes since 1916, and there is evidence of the existence of supermassive Black Holes."

16. Fred Adams and Greg Laughlin, "The Great Cosmic Battle," *Mercury* (January 2000), Vol. 29, p. 10. Article details the known history of the Universe and the underlying conflict between the attractive force of gravity and the tendency for physical systems to evolve toward more disorganized conditions (entropy). Adams is a physics professor at the University of Michigan in Ann Arbor, and Laughlin is a staff scientist at NASA Ames Research Center in Moffett Field, California.

17. <http://library.thinkquest.org/27585/large/large9.html> — "Black Holes and the Art of Trapping Light," *From Apples to Orbits, op. cit.*

18. James Glanz, "X-rays from afar bolster theory of a whirlpool in empty space," *The New York Times*, August 29, 2000, p. D-1.

19. Lawrence H. Ford and Thomas A. Roman, "Negative Energy, Wormholes and Warp Drive," *Scientific American* (January 2000), pp. 48–49.

20. "A Galaxy Less Far, Far Away," *The Sacramento Bee*, December 1, 2000, p. A-5.

21. <www.aiaa.org/public/index.hfm?pubp=0> — The White House National Science and Technology Council Fact Sheet on National Space Policy dated September 19, 1996.

22. <http://chandra.harvard.edu/edu/chandra1012.html> — "Chandra 101: Overview for Teachers and Students." Website is hosted by the Harvard-Smithsonian Center for As-

trophysics in Cambridge, Massachusetts. Educational tutorial includes class activities, background on the Chandra X-ray telescope project, interactive games, printable materials, links, and resources. Website includes general background information on X-rays and Black Holes.

[23] Mariette DiChristina, "X-Ray Vision," *Popular Science* (March 1999), Vol. 254 No. 3, pp. 70–74. Article about the Chandra X-ray Telescope: "A high-tech telescope that sees in X-rays will chronicle the Universe's dark secrets from its vantage point in space." The Chandra project is the third in NASA's Great Observatories series, and it bridges the part of the electromagnetic spectrum between the first two, the Hubble Space Telescope launched in 1990 and the Compton Gamma Ray Observatory launched in 1991.

[24] Ford and Roman, *op. cit.*, p. 49.

[25] Gravitation-oriented webpage for The Center for Space Plasma, Aeronomy, and Astrophysics Research (CSPAR) at the University of Alabama in Huntsville, Alabama. CSPAR is an internationally recognized center of excellence in space plasma, aeronomic, and astrophysical research. Established in 1986, it brings together faculty, research scientists, and students from the colleges of engineering and science at the University of Alabama – Huntsville. This webpage contains a link to Stanford University's Gravity Probe B homepage, which offers FAQ's such as "What is the purpose of Gravity Probe B?" and "Is Gravity Probe B worth doing?"

[26] <www.omicron-research.com> — Omicron particle space theory by James E. Tracy and Jurgen Schulz. Website explains Shifting Theory in which the gravity wave emanating from matter is simply a steady stream of gravitons emanating from all particles. Free gravitons, which move randomly through space, also randomly strike and are absorbed by particles of matter.

[27] Tom Bethell, "Rethinking Relativity," *The American Spectator* (April 1999), Vol. 32, pp. 20–24. Article abstract: "Albert Einstein's theory of special relativity is being reconsidered because of a research associate's claim that gravity is 20 billion times faster than light." Tom Van Flandern asserts the special relativity theory does not apply to binary stars with highly unequal masses.

[28] <www.altenergy.org/4/cofe/cofe.html> — Alternative Energy Institute, Inc., profile on Thomas Van Flandern's presentation at the 1999 Conference on Future Energy. Van Flandern explains how new findings regarding particle-gravity physics suggest that, in principle, free energy sources can be used for propulsion.

[29] <www.metaresearch.org> — Van Flandern's Meta Research website at the University of Maryland Physics, Army Research Laboratory. Van Flandern's article abstract titled "The Speed of Gravity — What the Experiments Say" states: "Standard experimental tech-

niques exist to determine the propagation speed of forces. When we apply these techniques to gravity, they all yield propagation speeds too great to measure, substantially faster than lightspeed. This is because gravity, in contrast to light, has no detectable aberration or propagation delay for its action, even for cases where sources of gravity accelerate significantly during the light time from source to target."

[30] <www.omicron-research.com/RecDocE/introE.html> — Project Omicron is the first serious gravity control research project being shared on the Internet. This website explains the work in progress of James E. Tracy and Jurgen Schulz.

[31] Alan Lightman, *Great Ideas in Physics* (New York: McGraw-Hill, 2000), pp. 124–125 and 128–129. Among the topics covered in this book are the Conservation of Energy; the Second Law of Thermodynamics, the Theory of Relativity, and Quantum Mechanics. Lightman is a physicist, writer, and educator. Currently he is John E. Burchard Professor of Humanities and senior lecturer in physics at MIT.

[32] *Mercury* (March-April 1996), Vol. 25, p.15. Short article. Abstract: "New ideas related to the zero-point electromagnetic field due to quantum vacuum would unify gravity and electromagnetism and also redefine mass. According to Harold Puthoff, charged particles oscillating due to the zero-point field emit electromagnetic waves. Other charged particles are drawn towards secondary field, causing gravitation. Based on Einstein's principle of equivalence of inertial and gravitational mass, inertia must be electromagnetic in nature too. Hence, mass is electromagnetic resistance. Accordingly, mass is not a property of matter but results due to charge."

[33] H. Hayashiya, D. Lizuka, H. Ohsaki, and E. Masada, "A Novel Combined Life and Propulsion System for a Steel Plate Conveyance by Electromagnetics," the Institute of Electrical Electronics Engineers *Transactions on Magnetics* (July 1998), Vol. 34, pp. 2093–2096. Article abstract: "A novel lift and propulsion system using alternating-current electromagnets is proposed. The system can simplify the configuration of a noncontacting conveyance system for a steel plate by magnetic force. The results indicate that the proposed system can produce not only the strong levitation force but also the propulsion force on a steel plate."

[34] M. Andriollo, G. Martinelli, A. Morini, and A. Tortella, "Electromagnetic Optimization of EMS-MAGLEV Systems," the Institute of Electrical Electronics Engineers *Transactions on Magnetics* (July 1998), Vol. 34, pp. 2090–2093. Article abstract: "In EMS-MAGLEV high-speed transport systems, devices for propulsion, levitation, and contactless on-board electric power transfer are combined in a single electromagnetic structure. The paper describes an overall optimization procedure, based on a suitable mathematical model of the system, which takes into account several items of the system performance."

[35] Mark Buchanon, "And God said…," *New Scientist* (July 26, 1997), Vol. 155. pp. 42–43. Article abstract: "The magnetic coil has the capacity to defy gravity, divide water and levitate frogs, grasshoppers and strawberries based on a process called 'The Moses Effect.' Initially reported by two scientists namely Ueno and Iwasaka of the University of Tokyo, the phenomenon remained unexplained until Kitazawa found out that water, being diamagnetic, has a mild repulsion to magnets. The magnet, being a coil, has its strongest pull at the center, thereby causing the parting of water at the middle. The possibility of levitating humans using the magnetic coil is also being studied."

[36] <www.padrak.com/ine/INE24.html> — "Electrogravitics Systems," United States Air Force examination of electrostatic motion, dynamic counterbary and barycentric control. 1956 paper by Gravity Research Group explores the possibility of "making efforts to discover the nature of gravity from cosmic or quantum theory, investigation and observation, with a view to discerning the physical properties of aviation's enemy." Site also includes a discussion about electrogravitics technology and its possible use as a means of space propulsion. Posted on the Institute for New Energy website; INE defines itself as a nonprofit technical organization whose foremost goal is "to promote research and educate society of the importance of alternative energy." The INE Internet address is <www.padrak.com>

[37] <www.newphys.se/elektromagnum/physics/Stirniman/Brown-Elektrokinetic-apparatus-patent> — On June 1, 1965, the United States Patent Office issued Thomas Townsend Brown patent # 3,187,206 for an "Electrokinetic Apparatus." The invention relates to an electrical device for producing thrust by the direct operation of electrical fields. Webpage posting of T.T. Brown's original patent application.

[38] <www.soteria.com/brown/> — Thomas Townsend Brown website includes an index of the family history, photographs, personal letters, research documents, patents, FAQ's, a website store, and related links.

[39] <http://antigravitypower.tripod.com/patents.html> — Antigravity Propulsion webpage with links to the United States Patent Office database from 1976 to present. Also contains links to the IBM Patent and European Patent Office databases with US patents from 1920. This website is dedicated to discovering the mechanism to construct an antigravity propulsion system. It is a clearinghouse for information concerning antigravity mechanisms, theories, concepts, ideas, experiments, and Internet links. The X Prize is a $10 million cash prize promised to the first private team that safely launches and lands a vehicle capable of transporting three people on two consecutive suborbital flights to 100 kilometers altitude. The X Prize Foundation was established in 1994 as an educational nonprofit organization dedicated to inspiring private entrepreneurial advancements in space travel.

40 Tim Beardsley, "The Way to Go in Space," *Scientific American* (February 1999), Vol. 280, No. 2, p. 83. Extensive collection of articles detailing the various technologies and ideas that scientists are currently investigating in order to make deep space travel possible.

41 <www.superconductivecomp.com/germandoc.htm> — Superconductive Components, Inc. (SCI), has been featured as a prominent superconductor manufacturer on a German television featuring gravity modification. SCI is headquartered in Columbus, Ohio, where they manufacture ceramic powders, laser ablation sources, single crystal substrates, and engineered ceramics for research and commercial applications of superconductors, solid oxide fuel cells, ceramic membranes, lithium ion batteries, and other products. The company sells product in the US and more than 40 other countries.

42 Eugene Mallove, "NASA — Antigravity Research Grant," *Infinite Energy* (Issue 25, 1999), pp. 29–30. Short article and contact information regarding SCI and comments by antigravity researchers.

43 <www.altenergy.org> — *op. cit.*, Type "Podkletnov" in AEI search engine window for his biographical information.

44 <www.inetarena.com/~noetic/pls/Papers/pc203.htm> — Copy of Podkletnov and Nieminen's 1992 preliminary paper titled: "A Possibility of Gravitational Force Shielding by Bulk Superconductor," Tampere University of Technology, Institute of Materials Science, Tampere, Finland, published in *Physica C 203*. Excerpt: "The sample was found to lose from 0.05% to 0.3% of its weight, depending on the rotation speed of the superconducting disk. Partial loss of weight might be the result of a certain state of energy which exists inside the crystal structure of the superconductor at low temperatures. The unusual state of energy might have changed a regular interaction between electromagnetic, nuclear and gravitational forces inside a solid body and is responsible for the gravity shielding effect."

45 <www.institute.ieee.org/INST/nov96/gravity.html> — 1996 article "Finnish Researcher Reportedly Discovers Gravity-change Effect," by Greg Gillespie, Assistant Editor. The Institute of Electrical and Electronics Engineers, Inc., helps advance global prosperity by promoting the engineering process of creating, developing, integrating, sharing, and applying knowledge about electrical and information technologies and sciences for the benefit of humanity and the profession.

46 <www.wired.com/wired/6.03/antigravity_pr.html> — Charles Platt, "Breaking the Law of Gravity," *Wired* (March 1998), p. 4. Veteran journalist Platt tracks down the elusive Eugene Podkletnov and interviews other researchers experimenting with gravity modification.

47 <www.gravity.org/msu.html> — Podkletnov's 1995 revised paper titled "Weak Gravitation Shielding Properties of Composite Bulk Superconductor below 70 $K$ under Electromagnetic Field." Italian theoretical physicist Giovanni Modanese provides a history and background to Podkletnov's research paper. Abstract states: "A high temperature bulk ceramic superconductor with composite structure has revealed weak shielding properties against gravitational force while in a levitating state at temperatures below 70 Kelvin."

48 <www.inetarena.com/~noetic/pls/Deltag.htm> — June 17, 1997, status report on NASA's gravity shielding experiment. L. Whitt Brantley, Chief of the Advanced Concepts Office at Marshall Space Flight Center in Huntsville, Alabama, released the document. The purpose of Project 96-07, High Temperature Superconductor Research, is to investigate an interaction reported in the peer-reviewed literature between a rotating-high-temperature superconductor in a magnetic field and the local gravity field.

49 Mallove, *op. cit.*, p. 30.

50 <www.inetarena.com/~noetic/pls/Papers/msfcuah1.html> — NASA Marshall Space Flight Center and University of Alabama, Huntsville 1997 *Physica-C* research paper by Ning Li, David Noever, Tony Robertson, Ron Koczor, and Whitt Brantley titled "Static Test for a Gravitational Force Coupled to Type II YBCO Superconductors." Recent experiments have reported that for a variety of different test masses, a Type-II, high temperature (YBCO) superconductor induces anomalous weight effects (0.05%-2% loss).

51 Corey S. Powell, "Zero Gravity," *Discover* (May 1999), Vol. 20, p. 31.

52 <http://chaos.fullerton.edu/Woodward.html> — James F. Woodward's homepage at California State University, Fullerton. Woodward is a faculty member at Cal State Fullerton in the physics and history departments. Includes his research interests: gravitation, Mach's principle, killing time, and propellantless propulsion. Also lists his recent publications.

53 <www.inetarena.com/~noetic/pls/woodward.html> — James F. Woodward: Mach's Principle Weight Reduction = Propellantless Propulsion. Contains a summary of Woodward's paper published in *Foundations of Physics Letters*, Vol. 9, No. 3, 1996, pp. 247–293. Presents an overview and basic tenets of this propulsion system as well as links, references, and comments by other physicists. "Mach's Principle explains inertia — the tendency of an object to resist acceleration — by the sum of the gravitational attractions of all objects in the Universe." This website, "Quantum Cavorite — The Physics of Gravity and Inertial Control: Is it Real?" created by Pete Skeggs, offers critical analysis of the various gravity modification experimenters, such as Podkletnov and Woodward. Quantum Cavorite is a central clearinghouse for both supporting and contradictory views of the issues presented, and facilitates both scientific and popular discussion. Pete Skeggs is a software developer and a pivotal figure in the gravity-enthusiast underground.

54 <http://chaos.fullerton.edu/~jimw/nasa-pap/> — James F. Woodward's "Mach's Principle and Impulse Engines: Toward a Viable Physics of Star Trek?" Paper presented at the NASA Breakthrough Propulsion Physics Workshop, held in Cleveland, Ohio, on August 12–14, 1997. Abstract: "Mach's principle and local Lorentz-invariance together yield the prediction of transient rest mass fluctuations in accelerated objects. These restmass fluctuations, in both principle and practice, can be quite large and, in principle at least, negative. They suggest that exotic spacetime transport devices may be feasible, the least exotic being 'impulse engines,' devices that can produce accelerations without ejecting any material exhaust. A scheme of this sort is presented and issues raised relating to conservation principles are examined."

55 Bynum, Browne, & Porter, *op. cit.*, p. 242.

56 <www.npl.washington.edu/AV/altvw83.html> — "Antigravity Sightings — The Alternative View" by physicist John G. Cramer. *Analog Science Fiction & Fact* article gives a critical overview of Woodward's experimental research with Mach's Principle. In a 1998 interview with *Wired*'s Charles Platt, John Cramer told the journalist that he did not believe that Podkletnov had found a way to block gravity.

57 <www.searleffect.com> — John R. R. Searl's website offers information regarding the historic background to the "Searl Effect." Searl is president of Direct International Science Consortium, Inc., a corporation dedicated to developing and marketing the Searl Effect Generator, a flying craft called an Inverse-G-Vehicle, as well as other related technologies. This website offers three levels of information: "free access," "members only," and "builders only."

58 John A. Thomas, Jr., *Antigravity: The Dream Made Reality* (London, England: Direct International Consortium, 1993), p. 39. This is the story of John R. R. Searl as told by his friend and partner — includes biography and an in-depth description of Searl's "the square technology." The book contains many photographs and diagrams of Searl's flying saucers. Thomas has a technical background in electricity and process control.

59 <www.altenergy.org/3/new_energy/magnetics_and_gravitics/gravitics/gravitics.html> — *op. cit.*, Alternative Energy Institute's biography on English inventor John. R. R. Searl. This information can be accessed by using the AEI search engine. Type "Searl" into the search engine window.

60 <www.searleffect.com/free/notice.html> — Public announcement by Direct International Science Consortium, Inc., disclosing the theft of magnetic coils and proprietary information by two former workers. Names and addresses are given for the two men. John Searl is the only rightful inventor of this technology; no one else has a license to build, use, or sell any device derived from his technology or from his company DISC. They do

encourage all who are interested to pursue experimentation on their own based on the information that has been released publicly by DISC.

[61] Paul S. Wesson, "Fundamental Unsolved Problems in Physics and Astrophysics," Prepared for the California Institute for Physics and Astrophysics, 366 Cambridge Avenue, Palo Alto, California. This 18-page paper is a list and discussion of what are arguably the top 20 unsolved problems in physics and astrophysics today. The list ranges from particle physics to cosmology and "opens the prospect that a solution to one or a few may lead to a significantly better understanding of modern physics." Paul S. Wesson is a member of the Department of Physics at the University of Waterloo, Waterloo, Ontario, Canada.

[62] Ken Croswell, "Interstellar Trekking," *Astronomy* (June 1998), Vol. 26, pp. 46–52. Article abstract: "Difficulties presented by interstellar travel are discussed. Enormous distances and inadequate vehicle speeds in addition to limitations imposed by physics are just a few obstacles to such a venture. Critics believe that the rocket should be eliminated as a source of energy for space vehicles."

[63] Paul S. Wesson, *op. cit.*, p. 17.

[64] Charles Platt, *op. cit.*, p. 64.

## SIDEBAR FOOTNOTES

[A] Peter Weiss, "Gravity gets measured to greater certainty," *Science News* (May 13, 2000, Vol. 157, Issue 20), p. 311.

[B] Richard C. Morrison, "Weight and Gravity — the need for consistent definitions," *The Physics Teacher* (January 1999, Vol. 37, Issue 1), p. 51.

[C] Levin Santos, "Weighing the Earth," *The Sciences* (July 2000, Vol. 40, Issue 4), p. 11.

[D] David Kestenbaum, "Gravity Measurements Close in on Big G." *Science* (December 18, 1998, Vol. 282, Issue 5397), p. 2180.

[E] "A Matter of Gravity," *The Economist* (December 25, 1999, Vol. 353, Issue 8151), p. 34.

[F] Martin Rees, "To the Edge of Space and Time," *Astronomy* (July 1998, Vol. 26, Number 7), p. 48.

[G] "Measuring Gravity," *The Economist* (May 6, 2000, Vol. 355, Issue 8169), p. 82.

[H] "Science Myths," Earth's gravity is still strong 200 miles up. As the shuttle circles Earth, it is continuously falling. But it is moving sideways so fast that Earth's curved surface drops away as quickly as the spaceship drops toward it. This "free fall" creates the sensation of weightlessness. A steep drop on a roller coaster produces a similar effect. *Discover* (May 2000, Vol. 21, Issue 5), p. 18.

I   Laura Tangley, "Found in space, Antigravity," *US News & World Report* (March 9, 1998, Vol. 124, Number 9), p. 8.

J   Phil Carpenter, David Hamilton, Dave Goodwin, et. al., *Breakthrough Energy Physics Research (BEPR) Program Plan* (Office of Energy Efficiency & Renewable Energy, Office of Science, and the Office of Nuclear Energy, Science and Technology — Joint Exploratory Research (October 2000 — *Draft for Agency Comment Only*), p. 9.

K   K. C. Cole, *Sympathetic Vibrations: Reflections on Physics as a Way of Life* (New York: Bantam Books, 1985), p. 95.

L   Charles Platt, "Breaking the Law of Gravity," *Wired* (March 1998) Issue 6:03, p. 13.

M   Tim Beardsley, "The Way to Go in Space," *Scientific American* (February 1999, Vol. 280, Number 2), p. 87.

N   James F. Woodward, "Rapid Spacetime Transport and Machian Mass Fluctuations: Theory and Experiment," an unpublished conference paper written by Woodward in early 2001, p. 2. This document is available at <http://chaos.fullerton.edu/~jimw/Jpcawf1.pdf>

O   W. E. Bynum, E. J. Browne, Roy Porter, *op. cit.*, p. 242.

P   Michael D. Lemonick, "Adventures in Antigravity," *Time* (August 7, 2000, Vol. 156, Issue 6), p. 74.

Q   Albert Einstein and Leopold Infeld, *The Evolution of Physics: From Early Concepts to Relativity and Quanta* (First published 1938. Renewed copyright by New York: Simon & Schuster, 1966), pp. 295–296.

# The Race for New Energy

## 16

The race to develop new, clean energy technologies is really heating up as we enter the 21st century. Revolutionary developments in pollution-free power generation will help break the industrial world's bond with gas, coal, and oil — which has negatively impacted our health, environment, and planet. New technologies must not produce carbon emissions, polluting our air and contributing to global warming. World carbon emissions have risen nearly fourfold since 1950, and the Intergovernmental Panel on Climate Change has stated that increasing average surface temperatures over the next century could cause "a wide array of dislocations to human and natural systems."[1]

SCIENTISTS, ENGINEERS, AND PROFESSIONAL RESEARCHERS are excited about a plethora of new emerging energy systems that, with sufficient financial and government support, might be developed and brought to market in time to help society make the transition from limited and polluting fossil-fuel-dependent energy systems to a future of clean, decentralized power generation. In Sacramento, California, a small group of retired rocket scientists has developed a steam-generation system that is fired by oxygen and fossil fuels but produces no emissions or pollutants. Clean Energy Systems' revolutionary innovation represents the world's only gas generation system that will economically sequester $CO_2$ when using fossil fuels

*A man with a new idea is a crank, until he succeeds.*
— Mark Twain

to produce zero-pollution, low-cost electricity.[2] The technology is so promising that, in 1999, CES received a $1.77 million grant from the Department of Energy to build its steam generator. The gas generator combusts clean fuel and oxygen to produce a high-temperature, highly pressurized gas composed of steam and carbon dioxide, which drives a multistage turbine and electrical generator. CES engineers still face several obstacles in developing this complex technology, including how to regulate precisely the steam-cooling process and balancing the most efficient mixture of fuel and oxygen. They must also prove that their system is durable and dependable, but CES officials are confident that these hurdles will be overcome.[3]

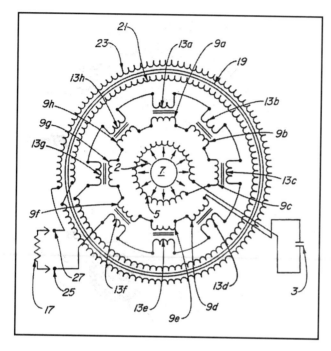

US Patent #4,835,433 issued to Paul M. Brown & Nucell, Inc., in 1989, for an "Apparatus for Direct Conversion of Radioactive Decay Energy to Electrical Energy." Image depicts a schematic of a nuclear battery.

Forward-looking scientists and engineers see a future of high-power energy packages, such as advanced batteries, inexpensive fuel cells, microgenerators of cheap electricity as well as decentralized power sources that will be affordable and environmentally clean. Modular, on-site heat and electricity generators will replace our current centralized utility-based energy system. This will be especially important in developing nations that cannot afford the extensive infrastructure of high voltage wires needed to distribute electricity. In fact, there are many new technologies in the theoretical or research and development stages that have the potential to convert energy into useful work.[4]

One advancement in battery technology has already been achieved by Paul Brown with his patented betavoltaic battery. This innovative device is powered by a benign nuclear power source called tritium (an isotope of hydrogen) that simply emits an electron over its half-life of 12.5 years. The tritium battery's useful life is estimated at about 25 years, making recharging unnecessary. When the long-lived battery finally goes dead after a quarter century's worth of service, disposal is safer than throwing out a home smoke detector. These inexpensive batteries have a wide range of applications and boast high-energy density as well as 25 years' worth of low-power generation. Paul Brown, a leader in the development of cutting edge nuclear technologies, is a member of the American Nuclear Society and holds several patents in the field.[5]

Brown is also the brains behind Global Atomics, Inc. (GAI), a company dedicated to the advancement of safe and economic scientific solutions to the problems surrounding the use of nuclear material and processes. Their proprietary *Photo-deactivation Technology*, which may offer a quick and inexpensive solution to the nuclear waste problem, is revolutionizing the nuclear industry. Photo-deactivation is the process of bombarding nuclear materials with X-rays, causing the target material to become nonradioactive in a short period of time. This new technology, which is based on long-standing and accepted principles of physics, is designed to decontaminate and eliminate large quantities of weapons-grade waste and secondary nuclear waste products. GAI claims that their photodeactivation technology is a safe, economical method for turning long-lived radioactive waste into benign and stable materials. Byproducts of the process include inert materials and also heat, which can be recycled using a conventional heat recovery/exchange system to generate commercial-grade electric power.[6]

Other, more exotic future energy technologies are being pursued in laboratories across the country and around the world. Peter Graneau has invented an apparatus that converts the chemical bond energy of water into kinetic energy. Graneau and his son, Neal, have been experimenting with the liberation of energy from ordinary water by means of an arc discharge since 1985, when they began measuring arc

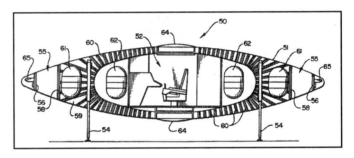

*Michael Sohnly was issued US Patent #5,211,006 in 1993 for his "Magnetohydrodynamic Propulsion System." This environmentally neutral propulsion system is fueled by hydrogen and oxygen derived from water through electrolysis to power aircraft and/or aerospace vehicles. The innovative system has possibilities for intercontinental, interplanetary or deep space travel.*

explosion forces at MIT.[7] The Graneau's use a high-voltage capacitor discharge to release the intermolecular bond energy stored in water. Normal rainwater is accelerated into a cold fog that produces low-grade heat and kinetic energy. These water explosions do not destroy individual water molecules, cause no air pollution, and reportedly generate an output of energy twice that of the input. This overunity condition could be favorable for reduction to a motorized conversion system and future propulsion source.

When it comes to developing novel propulsion systems, no one works harder than the National Aeronautics & Space Administration (NASA). Today's scientists and engineers realize that interplanetary travel is nearly impossible with current rocket technology. Extraordinary breakthroughs will be needed in fuel composition and engine design if humans are to explore space, a challenging technical feat NASA takes seriously. Engineers at the Johnson Space Center's Advanced Space Propulsion Laboratory (ASPL) in Houston, Texas, are developing what could become the next generation rocket engine for possible travel to Mars. This advanced propulsion system, called the Variable Specific Impulse Magnetoplasma Rocket and dubbed "VASIMR" for short, has been in development for 20 years and has cost $10 million, so far. Some consider this futuristic project a waste of time and money, but the $10 million invested in VASIMR over the last two decades is really quite small when compared to NASA's total financial appropriation of more than *$10 billion* in fiscal year 2001.[8]

The VASIMR plasma drive will propel rockets by expelling an ionized gas, or plasma, not by burning chemicals as today's rockets do. The propellant will consist of hydrogen, helium, or some other superlight element in the periodic table. The extraordinary light weight of the fuel will allow larger payloads, and, since the plasma drive is amazingly energy

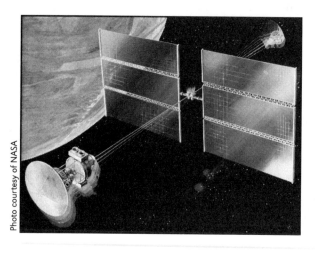

*A NASA artist's concept of a vehicle that could provide an artificial-gravity environment for Mars exploration crews. The piloted vehicle rotates around the axis that contains the solar panels. Levels of artificial gravity vary according to the tether length and the rate at which the vehicle spins.*

Photo courtesy of NASA

efficient, it can fire the engine continuously to generate constant acceleration and the higher speeds necessary for travel to a nearby planet. It also creates a weak pseudogravitational field that may help astronauts avoid the negative physiological effects of long-term weightlessness. There are serious engineering obstacles, primarily that the VASIMR's fuel must be heated to about 1 million degrees Celsius for combustion, compared to the much cooler 3,300 degree propellant temperature burned by NASA's current space shuttle engines. Due to the extreme temperatures of the plasma, no known substance can handle the gas without disintegrating. NASA engineers have solved this problem on a small scale by developing a system of magnetic fields to contain the superheated plasma. Despite the seemingly overwhelming technical obstacles involved in this research, NASA has successfully built a working — albeit small — prototype of this plasma drive. According to those developing VASIMR, it is funding, not physics, that is holding the program back.[9]

Funding has always been a serious problem for the men and women exploring the frontiers of science. Pioneers in the emerging new energy field are self-motivated and usually nonconforming; they prefer to learn and discover new things for themselves as opposed to being taught from the textbook. When it comes to scientific investigation into truly new energy frontiers, guidance from past records is often of little help and, in fact, may hinder efficient progress. Inventors, who work solo or in small teams, must choose a project or route of exploration that they can complete by themselves. Isolation often results as researchers seek to protect the proprietary information that gives their work commercial value and secures the inventor's share of future profits.

Theories are often of little value in new energy research; for inventors, the reality of success is the proper functioning of the machine they design and build. Many inventors believe that they don't need anything else for proof, let alone a theory for the machine. Working on their own, independent researchers struggle for financial resources, which tend to flow toward large established institutions. Ironically, the difference between research conducted using academic methodology and by creative, free-thinking inventors is probably nec-

> In fact, each new major advance in science starts with an anomaly that is unacceptable at first.
> — Beverly Rubik

> Fooling around with alternating current is just a waste of time. Nobody will use it, ever.
>
> — Thomas Edison

essary to foster invention in frontier science, but these complementary characteristics invite attacks from institutions that are striving to standardize information and maintain authority. Associating with an unorthodox researcher/inventor jeopardizes institutional stability and, therefore, the legitimacy for the all-important funding to that organization. More often than not, association with new-energy research by any scientist or researcher in a commercial or academic institution generally means professional excommunication.[10]

Money, however, has not been such a difficult problem for Dr. Randall Mills, the founder of Blacklight Power, Inc. Two utility companies have invested millions in Mills' claim that he has created a novel way of utilizing the hydrogen molecule to produce unprecedented levels of useable energy. Mills insists that lowering the energy state of hydrogen via catalytic reaction produces levels of energy 100 to 1,000 times the normal state achieved by burning the hydrogen. Mills' new atomic theory is purportedly based on accepted physical laws of nature and quantum mechanics. If Mills is right, his Blacklight Process (so-called because of the ultraviolet lightwaves emitted during catalysis), could generate nearly limitless clean energy, and has implications for traditional industries such as heating, electricity, transportation, and medical treatment.[11]

Dr. Mills, who attended classes in biotechnology and electrical engineering at MIT while obtaining his M. D. from Harvard Medical School, is not working out of his garage. The offices and laboratories of Blacklight Power, Inc., are located near Princeton, New Jersey, in a 53,000-square-foot research facility, staffed by 35 employees. Initial products currently in development are a direct plasma to electric power cell targeted at residential and other electric utility markets, and a high voltage battery to power portable electronics. De-

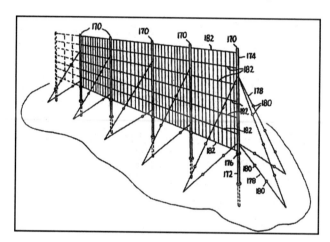

*Stephen Gregory and Alma Schurig were issued US Patent #4,146,800 in 1979 for their "Apparatus and Method of Generating Electricity from Wind Energy." When wind blows through the fence's electrostatic field the device utilizes the charged particles and increased potential to produce electricity.*

spite recent conflicts with investment groups, the US Patent and Trademark Office, and vocal opponents who accuse Mills of making unsubstantiated claims, Blacklight Power continues to develop its technology and receive additional support.

One simple and untapped method to generate decentralized, pollution-free electricity in good weather has been around since the 18th century when Benjamin Franklin improvised the first electrostatic motor. This energy system is based on atmospheric conditions in which the physics of high-pressure and fair-weather create an abundance of positive electric charges that diminish rapidly with altitude. Specialized antennas have been designed to tap this charge difference to generate electricity, but research scientist Oleg Jefimenko discovered that by adding a small radioactive source (alpha or beta) to the aerial, he can avoid the clumsy height requirements associated with the traditional antennas. Jefimenko's innovative design creates larger currents because the radioactive source ionizes the air in the vicinity of the antenna. The positive-negative differential generates enough charge to power an electrostatic motor, which can operate on a variety of sources and range of voltages.

It is estimated that a small desk-sized electrostatic motor can generate one megawatt of power. (One megawatt provides enough electricity for 1,000 American households.) Jefimenko has already succeeded in constructing demonstration motors that run continuously off atmospheric electricity. The atmospheric potential of the planet is estimated at not less than 200,000 megawatts, an abundant source of free energy. With design improvements and a power upgrade, Jefimenko's electrostatic motor has the potential to tap an inexpensive and renewable source of electricity, much like wind and solar power.[12]

Two other new technologies deserve second mention because they represent two important stages of research and development. First, David Wallman's patented carbon-arc gasification process produces a "clean" carbo-hydrogen gas from biomass waste solution, and boasts an overunity efficiency of about 125% to 150%.[13] The carbo-hydrogen gas burns cleanly, producing water vapor and only the amount of $CO_2$ that was originally absorbed by the biological mass

> Science...avoids the most fundamental questions. It is characteristic of physics as practiced nowadays, not to ask what matter really is, for biology not to ask what life really is, and for psychology not to ask what the soul really is.
> — C.F. von Weizsaecker

when it was recently growing. In contrast to burning fossil fuels like coal, oil, and natural gas that release enormous quantities of ancient buried carbon into our atmosphere, Wallman's biomass gas recycles only carbon dioxide that would be released into the atmosphere when the plant died and decayed anyway. This technology boasts excellent energy efficiency and can generate electricity from the millions of gallons of farm-produced liquid biomass that go to waste every year. The process is scaleable for commercial use, and demonstrations of pilot plant designs are already available from Wallman's company, DW Energy Research.

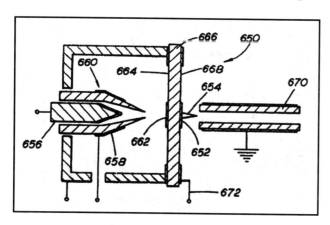

*Inventor Ken Shoulders' "Method of and Apparatus for Production and Manipulation of High Density Charge" was granted Patent #5,054,046 in 1991. This image shows a schematic diagram of a cathode and anode designed by Shoulders to produce high electrical charge density entities.*

The energy produced by the atom is a very poor kind of thing. Anyone who expects a source of power from the transformation of these atoms is talking moonshine.
— Ernst Rutherford

Though David Wallman's biomass-to-gas process is fine-tuned and ready for market, Ken Shoulders' amazing electron charge cluster experiments (EV) are in the infant stage and still knee deep in R&D. Shoulders' patented process produces electron clusters with such high energy density, they equal processes exceeding 25,000 degrees Celsius upon impact, yet he uses only 20 microjoules to produce the effects.[14] Power is measured in watts, kilowatts, and horsepower. Energy is measured in joules (watt-seconds) or kilowatt-hours. A power level of one watt that continues for one second equals one joule; the integrated energy from a 100-watt light that runs for 60 seconds equals 6,000 joules. Shoulders' EV clusters travel at a maximum of one-tenth the speed of light and penetrate any substance with accuracy and sharp precision. Charge-cluster bore holes in refractory dielectric or semiconductor material exhibit an electrified, melted look with physical evidence of thermal processes in the range of 25,000 degrees C. Despite the obvious physical evidence that the targeted material melted, it occurred at temperatures so cool that it did not melt wax or harm the sensitive photographic emulsion.

Shoulders has observed low energy nuclear transmutation in some target materials, but the process is not very efficient because the basic mechanism is still unknown and has not been optimized. He thinks that the low efficiency is likely due to a multistep process involving shock wave interaction between rare dual EVs. Shoulders believes that the enigmatic EV action is the root process for all forms of what is commonly called "cold fusion." He postulates:

> "The problem with the cold fusion process understanding and control is that the interaction takes place on a very small size scale and is not amenable to either easy fabrication or inspection. The size scale is approximately 1 micrometer in total extent and has requirements for even greater detail in material handling. In fact, the structures of interest are seen by most investigators as contamination sites. The gross methods currently used in the technology are no match for the requirements."[15]

The new physics of like charges clustering in bundles under low power conditions opens a wide range of applications including spacecraft maneuvering microthrusters, but Shoulders has no illusions that his EV technology will be powering anything anytime soon. This realistic researcher has stated that energy is not likely to be produced as much as converted from one state to another. The two primary methods for performing this conversion with EVs are thermal and electrical, but, after years of experimentation, Shoulders sees no final resolution in the form of a working machine to validate any one solution. He believes that the thermal processes encountered thus far all produce disruption of the materials used during the process, and no satisfactory engineering solution to preserve or economically rebuild the materials has been found. Electrical energy production using these clusters of pure electronic charge does not necessarily destroy the energy conversion device, as it does with thermal processes, but there are inherent disadvantages that have not yet been overcome in an engineering sense. Because the small size of the reaction

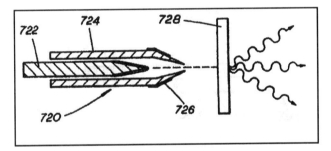

The new scientific revolution is still in its early phases, a full-fledged theory is but a hopeful prospect.[A]
— Ervin Laszlo

*Shoulders has named the high-density state of matter produced by charge-cluster technology* Electrum Validum, *abbreviated* EV, *from the Greek* elektron *for electronic charge and from the Latin* valere *meaning to have power, to be strong, and having the ability to unite.*

> Don't sell your oil shares yet — but don't be surprised if the world again witnesses the four stages of response to any new and revolutionary development: 1) It's crazy! 2) It may be possible — so what? 3) I said it was a good idea all along. 4) I thought of it first.[8]
>
> — Arthur C. Clarke

site is very difficult to work with for either fabrication or analysis, large-scale energy applications are not yet practical.[16]

New energy research poses technical difficulties and funding problems for big government projects as well as the independent scientist. The US has spent billions of dollars since 1986 in an international collaboration with Russia, Europe, and Japan to design and build a self-sustaining fusion reactor. Skeptical of the design and concerned with the great cost and lack of progress associated with the International Thermonuclear Experiment Reactor (ITER), the US pulled out in 1998. Generating electricity from fusion still appears to be decades away.[17]

Developing new technologies is neither cheap nor easy. The ill-fated Superconducting Supercollider was a bold idea that captured the imagination and generous funding by officials at the Department of Energy, only to run into complex technical difficulties that dried up financing and killed the project. The United States' effort to construct the world's most powerful laser is at least $1 billion over budget and four years behind schedule. Scientists insist that the stadium-sized National Ignition Facility at the Lawrence Livermore National Laboratory can help maintain the US nuclear arms stockpile and increase their knowledge of astrophysics and the behavior of materials at very high pressure and densities. After visiting the Lawrence Livermore laboratory in 1999, Energy Secretary Bill Richardson said the laser was on time (scheduled for completion in 2002) and on budget (originally $1.2 billion). But the following year the DOE revised its estimates and now expects to spend at least $195 million more than budgeted through fiscal 2002. The estimated completion time has been rolled back to 2008. Despite the tremendous financial cost and pooled intellectual brain power behind the project, Congress was told that the odds of a successful ignition were only 50%.[18] US Senator Tom Harkin, who seeks to withhold extra funding for the project, has said, "We've seen this pattern before; massive cost overruns, schedule delays, poor management, and unresolved technical problems."[19]

Research into breakthrough energy systems is challenging, expensive, time-consuming, and, since most of the new technologies will not prove successful in commercial application, often frustrating. A major problem in the alternative energy field is that most conventional physicists and academics believe this kind of research violates the laws of physics and, therefore, is a waste of time. Many individuals working on their own feel that they are a victim of a "conspiracy of suppression" when, in fact, they are only battling the inertia of the established scientific paradigm. Some men and women working at the periphery of known science have not helped their cause by making outlandish promises and overstated claims of experimental success.

It has been said that in the heat of battle, truth is often the first casualty. In fact, most overunity energy claims usually fall into these categories: 1) The inventor has made honest but erroneous measurements that are discovered when the device is built and tested by others. Most overunity claims fall into this category. 2) The inventor observes an anomaly or effect but refuses to allow any independent testing or duplication out of fear that someone will steal his idea. Some researchers refuse to accept the fact that their experimental results can't be reproduced. 3) The "inventor" is really a con-artist, and the purpose of the "overunity device" is to attract "true believer" investors in their company in the form of "partnerships" or "franchises" that will become profitable when the new invention or device becomes available. Of course, it never does, and the careless, overly optimistic "partners" never recoup their investment. The line between foolishness and fraud is thin, and it is usually these fraudulent scam artists who make media headlines and taint the legitimate professional researcher struggling in the new energy field.[20]

The search for 21st century carbonless energy systems will continue despite political and financial obstacles. In April 1998, the United States Department of Energy issued its *Comprehensive National Energy Strategy* (CNES). Included among its five goals was: "Goal IV: *Expand future energy choices* — pursuing continued progress in science and technology to provide future generations with a robust portfolio of clean and reasonably priced energy sources." Included as two ob-

> Imagination is more important than knowledge. Knowledge is limited. Imagination encircles the world.
> — Albert Einstein

> Science is not just a collection of laws, a catalogue of unrelated facts. It is a creation of the human mind, with its freely invented ideas and concepts.[c]
> — Albert Einstein & Leopold Infeld

jectives are: 1) "Maintain a strong national knowledge base as the foundation for informed energy decisions, new energy systems, and enabling technologies for the future; 2) Expand long-term energy options."[21] An independent study was performed by Alternative Energy Institute, Inc., on the DOE's progress two years into the CNES declaration, and it is clear that the DOE has not yet engaged in developing, much less maintaining, a robust knowledge base of future energy choices. There has been no expanded research into new energy systems or long-term energy options, mainly due to upper management decisions. Instead of investigating potential clean carbonless energy technologies, the DOE has endorsed natural gas use for future generations (it burns cleaner than coal or oil but is also a limited, nonrenewable resource), and supports using coal and nuclear energy to provide much of the nation's electricity.[22]

Optimists claim that there is enough oil worldwide to last another 40 years.[23] Some experts consider it to be much less. What then? *Now* is the time to make the decisions that will facilitate the transition from the pollution-ridden fossil fuel age to a future of clean and virtually limitless energy technologies. At this time, there is no new technology or renewable resource that can single-handedly replace oil as the energy product of choice. It will take a combination of well-designed, reliable, environmentally friendly, renewable energy systems to take the place of cheap petroleum. It is time to wholeheartedly support and implement renewable energies like wind, biomass gasification, and solar power, as well as the search for future power systems that will tap the sea of energy that surrounds us all. The 19th century philosopher Henry David Thoreau got it right when he said, "If you have built castles in the air, your work need not be lost; that is where they should be. Now put foundations under them."

## FOOTNOTES

1. Lester R. Brown, Michael Renner, and Christopher Flavin, *Vital Signs 1998* (New York: Norton/Worldwatch Books, 1998), p. 66.

2. <www.cleanenergysystems.com> — The founders of Clean Energy Systems have advanced a superior power generation concept. Using technology from the aerospace industry, they have developed a high-energy gas generation system that powers turbines, which in turn produce electricity with no emissions or pollutants. Check out the CES technical papers on their website.

3. Mark Glover, "Rocket Science," *The Sacramento Bee*, May 1, 2000, p. D-1.

4. Thomas Valone, "The Right Time to Develop Future Energy Technologies," Integrity Research Institute, 2000. Paper prepared for Senator Bob Smith, Senate Committee on the Environment and Public Works. An appendix to this paper, "Future Energy Technologies," was published in *FutureFocus 2000*, The Proceedings of the Annual Conference of the World Future Society and *Energex 2000*, Proceedings of the International Energy Foundation Annual Conference/Exposition.

5. <www.altenergy.org/4/cofe/cofe.html> — Paul Brown presented information on his nuclear battery and his nuclear waste remediation process at the First International Conference on Future Energy (COFE) held in Bethesda, Maryland, in 1999. In the US alone, there are currently 34,000 *tons* of nuclear waste growing at an estimated 3,000 tons per year.

6. <www.globalatomics.com> — Global Atomics, Inc. (GAI), website. GAI develops and implements advanced solutions for nuclear applications. The technologies are intended to eliminate the threat of nuclear waste production and contamination. Their proprietary photodeactivation technology facilitates the neutralization of radioactive waste.

7. <www.altenergy.org/4/cofe/cofe.html> — COFE profile on the arc-discharge research of Peter and Neal Graneau. The Graneaus believe that the original source of this stored energy [in water] must be kinetic energy of the free particle that is given up when it forms the bond.

8. <www.aaas.org/spp/cstc/stc/stc01/01-01/table.htm> — *Science & Technology in Congress* electronic newsletter. "January 2001 Special update and review of the Congressionally approved research and development final appropriations for fiscal year 2001," written by Kei Koizumi, Director, AAAS R&D Budget and Policy Program.

9. Ed Regis, "Zip Drive," *Wired* (January 2001), pp. 96-108. "NASA scientists are building a hot little ride: VASIMR, a rocket that runs on million-degree plasma and could someday fuel a fast-track trip to Mars."

[10] Ken Shoulders, "Exploring," December 16, 2000. Short 2-page paper defining obstacles to new-energy funding.

[11] <www.blacklightpower.com> — Blacklight Power, Inc., has designed and tested cells for generating thermal energy for heating, electrical power generation, and motive power. The company plans to commercialize its energy and chemical technologies by forming strategic alliances. Morgan Stanley is serving as the company's investment banking firm.

[12] Oleg Jefimenko, "Electrostatic Energy Resources, Electrostatic Generators, and Electrostatic Motors," *Proceedings of COFE*, p. 195.

[13] <www.altenergy.org/4/cofe/cofe.html> — AEI's COFE profile on David Wallman's Biomass Gasification process.

[14] Ibid, COFE.

[15] Ken Shoulders, "Permittivity Transitions," September 4, 2000, p. 7. Shoulders' excellent 18-page paper explaining the anomalous effects of his research with highly organized micron-sized clusters of electrons (EVs). In his prologue, Shoulders writes, "Many of the observations of EV phenomena are totally without explanation. It would seem that EVs live in an entirely different and bizarre world compared to ours." The paper is Shoulders' attempt to connect others who use different methods for viewing the same basic thoughts.

[16] Ibid., p. 2.

[17] Editor, "Burning Times for Hot Fusion," *Scientific American* (March 2000), p. 19. Despite serious problems with the International Thermonuclear Experiment Reactor (ITER), scientists involved with the project remain determined to take the next step in hot fusion.

[18] Carrie Peyton, "High Price, Odds Don't Discourage Laser Backers," *The Sacramento Bee*, May 4, 2000, p. A-1.

[19] Robert Lee Hotz, "Huge Livermore Laser Project Dimmed by Overruns, Doubts," *Los Angeles Times* article, reprinted in *The Sacramento Bee*, November 24, 2000, p. B-5.

[20] Robert L. Park, *Voodoo Science: The Road from Foolishness to Fraud* (Oxford/New York: Oxford University Press, 2000), p. 10.

[21] US Department of Energy's *Comprehensive National Energy Strategy* (#DOE/S-0124), April 1998.

[22] Alternative Energy Institute, Inc., "An Independent Follow-up Evaluation of the US Department of Energy Comprehensive National Energy Strategy." In-depth analysis of the DOE's CNES, published by AEI in January 2001.

[23] Seth Borenstein, "Oil Crunch Isn't a Major Worry Yet, Experts Say," *Knight Ridder Newspapers* article, published in *The Sacramento Bee*, October 1, 2000, p. D-1. Many experts

and two recent comprehensive US Government studies insist that there is *plenty* of petroleum in the world, enough to last at least another *forty* years.

SIDEBAR FOOTNOTES

[A] Ervin Laszlo, *The Whispering Pond: A Personal Guide to the Emerging Vision of Science* (Boston, MA: Element Books, Inc., 1996), p. 126.

[B] Arthur C. Clarke, "Space Drive: A Fantasy That Could Become Reality," *Adastra*, November/December 1994, p. 38.

[C] Albert Einstein and Leopold Infeld, *The Evolution of Physics: From Early Concepts to Relativity and Quanta* (New York: Simon & Schuster, 1938), p. 194.

# Section IV

# Arcs, Sparks, & Electrons: Accelerating Into the Future

In the 21st century, the energy trend will be toward technologies that can best serve the greatest population with the least use of valuable resources, while also having the smallest adverse impact on the environment. Changing from a global system in which more than 85% of the energy used produces carbon, to a system where very little carbon is released, requires fundamental changes in technology and major investments in capital equipment turnover or replacement. As a result of steady population growth and needed economic expansion, projections indicate that the world will require 50% to 100% more energy in 2050 than it does today. By mid-century, 85% of the world's population will be living in developing countries, and those countries will account for the major part of the world's greenhouse gas emissions. Even with continuous efficiency improvements, the stabilization of atmospheric $CO_2$ concentrations will require a serious long-term commitment to zero-carbon alternatives for large-scale global deployment. Solving the energy/carbon conflict represents a critical milestone toward achieving a more holistic approach to environmental progress.[1]

As scientists, researchers, and engineers search for answers to the looming energy crisis, politics and government energy policies will undoubtedly follow established technological courses. They will be heavily influenced by short-term gains at the expense of long-term energy sufficiency

> If pickups and minivans were lumped with SUVs and replaced by more-fuel-efficient vehicles, the potential reduction in demand would be 690,000 barrels per day....^A
>
> — *The Philadelphia Inquirer*

and independence. US energy policy currently supports a supply-sided effort to further subsidize and exploit environmentally dangerous fossil fuels and uneconomical nuclear power to satisfy America's gargantuan appetite for transportation fuels and electric energy. President George W. Bush's 2002 National Energy Plan follows the traditional approach by recommending greater investment in coal and hazardous nuclear power, at the expense of already-available efficiency measures as well as proven renewable, carbonless technologies like wind, biomass gasification, and solar power.

For three decades now, American presidents have promised comprehensive energy policies that would make the United States less dependent on imported oil, but, in 2001, foreign petroleum accounts for more than 50% of domestic consumption. Most geologists agree with the estimate that the peak in global oil production will probably occur within the next 20 years, to be followed by diminishing supplies. Because petroleum is used principally as transportation fuel, it is imperative that the US upgrade antiquated fuel efficiency standards in its automotive fleet, especially for light trucks and SUV's. Revising the tax structure to penalize low mileage vehicles and reward fuel efficient models will help jump-start the market for new high-mileage electric/gasoline hybrid vehicles that significantly reduce fuel consumption and harmful emissions.

Fossil fuels should be taxed for their true cost to society, not subsidized. If the social costs, such as air pollution and sprawl, were included in the market price of gasoline, the price at the pump would be much higher. A carbonless energy future necessitates a reversal in tax policies that currently encourage the use of fossil fuels, as well as the phase-out of inefficient and dirty technology that causes pollution. There is no way around the fact that the glory days of cheap and abun-

*EVs run much smoother and quieter than gasoline- or diesel-powered vehicles.*

Photo: Alternative Energy Institute, Inc.

dant oil are numbered, and that competition for this dwindling resource will strain international cooperation in the future. The inevitable peak in world oil production will be a major challenge for the United States, a nation that consumes 25% of all the oil produced.

In order to meet environmental goals aimed at reducing carbon in the energy economy, the developed world over the next two to three decades will largely meet the need for additional capacity with natural gas-fired combustion generators. Small modern microturbines utilize a wide range of fuels, from methanol and propane to natural gas. Compared to the traditional huge, central-generation power plant that must send electrons long distances through a massive, aging, and inefficient transmission grid, these new hi-tech generators can be located closer to the consumer and scaled to the load, which reduces energy waste. On average, 10% of the electricity generated by a large centralized power plant is lost as waste heat in transmission from plant to consumer. Each microturbine, however, serves small groups of consumers on a small portion of the grid. Another advantage in moving the power source closer to its market is that, instead of wasting the thermal energy, these turbines produce, they can provide heat, run air-conditioning systems, or be used to boil water, making steam that generates even more electricity. This type of *cogeneration* system is being employed in Europe, Australia, and Asia because it can result in a total energy efficiency of 70% or higher and cuts $CO_2$ emissions in half, as opposed to a conventional gas turbine, which wastes two-thirds of its energy input into the atmosphere.[2]

As increasing numbers of gas-fired microturbines are connected to the existing transmission grid, the more complex operation of the grid will require better monitoring and control of energy flows. Power management will improve as information technology drives gains in productivity and efficiency of engineering, design, operations, and maintenance. But there is an important reason why President Bush's strategic energy plan is pushing coal and nuclear systems for future power generation in the US, and not relying on efficient gas-driven turbines. North American

> If Ford really wanted to help the environment, it could try harder to change consumers' buying habits by spending more money to advertise fuel-efficient vehicles and less to market SUVs.[B]
>
> — *The Wall Street Journal*

natural gas reserves and distribution are strictly limited, which means this relatively clean-burning fossil fuel cannot logistically, strategically, or economically meet the dual role of replacing coal for cleaner electricity generation, while simultaneously substituting for oil products as a transportation fuel for cars, trucks, and buses. Increasingly price-sensitive, natural gas reserves, production, and depletion profiles closely shadow the demise of domestic conventional petroleum.

The Bush administration has touted new *Clean Coal Technologies* as a panacea for burning coal, but so far the CCTs have failed to solve the energy efficiency and environmental problems associated with this hydrocarbon — the dirtiest of all the fossil fuels. Engineers are developing advanced high-temperature materials for supercritical steam cycles and technologies such as pressurized fluidized-bed combustion, which have the potential to achieve 50% efficiency in terms of higher heating value. Current estimates suggest that these technology advances in coal-firing have the potential to make clean-coal power generation competitive with natural gas on a cost-of-electricity basis sometime within the 2010 to 2020 time frame. The US Department of Energy envisions that in addition to generating electricity, the future coal refinery also will provide hydrogen separation, chemical production, and carbon dioxide sequestration; but these potential technologies will also require major infusions of R&D funding to achieve commercial viability before 2020.

Odds are that nuclear power will reappear in the 21st century as a technology of necessity, primarily because of its lack of greenhouse gas emissions. Nuclear engineers are designing more reliable gas-cooled reactors and have also developed a modular reactor that is fueled by 0.5-mm uranium oxide granules sealed in tennis-ball-sized "pebbles" made of graphite and silicon carbide steel. Both designs are an improvement in safety compared to conventional nuclear power plants. To succeed socially and

*Germany is shutting down its nineteen power plants, making it the world's largest industrialized nation to willingly forego the technology.*

economically, the troubled nuclear industry will have to address public concerns of safety, shorten plant construction times, and lower initial capitalization costs. Emerging technologies in the large-scale transmutation and stabilization of lethal radioactive waste could make nuclear waste disposal immaterial as a serious environmental issue. The transmutation of high-level radioactive waste is another developing technology that needs significant R&D funding to become viable. If any new nuclear power plants are built in the US, they will be the first ones in nearly 30 years.

The global demand for electricity is expected to grow rapidly in the 21$^{st}$ century. Both industrialized and developing nations will utilize this type of energy, which can be applied with great precision, efficiency, and cleanliness at the point of use, is compatible with the streamlined infrastructure of modern economies, and can be generated from a wide variety of sources. Energy conservation and increased resource efficiency are logical first steps in reigning in the United States' runaway energy consumption and rising greenhouse gas emissions. In order to meet projections of future demand for electricity, the US Department of Energy has estimated the need for 1,300 new power plants by the year 2020 (at an average size of 300 megawatts). But a new study shows that improved energy efficiency measures would reduce that estimate to only 170 new power plants.[3] Installing energy efficient household appliances like clothes washers, air conditioners, and water heaters would save the equivalent energy generated by 127 power plants. Programs to reduce energy use in new buildings, such as building energy codes, tax credits, and public benefit programs, would avoid another 170. Everyone can help: if each household in the United States replaced four 100-watt bulbs with compact and long-lasting fluorescent bulbs, it could eliminate the need for thirty new 300 megawatt power plants.

The United States, a country with less than 5% of the world's population, is the world's least efficient major economy, uses a disproportionate amount of the planet's resources, and contributes 25% of all greenhouse gas emissions. Other countries are alarmed by the fact that the

> If Accelerator Transmutation of [nuclear] Waste (ATW) technology could be successfully implemented to overcome all technical issues, it could potentially facilitate the long-term management of a repository system.[D]
> — US Department of Energy

> A California poll shows that 59% of those surveyed support building new nuclear power plants in the state — compared with 36% who oppose the idea and 5% undecided.[C]
> — San Francisco Chronicle

increase in US emissions over the last ten years equals the combined emissions increase from China, India, and Africa, which together have a population more than ten times that of the United States.[4] Despite the alarming trend in foreign oil imports, US public expenditures on energy R&D have declined by a third over the last 20 years, and less than 10% of that funding goes to energy-efficiency improvements. Unfortunately, the US 2002 Federal Budget includes more cuts: energy efficiency research funding will drop by 25%; solar energy research is cut in half to $42.9 million; and research on geothermal, hydrogen, and wind energy is down 48%. Federal funding for functioning renewable energy programs will be slashed 40%, and the budget for the National Oceanic and Atmospheric Administration science office (which studies climate change) will be trimmed by 10%. In contrast, research into cleaner coal technology will be funded to the tune of $2 billion over the next ten years.[5]

*Treat every day like Earth Day — act locally, think globally.*

Reversing these disturbing policy trends will require strong leadership by the US Congress and an informed, politically proactive voting public, whose children and grandchildren will ultimately pay the price for not making timely decisions concerning our energy future. We will somehow make the transition from a hydrocarbon-based economy to a more sustainable one dependent largely on energy efficiency, and traditional renewable energy systems like wind and solar power, which can cleanly produce electricity and hydrogen. The only question is when and how. The US must implement new policies that offer renewable and alternative energy technologies a level playing field with the entrenched fossil fuel and nuclear industries, including a fair price that reflects their low environmental impact. It is time to cut subsidies to industries that harm our health and environment, and initiate renewable energy system procurement policies to stimulate growth in the market. Market-based investments in renewable resources will increase system efficiency and drive down costs. Cost-competitive renewable energy generation technologies will displace increasing amounts of fossil fuel through increased efficiency, low-cost mass production, and wider application.

Solar panels and wind generators are reliable low-maintenance, distributed electric-power systems for rural and remote locations, but contrary to popular perception, traditional renewable energy is not a Golden Goose that will quickly or easily replace the energy densities we currently derive from oil and natural gas. Although both solar and wind power technologies are growing at double digit rates, they still contribute less than 2% of the world's electricity — hydro and biomass contribute 8% — and all are a long way from a majority share of power generation. Electricity is the key to diversifying transportation fuels beyond petroleum, with enormous strategic advantages for global security, but electricity is not a primary energy source. Electricity is an energy carrier that has zero mass, travels near the speed of light and, for all practical purposes, can't be stored efficiently. For comparison, one gallon of gasoline weighing about 8 pounds has the same energy as 2,000 pounds of conventional lead acid storage batteries. In other words, the 15 gallons of gasoline sloshing around in your car's gas tank represents the equivalent of 15 tons of storage batteries.

Wind and solar power can significantly improve the standard of living for the two billion impoverished humans that have no electricity at all, but it is highly unlikely that these diffuse and intermittent energy sources will single-handedly sustain affluent societies like the United States and Europe, which rely on high levels of energy use. It is commonly assumed that hydrogen will

> Next month the largest wind farm in the world, which will generate enough electricity to power 70,000 households, will start construction along the Oregon-Washington border, and similar projects are under way in California, Texas, Minnesota and Iowa.[E]
> 
> — *Time* (January 2001)

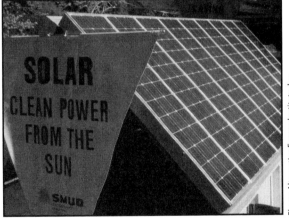

*Some utilities argue that solar panels don't work at night, but they produce power during the precise hours when air conditioners are roaring and the most power is consumed.*

replace oil as the principal energy carrier of the 21st century (along with electricity), but major problems become evident when basic consideration of volume, conversion, and transport losses are considered.[6]

In addition to mitigating rapid climate change by reducing greenhouse gas emissions, another major challenge in the 21st century will be providing food, housing, and energy for the developing world where population growth will be greatest. Overpopulation is the root cause of many kinds of environmental deterioration. Climate change, rain forest destruction, desertification, and most types of pollution are linked to an expanding world population and increased pressure on limited resources. By 2050, the US population will jump to 400 million from 238 today, and will be the *only* developed country among the world's most populous nations. Historic world population growth and/or growth in the rates of consumption cannot be sustained. Since 1960, nearly one-third of the world's arable land has been lost due to urbanization, highways, soil erosion, salinization, and water logging of the soil.[7] Ominous trends are developing in response to the burgeoning human population. World energy production per capita peaked in 1979 and has been declining ever since. Grain production per capita started declining in 1984, and irrigation per capita started declining in 1989. In order to reduce the huge human population that is negatively impacting the planet's biosphere, the US and United Nations should initiate major comprehensive educational, technical, and outreach programs in the areas of social responsibility, contraception, and family planning.

It's no secret that to replace fossil fuels with robust and carbonless energy systems, new technical solutions will be required. To be viable as a future energy source, emerging energy technologies must not produce carbon emissions nor contribute to climate change. Engineers are making strong advances in exotic conductive materials and

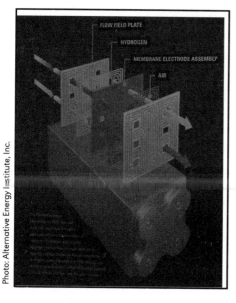

*Fuel cells work so well that they could eventually nail the lid on the coffin of gasoline and diesel engines.*

computer technology in all phases of power generation and distribution. Fuel cells utilize electrochemical combustion of hydrogen with oxygen to generate electricity. Fuel cells produce reliable streams of current and emit only water vapor and heat when fueled with hydrogen. They are quiet, require little maintenance, and, when hooked up to water electrolyzers, can also store electricity as hydrogen for energy that can be fed back into the system during peak demand. Solar and wind power can be used to produce hydrogen to feed the fuel cell resulting in a clean and renewable system for sustainable energy production. However, for sites without solar and wind power, a whole new infrastructure to produce and distribute the hydrogen will be required.

The growth in global energy demand will lead to soaring carbon dioxide emissions through 2020. US carbon emissions alone are projected to increase 33% over 1990 levels by 2010 and 47% by 2020. The US Energy Information Administration estimates that worldwide oil demand will increase from more than 70 million barrels per day to more than 110 million barrels per day in 2020. Whether petroleum producers — mainly OPEC — will be able to keep up with, let alone sustain that projection level, seems unlikely.[8] International concern regarding the potentially disastrous impacts associated with rapid climate change due to elevated greenhouse gas levels in the atmosphere initiated the United Nation's Framework Convention on Climate Change. The controversial international treaty known as the Kyoto Protocol that emerged from the Framework Convention requires an aggressive shift to low- and zero-carbon energy technologies on a global basis in the 21$^{st}$ century — and application of methods for capturing and sequestering carbon during the transition. In 2001, the United States, which had signed the treaty but not yet ratified it, pulled out of the Kyoto Protocol process because of a perceived risk to the US economy.

Politicians in Washington, D.C., may consider the Kyoto Protocol a flawed contract, but as the greatest polluter and energy consumer in the world, the US should not unilaterally walk away from this international agree-

> Industry officials say they could easily double and possibly triple the average number of miles per gallon obtained by conventional internal combustion engines for both cars and light trucks.[G]
>
> — *Washington Post*

ment to curb global greenhouse gas emissions. The troubles plaguing the controversial Kyoto Protocol do not preclude the need for a strong catalyzing action among the world's worst contributors to emissions and air pollution. The present Kyoto agreement will not reverse climate change or even result in the stabilization of atmospheric greenhouse gases, but it was never meant to. The treaty was intended to be the only first step toward an international effort and commitment to address a serious global problem. If the spirit of the dying Kyoto Protocol is be to revived, the impossibly strict timetable for emission reduction mandates should be adjusted to allow large-scale green house gas contributors like the US more time to meet the treaty's goals.

Technological innovation has emerged as the primary driver for economic growth. Because existing renewable energy systems will not be sufficient to sustain the industrial world's present levels of energy consumption, new, exotic technologies must be developed that can tap other sources of energy. Emerging technologies currently under experimental development are unlikely to translate into robust energy or propulsion systems in time to replace diminishing oil production. If history is any guide, once a new energy system is discovered, it will take several decades to develop and implement the energy resource. Under-investment in energy technology R&D is detrimental to both long-term energy security and global sustainability. Further, it could foreclose the technology options that the global community will need to address systematically the environmental impacts of energy.

In order to improve energy-conversion efficiencies in conventional renewable energy systems, as well as accelerate the development of alternative breakthrough energy/propulsion research, funding must be provided, and the institutional, political, corporate and financial obstacles must be eliminated. At this time there is no DOE mechanism for the support, evaluation, testing, and dissemination of information about new scientific discoveries. By contrast, older energy systems that have been supported with billions of dollars of government funding serve to

> Two main sources of energy are at the forefront of consideration for mainstream future transportation: low-cost natural gas and onboard fuel-cells for electric-power generation. [F]
> 
> — *The Oil and Gas Journal*

perpetuate government bureaucracies and sustain federally funded national laboratories and academic projects. It is time for the DOE to fund and coordinate exploratory R&D in potential energy and transportation breakthroughs, such as low energy nuclear reactions, zero-point energy extraction, and anomalous gravitation/inertia mass effects. When it comes to new energy technology, the US government should encourage teaming, partnership, and collaboration with the private sector, the states, and other nations.

For the most part, the organized scientific community varies from highly resistant to openly hostile toward novel scientific research. However, each new major advance in science starts with an anomaly that is unacceptable at first. The anomalies are important because they inspire new ways of thinking. The energy crisis may eventually be solved by a dramatic change in perspective, a paradigm shift in which upcoming scientists are encouraged to challenge the conventional laws of physics as they seek answers to the global energy crisis. Future generations are relying on our efforts, so they will inherit a healthy, sustainable environment and economy.

# FOOTNOTES

[1] <www.epri.com/corporate/discover_epri/roadmap/index.html> The Electricity Technology Roadmap Initiative prepared by the Electric Power Research Institute, p. 88. This "roadmap" explores a period of immense technological and institutional change in the electricity enterprise and in the society it serves. The choices made in such periods of change can have profound consequences on whether future opportunities are opened or foreclosed, and whether threats are eliminated or realized. A broad range of benefits for society can be realized by a strong and continuing commitment to scientific and technological innovation. The essential activity of the roadmapping process is the evaluation of various technological opportunities and their potential benefits.

[2] Steve Silberman, "The High Energy Web," *Wired* (July 2001), pp. 119 and 126.

[3] <www.ase.org/media/factsheets/facts1300.htm> Study by Bill Prindle, a program director at the Washington, D.C.-based Alliance to Save Energy, The Alliance <www.ase.org> is a coalition of prominent business, government, environmental, and consumer leaders who promote the efficient and clean use of energy worldwide to benefit consumers, the environment, economy, and national security.

[4] Christopher Flavin, "America's Energy Problem," *World Watch* (July/August 2001), p. 11. Christopher Flavin is president of the Worldwatch Institute <www.worldwatch.org>

[5] *World Watch* (July/August 2001), p. 18.

[6] F. E. Trainer, "Can Renewable Energy Sources Sustain Affluent Society?" *Energy Policy*, 1995, Vol. 23, No. 12, p. 1016. In this professional research paper, F.E. Trainer concludes that "renewable energy sources will not be able to sustain present rich world levels of energy use and that a sustainable world order must be based on acceptance of much lower per capita levels of energy use, much lower living standards, and a zero-growth economy."

[7] Andrew R. B. Ferguson, "Perceiving the Population Bomb," *World Watch* (July/August 2001), p. 38. Andrew R.B. Ferguson is research coordinator at Optimum Population Trust in the United Kingdom. For more information: <www.globalideasbank.org/1993/1993-177.HTML>

[8] United States Energy Information Administration report. Article "Grim future for energy costs," published in *Resource: Engineering & Technology for a Sustainable World* (February 2000), Vol. 7, p. 16. Copyright 2000 American Society of Agricultural Engineers.

## SIDEBAR FOOTNOTES

A  Joseph A. Gambardello and Kristen A. Graham, "SUVs now a symbol of plight," *The Philadelphia Inquirer*, May 11, 2001, p. A-1.

B  Jeffrey Ball, "Warming Trend: Auto Makers Juggle Substance and Style in New Green Policies," *The Wall Street Journal*, May 15, 2001, p. A-1.

C  Carla Mainucci, "Nuclear Power's California Comeback," *San Francisco Chronicle*, May 23, 2001, p. A-1.

D  *The OCRWM Enterprise*, A Publication of the Office of Civilian Radioactive Waste Management (OCRWM), November 1999, p. 8.

E  Daniel Eisenberg, "Which State is Next?" *Time* (January 29, 2001), p. 48.

F  Louella E. Bensabat, "US Fuels Mix to Change in the Next Two Decades," *The Oil and Gas Journal* (July 12, 1999), Vol. 97, p. 46.

G  Juliet Eilperin, "SUVs Likely Face Stricter Fuel Rules," *Washington Post* article reprinted in *The Sacramento Bee*, June 22, 2001, p. A-1.

# Transition:
# Life Beyond the Oil Patch

## 17

A major energy revolution is just around the corner, and, with adequate assistance, guidance, and a bit of luck, it will arrive in time to stave off a potentially disastrous energy crisis. The slow transition from our historic reliance on nonrenewable fossil fuels to renewable, decentralized, and carbon-free energy has already begun. It will require a panoply of new alternative energy systems, coupled with strong public demand for clean and green power, to wean ourselves from polluting fossil fuels. This transition to alternative energy won't be quick or easy, but the well-being of every future generation of Americans will rely on the decisions we act on today.

THE UNITED STATES WAS BLESSED with good geology and endowed with vast primordial reserves of coal, oil, and natural gas, but after more than 100 years of heavy consumption, that scenario has markedly changed. These fossil fuels account for the vast bulk of global energy supply, but they formed over millions of years and are finite and nonrenewable. There are still substantial coal reserves buried in the continental US, but the environmental and health costs associated with using this carbon-rich energy source as a primary fuel are prohibitive. Once-abundant petroleum and natural gas fields are being rapidly depleted and becoming increasingly expensive to produce. The US once led the world in oil output, but is now a densely drilled, mature producing region in which oil production has been declining for 30 years. As recently as 1950, the United States was producing half the world's oil, but now its proven reserves amount to only 3% of global petroleum assets. Indicative of declining domestic production in the US, during the 1950s, oil producers discovered about 50 barrels of oil for every barrel invested in drilling and pumping. Today, the world finds only one new barrel of oil for every four it consumes.

This alarming trend of diminishing returns on oil and natural gas exploration is occurring all around the world, and fewer and fewer countries can produce enough petroleum for their own use, let alone for export. The critical factor for meeting global crude oil needs is not the world's total oil production but the amount of net global production that is available for export. The world's 370 known giant oil fields contained 75% of all oil found through 1985, but few "elephants" have been discovered since then. As global demand for petroleum increases, the competition for that exportable oil will become fierce. In 1998, Australia was 80% self-sufficient in oil, but, by 2010, the country will be able to supply only about 45% of its growing demand for liquid petroleum.[1] Within 10 years or so, the bulk of the world's exportable oil will come from a handful of OPEC countries located in the Persian Gulf region. Petroleum is more than just gasoline: diesel fuel, herbicides, pesticides, fertilizer, medicines, jet fuel, plastics, and a huge

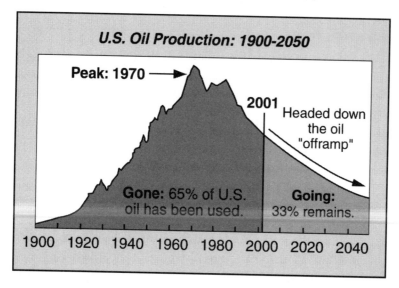

*US domestic oil production peaked in 1970.*

number of other useful products are made from this vital, nonrenewable resource.

Similar supply-and-demand scenarios are playing out in the natural gas arena. North American natural gas production has no excess capacity and, in fact, is suffering severe depletion rates in almost every key production basin. Canada currently provides about 15% of US natural gas needs, but geologists warn that Canada's production will eventually decline also. Mexico's gas production reached a plateau in 1998 and has had a downward slope of around 2% ever since. Natural gas exports from Mexico to the US fell from a total of more than 54 billion cubic feet (bcf) in 1999 to just 4.71 bcf for the first four months of 2000, and then to nothing. Increasing domestic demand in Mexico for natural gas no longer allows for exports. Despite diminishing natural gas reserves in North America, more than 275 gas-fired electrical generation plants are planned to begin operating through 2006, up from 158 in 1999, which will increase gas consumption by more than 8.5 *trillion* cubic feet. Forecasts show gas demand could outstrip supplies from traditional sources by as much as 4 billion feet a day within a decade.[2] After that, where will the supply come from, and what happens when it's gone?

As we enter the 21st century, a large number of the giant older fields that anchor the world's hydrocarbon production base (including the North Sea — a key non-OPEC producing region), have now started to decline. Petroleum geologists have warned for 50 years that global oil production would "peak" and begin its inevitable decline within a decade or two of the year 2000.[3] As the inevitable apex of world petroleum production looms ever closer, policies

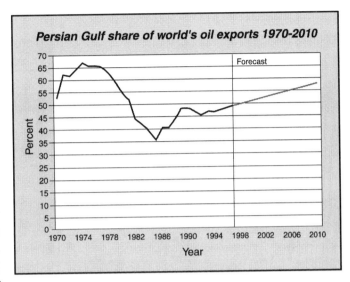

*OPEC's share of world oil production continues to increase.*

We are now in an energy crisis. I'm worried about the economy; I'm worried about the lack of an energy policy; I'm worried about rolling blackouts in California.[A]

— US President George W. Bush

encouraging conservation, more efficient energy use, and the development of alternative energy sources are urgently needed. A major obstacle is that renewable energy systems can generate only a fraction of the power now being produced by fossil fuels. Hard choices will have to be made by government, industry, and the American public if the US economy is going to survive the inevitable transition from a world swimming in inexpensive oil to one without enough to meet global demand.

Oil is the most important form of energy we use, especially in transportation. The intrinsic qualities of liquid petroleum seem irreplaceable: its extractability, transportability, versatility, and low cost. This viscous black magic makes up about 40% of the world's energy supply, and there is serious concern as to whether America's highly consumptive standard of living can survive in the period after the Carbon Era. Americans, who represent only 4% of the world's population, consume 26% of the world's oil, importing nearly 60% of it at a cost of about $100 billion.[4] Driving industrialized economies with fossil fuels has polluted the biosphere and contributed significantly to global warming, but the Age of Carbon Man is ending, and whether we want it or not, a new energy future awaits us.[5]

The long, potentially disruptive journey to carbon-free energy has already begun with incremental baby steps such as the scattered implementation of wind, solar, biomass gasification, geothermal, and other renewable energy systems. Energy-producing technologies that draw on renewable sources avoid the severe environmental impacts of the fossil fuel cycle. Wind power has really taken wing in the last decade, boasting double-digit growth. Western Europe has installed the bulk of the world's wind turbines so far, but, between 1990 and 2000, global generating capacity posted an impressive average growth rate of 24% per year.[6] In Japan, five companies that generate electricity from wind have formed an industry association to encourage and promote wind-power generation in that country. Japan is the world's fourth largest energy consumer and second largest energy importer after the United States. The island nation lacks significant domestic sources of energy and must import substantial amounts of crude oil, natural gas, and other energy resources, such as uranium.

Last year, the world's first commercial wave-power station was installed on the Scottish island of Islay, already famous for its malt whiskies. Jointly developed by WAVEGEN, the world's leading wave power company, and Queen's University Belfast, this innovative renewable energy system dubbed LIMPET took 20 years of collaborative academic research to develop commercially. The power plant is rated at 500 kW and generates enough electricity for about 400 local homes. Due to its modular construction and simple operation, the new system can offer some small coastal communities a choice between dirty diesel generation and clean indigenous power. The station's operators secured a 15-year power purchase agreement with major public electricity suppliers in Scotland, thereby encouraging more funding and development for the LIMPET project.[7]

Dependable, quality electricity is more important than ever in the new information-oriented, computer-based economy. Companies relying on computer systems or who have global 24/7 operations increasingly are recognizing that they are at risk to serious or possibly fatal business interruptions tied to poor quality power. All companies now have a heightened awareness to everything from minor surges or dips in quality to major blackouts that can negatively affect their business.

> [Data centers], which house shelf after shelf of computers, sometimes with office space and other hi-tech services alongside, are forecasting "unheard of" leaps in electric demand.[8]
> — Pacific Gas and Electric Company

*Dependable, quality electricity is more important than ever in the new information-oriented, computer-based economy.*

*Royal Dutch/Shell is investing $500 million in renewable energy sources such as wind, biomass, and solar power.*

During the first six weeks of 2001, California endured blackouts followed by 32 consecutive days of Stage 3 power alerts (when electricity reserves fell below 1.5%), the highest stage possible before rolling blackouts sweep the grid. Concerns about power outages and spiking energy costs have inspired some concerned California companies to consider other more reliable sources of electricity in this volatile market. On-site generation systems, such as solar PV, microturbines, natural gas reciprocating engines, fuel cells, and other alternatives, are being installed around the state. These units can generate electricity and heat, displacing energy consumption from central power plants during peak daytime hours when electricity demand is greatest, most expensive, and most susceptible to spikes, surges, brownouts, and blackouts.

Two major Southern California solar-energy projects, one located in Carlsbad, the other in Fountain Valley, went online in December 2000. The Fountain Valley project consists of a one-acre, 230-kilowatt photovoltaic array, which can supply about half the electricity demand for a 110,000-square-foot office building. The heavily subsidized project cost $2 million, of which more than $880,000 came from the state of California. Touted as the largest private commercial photovoltaic system in the Western Hemisphere, the successful project represents a positive trend towards clean renewable energy. But it also offers a glimpse into the ongoing war of competitive pricing in the marketplace between renewable energy sources versus the coal, oil, and natural gas industries.[8]

For decades, proponents of renewable energy systems such as wind and solar power have been fighting an uphill battle against an entrenched fossil-fuel industry. Like David versus Goliath, oil companies and electric utilities that have billions of dollars already invested in infrastructure and inventory have been unwilling to switch to buying from alternative energy providers, which are not yet cost-competitive on the open market. Fossil fuels are cheaper only when environmental and health costs are ignored, but an energy market that considered the total cost to society of our energy choices would greatly encourage the deployment of renewable energy technologies. Big Oil can read the writing on the wall and is transforming itself into the energy business. Although real profits with renewable energy are some way down the road, corporations that sell hydrocarbons know that a long-range strategy that includes environmental policies is good for business.

In February of 2001, Jeroen van der Veer, President of Royal Dutch/Shell Group, the second largest oil company in the world, stated that his company is boosting natural gas production and is investing in renewables in response to consumer demand for cleaner energy. He admitted that coal, oil, and natural gas currently supply about 85% of the world's energy needs and that would not change overnight. But, in response to public concern about air pollution and global warming, Shell is investing $500 million over the next five years in renewable energy sources such as wind, biomass gasification, and solar power.[9] The Anglo-Dutch giant established its Shell International Renewables division in late 1997 as a fifth business unit. Half-a-billion dollars is a step in the right direction but only a minor financial investment compared to the billions Shell spends in other business areas and not yet commensurate with the risks of potential climate change. British Petroleum is considered number one among the "super major" oil companies in anticipating the profits of environmental policies; its Solarex division is a world leader in photovoltaic R&D.[10]

Lacking serious government and industry support, renewable energy entrepreneurs often rely on the private sector for funding, but, in the US, less than 1% of private equity investment flows to the energy and environmental sectors. Strong capital investment in renewable energy systems has been slow to develop, primarily because venture

*As the price of fossil fuel-generated electricity increases and becomes less reliable, more businesses and homeowners are converting to dependable on-site solar power.*

> [The US] spends as much as $100 billion annually on costs related to the effects of polluted air.[c]
> — *Natural Capitalism*

capitalists have not seen any major success stories yet. The production cost of mass-produced photovoltaic panels is decreasing, but the wholesale price remains relatively expensive as companies try to recover some of their billion dollar plus investments in PV technology and production facilities. Financiers are not usually attracted to new businesses that rely heavily upon government subsidies or mandates to be competitive, as solar ventures often do. Solar photovoltaics are the most expensive form of renewable energy in terms of cost per hour of electricity generated, at about 15 cents per kilowatt-hour to produce compared to fossil-fuel generated electricity, which can run as low as 3 cents. Another perceived drawback cited by private investors is the lack of experienced and successful management talent in the solar field. These fundamental inequities in costs and investment will be resolved over time as oil and natural gas become more expensive to produce, making solar-derived energy more competitive.[11]

Only recently have the renewable energy units of national laboratories and academies of science in many countries begun to examine the commercialization of small-scale and decentralized technologies. Consequently, the application of these technologies to rural development, and the methods needed to finance their dissemination have been left to institutions and individuals lacking the necessary resources and training.[12] Japan and member countries of the European Union have begun investing aggressively in renewable energy systems by promoting and subsidizing these technologies. Critics of government support in private industry should bear in mind that the United States provides more than $20 billion in federal subsidies to the oil, coal, and natural gas industries every year. It is sad and bitter irony that tens of billions of American tax dollars go to bolster the fossil-fuel industry, which pollutes the environment and causes serious health problems, both of which cost taxpayers more billions every year. Offering equivalent funding for renewable energy technologies would simply level the playing field and give the clean energy industry the jumpstart it needs to become commercially competitive.[13]

An examination of US spending policy on energy technologies is very revealing. One study, conducted for the Renewable Energy Policy Project (REPP) and Center for Renewable Energy and Sustainable Technology (CREST), was released in a July 2000 report "Federal Energy Subsidies: Not All Technologies Are Created Equal."[14] The detailed study indicates that, during the last 50 years, the US government has spent approximately $150 billion on energy subsidies for wind, solar, and nuclear power, but nearly all of the money (96.3%) was funneled into nuclear energy. Since 1947, cumulative subsidies to nuclear power has had an equivalent cost of $1,411 [1998 dollars] per US household, compared to $11 for wind energy research and development. "It requires a great deal of money to establish an energy technology," says Adam Serchuk, Research Director for REPP-CREST, "and very few have reached maturity without public investment."[15]

Despite the hefty financial subsidies supporting the nuclear industry, generating electricity through atomic fission has never been economic in any sense of the word.[16] Proponents of nuclear energy, which supplies about 20% of the US demand for electricity, have claimed that although the capital and operating costs for nuclear power stations are higher than those built to burn fossil fuel, their fuel costs are lower than for coal or oil. But similar to coal, oil, and natural gas, uranium is also a finite nonrenewable resource whose price has risen as have the costs associated with reprocessing the fuel as well as those associated with radioactive waste storage and disposal. The serious dangers inherent in dismantling and safeguarding a decommissioned nuclear power plant add hundreds of millions of dollars to the tab. Those who support using nuclear energy to generate electricity correctly point out that atomic power releases no greenhouse gases, but public and political resistance to the industry is so formidable that it has been 23 years

*The high costs of storing dangerous and long-lived radioactive wastes make nuclear power unpopular and uneconomical.*

TRANSITION: LIFE BEYOND THE OIL PATCH    313

since anyone in the United States has even applied to build a nuclear unit.[17]

Although the US is by far the number-one user of nuclear energy in the world in numbers of units and electrical output (103 of the world's 433 nuclear power plants are in the US), 19 other countries derive a greater percentage of their total electricity from nuclear generators. But at what cost? In 1999, German taxpayers revolted against the price tag of US$17 billion needed to subsidize their nuclear power industry for that year, and elected a government that has pledged to abandon nuclear energy over the next few decades. Germany currently produces 31.2% of its electricity with nuclear power, but the country intends to shift those funds into more cost-effective and ecologically sustainable renewable technologies.[18] Despite the many compelling arguments against nuclear power, the Energy Information Administration projects a 40% increase in US demand for electricity by 2020, sparking interest in a possible resurgence of nuclear-generated power in the United States. The revival of nuclear power faces many hurdles, especially in the US, but it may become more palatable in the future if: 1) fission technology is made safer: for example, the South African-designed Pebble Bed Modular Reactor; 2) radioactive waste disposal issues are ameliorated by transmutation that stabilizes the dangerous nuclear byproducts. Both the reactor and promising transmutation technology, which converts high-level radioactive liquids and spent-fuel pellets into stable elements, are already under development.[19, 20]

During the 1990s, the world increased its consumption of natural gas and oil, but depressed prices of these commodities decimated industry profit margins and forced a downsizing in exploration and development of more oil and gas fields, especially in the US. At the same time, the

*In order to manage demand and encourage energy conservation, University of California economists want the standard utility meters of California's big and midsized power users changed to "real-time metering" under which the exorbitant cost of peak power — delivered at the hours of highest demand — is billed at a higher rate.*

United States did not develop its electric power structure and is now having difficulty providing quality peak power demand in certain regions of the country. The only immediate solution for the US is to reduce systematically the consumption of electric power, natural gas, and oil because it will take time and money to overcome the physical constraints associated with the current regional energy shortfall. The most realistic short-term solution for the average citizen is to reduce personal consumption of both electricity and natural gas. Americans faced with steeply rising energy costs have two key weapons at their immediate disposal: energy conservation and energy efficiency. Both begin at home, in the car, and at the office.

New technologies promise to bring clean, renewable energy in the future, but you can take action today to help save both the environment and your hard-earned dollars at the same time. On average, Americans waste as much energy on a daily basis as two-thirds of the world's population *consume*. Gas-guzzling vehicles, inefficient furnaces, leaky appliances, and poorly insulated buildings all contribute to wasted energy.[21] In California, consumers are trying to save money by conserving their use of electricity while waiting impatiently for more natural gas-fueled power plants to be built. Outraged rate payers should be aggressively petitioning their legislators and regulators for greater investment in time-of-use meters and for more decentralized, on-site energy systems such as solar panels and fuel cells, technologies which are already available today. Time-of-use meters allow utilities to charge "real-time" prices to industry and other big power users. Prices are cheap during nonpeak hours and climb when demand skyrockets. The hour-to-hour pricing of electricity, which can vary greatly and is normally invisible to the consumer, will allow customers to curtail usage on hot summer afternoons when California is forced to pay whopping last-minute prices for power. Currently, most homeowners and businesses pay the same price for electricity throughout the day, regardless of the wholesale costs of that power. As a result, customers have little incentive to curb their appetites for expensive energy.

> As squads of lobbyists relentlessly remind state lawmakers, there's more money invested in the half-trillion-dollar American electric utility industry than in banking, telecommunications, and airlines combined.[D]
> 
> *— Harper's*

Most modern US houses incorporate energy-efficient thermal windows and adequate insulation, but today's larger, gadget-filled American homes need more juice than ever before. Computers, televisions, VCRs, and other appliances drain electricity 24 hours a day and are stealthily boosting the nation's electricity demand. In a recent study of San Francisco Bay Area households, researchers at UC Berkeley and the Lawrence Berkeley National Laboratory found that, on average, 10% of all electricity went to supply standby power, costing each household about $80 a year.[22] The leakage from TVs and VCRs alone costs American consumers $1 billion a year. In 1980, the average California household annually used about 6,000-kilowatt hours of electricity. Twenty years later, the rate climbed to 6,700 kilowatts and is projected to reach 7,000 kilowatt hours by 2009.[23] California is one of the most successful states when it comes to energy efficiency. The Golden State ranks second to Texas in total energy consumption, but, when the data are converted to energy use per capita, only three states consume less per person. Consumers can fight back by buying appliances and electronic devices that display the federal Energy Star label. Units with this label contain an embedded microchip that reduces the standby power drain by 75% to 90%.[24]

*Southern California has the largest number of electric cars and publicly available battery chargers in the nation. Electric-car drivers are test pilots for the 21st century.*

One of the most dramatic aspects of the 21st century energy revolution will soon be parked in your driveway. Within a decade, a whole slew of new fuel-efficient, low-emission hybrid cars, trucks, and SUVs will be rolling off assembly lines and coming to a showroom near you. These vehicles will boast cutting-edge technology such as an electric drive at the wheels, will be built for safety with new, strong but lightweight materials, and will be powered by fuel cells, advanced batteries, and electrically-assisted gasoline engines. The innovative technology is designed to improve gas mileage radically, thereby reducing air pollution and our demand on petroleum, a dwindling resource. The challenge for the auto industry is to design and build a low-emission vehicle (LEV) that can accelerate from 0 to 60 mph in less than 10 seconds, get better than 25 miles per gallon of gasoline, and run for 300 miles or so between refueling. To be commercially successful, the vehicles must be produced in large quantities like today's standard automobile, and be sold at an average price of about half the median US annual household income.[25]

Cutting gasoline consumption is a critical first step in trying to steer our society toward a cleaner, more sustainable lifestyle. In 1996, there were 600 million cars and trucks on the world's roadways (nearly 200 million of them in the US); by 2010 there could be nearly 2 billion. Gasoline and diesel provide 97% of the world's transportation energy needs, and Americans alone consume more than 120 billion gallons of these two petroleum products every year. Gasoline and diesel are poisonous to humans, plants, and animals, and their vapors are toxic. The environmental impact associated with petroleum production lays waste to many fragile ecosystems around the world, and oil spills contaminate oceans, coastal waters, rivers, and ground water. With the exception of hydrogen, when gasoline, diesel, or other fuels are burned in car engines, combustion is never perfect, and a noxious mix of hazardous pollutants is released in the exhaust. Both gasoline and diesel fuel emit volatile organic compounds (VOCs) and nitrogen oxides ($NO_x$), which contribute to ground-level ozone (smog). Transportation causes more than half of urban smog and produces one-third of US emissions linked to global warming. It is also the fastest growing pollution source.[26]

Because of the ongoing effort to fight air pollution in the United States, all new vehicles must meet either the emissions standards set by the US Environmental Protection Agency (EPA) or those set by the California Air Resources Board (CARB). Generally, California standards are more stringent than the Federal standards. A number of

If the automobile is to change radically in the new century, it must start now. Cars in today's showrooms will still be on the road when the world's oil supplies begin to diminish.[E]
— *Popular Science*

Dismayed by sluggish advances in battery-powered cars, the California Air Resources Board voted unanimously [January 2001] to reduce by nearly 80% the volume of zero-emission vehicles that automakers must offer for sale within two years, from 22,000 to 4,650.[F]
— *The Sacramento Bee*

northeastern states have adopted the California standards, and vehicles meeting these standards are becoming more and more common nationwide. During the 1990s, CARB mandated that by 2003, at least 4% of cars sold in California by auto manufacturers must be powered by pollution-free, zero-emission engines. Currently, the only zero-emission vehicles (ZEVs) commercially available are battery-powered; they run on the electricity stored in on-board batteries. Despite waiting lists of buyers who want an electric car but can't get one, in January 2001, CARB voted unanimously to revise their mandate and reduce by nearly 80% the volume of ZEVs that automakers must offer for sale by 2003, from 22,000 to 4,650. Under the old rule, Californians would have been offered 40,000 ZEVs for sale by 2007. Under the new rule, that level won't be reached until 2015 but will include sport utility vehicles. This was the third such rollback in a decade, and board members were not happy about cutting back on quotas once again; but expected advances in battery technology have not been realized. The board realized that the average consumer is not interested in an electric car that is too expensive and doesn't travel far enough between charges.[27]

> A new theory is always announced together with applications to some concrete range of natural phenomena; without them, it would not be even a candidate for acceptance.[G]
> —Thomas S. Kuhn

The auto industry is excited about the new hybrid-electric cars, which are extremely fuel-efficient. The hybrid-electric gasoline engines will dramatically slash greenhouse gas emissions by 50%. The Toyota Prius and Honda Insight represent the first generation of hybrid designs that may one day power all vehicles. These rigs link a combustion engine to a motor/generator, power electronics, and a modest battery pack in a hybrid configuration. Innovative features include a hybrid-electric drivetrain, regenerative braking that reclaims kinetic energy, and idle-off. American manufacturers are also developing hybrid-electric SUVs within the next few years.

The most exciting development in new transportation technology is the electric fuel cell, which can use hydrogen as a pollution-free fuel. Unfortunately, this system faces serious hurdles before it can hope to join the transportation energy revolution any time soon. Fuel cells generate electricity through a process of low-temperature oxidation, which is a kind of reverse hydrolysis whereby electrons left behind when hydrogen molecules slip through a membrane join the oxygen on the other side to become a source of power. Fuel cells can operate on hydrogen derived from fossil fuels, but the process to free the hydrogen from the chemical bonds with carbon requires some other energy source. The process produces carbon dioxide, which contributes to global warming. It is important to remember that neither hydrogen nor fuel cells are primary energy sources. (Electricity is not a primary energy source either but rather an "energy carrier," which has zero mass, travels near the speed of light, and, for all practical purposes, can't be stored.)

Fuel cells are being developed for use in cars, trucks, buses, etc., as a possible substitute for the internal combustion engine, but they are expensive and must be fueled with hydrogen to be pollution-free. Hydrogen is also difficult to handle and highly explosive and must be compressed or cooled to a liquid at minus 253 degrees Celsius in order to be carried in significantly useable amounts. Any radically new car running on an alternative fuel must coexist with older cars that will be on the road for another 10 to 20 years. In order to replace gasoline with a fuel like natural gas or hydrogen, another extensive system of distribution (think pipelines and gas stations) will have to be built and coexist for several decades before conventional fuels like gasoline and diesel are totally replaced.

There are many optimists who believe that America can quit guzzling oil and natural gas and switch to clean and renewable sources of energy like solar and wind power without disrupting our high material living standard. One thing is for sure, the 21st century will be a world with many different energy systems combined to help us make the transition to sustainable energy use. Back in 1994, British scientist and statesman Sir Crispin Tickell wrote, "We have done remarkably little to reduce our dependence on a fuel [oil] which is a limited resource and for which there is no comprehensive substitute in prospect."[28] It's time we did something about that. If not now, when? If not us, who?

> If this threshold is crossed, changes are likely to come at a pace and in ways that we can only begin to anticipate. The overall effect could be the most profound economic transformation since the Industrial Revolution itself. If so, it will affect every facet of human existence, not only reversing the environmental declines with which we now struggle, but also bringing us a better life.[H]
>
> — *World Watch*

## FOOTNOTES

[1] Environment News Service, "Protests Grow at Australia's Shale Oil Project," November 30, 2000, pp. 1–4.

[2] <www.dieoff.org/synopsis.htm> — Abstract: "Petroleum geologists have known for 50 years that global oil production would 'peak' and begin its inevitable decline within a decade of the year 2000. Moreover, no renewable energy systems have the potential to generate more than a tiny fraction of the power now being generated by fossil fuels. *In short, the end of oil signals the end of civilization, as we know it.*" Compelling 12-page synopsis by Jay Hanson, December 30, 2000. Hanson's "Dieoff" website showcases an excellent assortment of articles and graphs concerning declining per capita oil production, population growth, and the approaching energy crisis.

[3] <www.altenergy.org/2/nonrenewables/fossil_fuel/depletion/denver_energy_meeting/denver_energy_meeting.html> — or use the AEI search tool by keying in "World Oil Forum" to read the AEI profile of the October 30, 1998, World Oil Forum, held in Denver, Colorado. Two hundred geologists, oil and gas professionals, and US think-tank policymakers attended this conference to brainstorm the future availability of the planet's conventional oil supply.

[4] <http://groups.yahoo.com/group/dieoff/message/5> — "USA's Triple Energy Whammy in Electric Power, Natural Gas & Oil" — Position paper by Brian J. Fleay, dated January 10, 2001. Overview: "The USA is in a major electric power crisis. Peak power demand is exceeding reliable generation and transmission capacity, especially in California and on the West Coast. The industry is trying to overcome the crisis by installing gas turbines just when the supply of natural gas in North America has reached a peak, creating in turn a natural gas supply crisis. The only short term solution is to reduce consumption of both electricity and natural gas, bringing energy efficiency back on the agenda." Brian Fleay is an Associate of the Institute of Sustainability and Technology Policy at Murdoch University in Western Australia. His book *The Decline of the Age of Oil* was published in 1995, and he has authored several papers for conferences on the future of petroleum supplies and the consequences for transport.

[5] <www.dieoff.org/page224.htm> — "The Peak of World Oil Production and the Road to the Olduvai Gorge" — Keynote Address to the Geological Society of America by Richard C. Duncan. The conference "Summit 2000" was held in Reno, Nevada, on November 13, 2000. Duncan's research indicates that world energy production per capita peaked in 1979 and has been in decline ever since.

[6] Michael Renner, "Going to Work for Wind Power," *World Watch* (January/February 2001), p. 26.

7. <www.futureenergies.com> — "First Commercial Wave Power Station" Article posted on the Future Energies website on November 20, 2000. The Managing Director of WAVEGEN, Allan Thomson, said, "Wave power has joined the important group of commercially viable, competitive, and clean forms of sustainable energy; this is the launch of a new global market."

8. Associated Press article, "Power Crunch has Companies Thinking Solar," published in *The Sacramento Bee*, December 7, 2000, p. A–3.

9. <http://evworld.com> — Reuters article, "Shell Sees Growing Role for Natural Gas & Renewables," posted February 20, 2001. Sign up for the free EV World electronic newsletter.

10. Jane Collin, "Firms Come to Believe There's Profit in Being Pro-Environment," *The Oil Daily* (October 20, 1999), Vol. 49, Issue 200.

11. Richard T. Stuebi, "Attracting Equity for Solar Businesses," *Solar Today* (March/April 1999), p. 66. Stuebi is President of NextWave Energy based in Denver, Colorado. This article is an edited excerpt of a longer piece that is available on the NextWave website at <www.nextwave-energy.com>

12. Daniel M. Kammen, "Promoting Appropriate Energy Technologies in the Developing World," *Environment* (June 1999), p. 15.

13. Ross Gelbspan, "Putting the Globe at Risk," *The Nation* (November 30, 1998), pp. 20–22. Gelbspan, a 30-year journalist with stints at *Philadelphia Bulletin*, *Washington Post*, and *Boston Globe*, is author of *The Heat Is On: The Climate Crisis, the Cover-up, the Prescription* (HarperCollins).

14. <http://solstice.crest.org/index.shtml> — This website is the Internet information service of the Renewable Energy Policy Project and the Center for Renewable Energy and Sustainable Technology (REPP & CREST).

15. <www.repp.org/articles/resRpt11/subsidies.pdf> — Renewable Energy Policy Project, research report by Marshall Goldberg entitled "Federal Energy Subsidies: Not All Technologies are Created Equal," July 2000, No. 11. Wind, solar, and nuclear power received approximately $150 billion in cumulative federal subsidies over roughly 50 years, some 95% of which supported nuclear power. Goldberg is the principal of MRG & Associates, an environmental and economics consulting firm in Madison, Wisconsin. He is also resource planner specializing in energy and environmental policy analysis.

16. Peter Bunyard, "Nuclear Power — A Dead Loss," *The Ecologist* (November 1999) Vol. 29, Issue 7, p. 412. The author examines the true cost of nuclear power. Topics include prices and rates, energy policies, and power plant operating costs.

[17] Edie Lau, "Crisis Fuels Nuclear Talk," *The Sacramento Bee*, February 19, 2001, p. A–1.

[18] Stewart Boyle, "Making Progress Towards a Fossil Free Energy Future," *The Ecologist* (March/April 1999), Vol. 29 Issue 2, pp. 129–133. Abstract: "The shift from fossil fuels in the global energy system requires massive changes in energy and economic systems. A fossil-fuel-free energy future necessitates a change in current tax policies to stop encouraging the use of polluting energy and promote the phase-out of inefficient and dirty technology that causes pollution."

[19] <www.globalatomics.com> — Paul Brown's Global Atomics Corporation utilizes a proprietary photodeactivation process to stabilize radioactive materials.

[20] <www.padrak.com/ine/JNEV4N4.html> — Institute for New Energy website. Editorial comments on The New-Energy Educational Challenge and sample titles of content published in the *Journal Of New Energy, An International Journal of New Energy Systems*. Fox is the president of Emerging Energy Marketing Firm, Inc. (EEMF), which publishes this peer-reviewed quarterly journal. EEMF has acquired the patents to the new HDCC charge-cluster technology that produces thermal and direct electrical energy to stabilize high-level radioactive liquids and spent-fuel pellets into stable elements.

[21] <www.altenergy.org/3/solutions/solutions.html> — Alternative Energy Institute's energy conservation webpage. Offers practical solutions for saving energy and money at home, on the road, and at work.

[22] Frank Jossi, "Standby Sucks," *Wired* (February 2001), p. 80. Brief article reveals how electronic devices can consume almost as much electricity when they're turned off as when they're on. San Jose-based Power Integrations <www.powerint.com> has developed a microchip that can reduce standby power drain by 75% to 90%. Power Integrations leads the industry in high-voltage analog integrated circuits, providing cost-effective AC to DC and high-voltage DC to DC power conversion solutions.

[23] Editorial opinion, "Surviving the Heat," *The Sacramento Bee*, August 1, 2000, p. B–6.

[24] <www.energystar.gov> — The Energy Star website offers information and links to energy conservation solutions such as efficient appliances and electronic devices.

[25] Dan McCosh, "Hydrogen on Wheels: Space-Age Technology Comes Down to Earth to Power the Car of the Future," *Popular Science* (May 2000), pp. 53 – 61.

[26] <www.greenercars.org/autoenviron.html> — Most of the environmental impact associated with motor vehicles occurs during the operation of the vehicle due to exhaust pollution associated with fossil-fuel combustion. The GreenerCars' website offers information on low emission vehicles and rates vehicles according to their environmental impact.

27 Chris Bowman, "State Eases Pollution-free Car Mandate — Again," *The Sacramento Bee*, January 26, 2001, p. A–4.

28 Sir Crispin Tickell, *The Future and Its Consequences: The British Association Lectures 1993*, The Geological Society, (London, England, 1994), pp. 20–24.

SIDEBAR FOOTNOTES:

A Sacramento Bee staff and News Services, "Bush Says Energy is Paramount Concern," *The Sacramento Bee*, March 30, 2001, p. A–1.

B Carrie Peyton, "Data Servers Crave Power," *The Sacramento Bee*, November 26, 2000, p. A–1.

C Paul Hawken, Amory Lovins, and L. Hunter Lovins, *Natural Capitalism: Creating the Next Industrial Revolution* (Boston, Massachusetts: Little, Brown and Company, 1999), p. 58.

D Alan Weisman, "Power Trip: The Coming Darkness of Electricity Deregulation," *Harper's* (October 2000), p. 80.

E Dan McCosh, "Hydrogen on Wheels: Space-age technology comes down to Earth to Power the Car of the Future," *Popular Science* (May 2000), p. 52.

F Chris Bowman, State eases pollution-free car mandate — Again," *The Sacramento Bee*, January 26, 2001, p. A–4.

G Thomas S. Kuhn, *The Structure of Scientific Revolutions* (Chicago: University of Chicago Press, 1962), p. 46.

H Lester R. Brown, "Crossing the Threshold: Early Signs of an Environmental Awakening," *World Watch* (March/April 1999), p. 13.

# Twenty-first Century Solutions: Turning the Corner

## 18

Technological optimism is a tenant of faith in Western science, but relying on major advances in new energy research as a quick and painless solution to the environmental/energy crisis would be a mistake. Energy systems take years, if not decades, to develop. Conversely, funding and coordinating a large-scale research effort along the lines of a "Manhattan Project" could yield dramatic progress in developing breakthrough energy and/or propulsion systems. The challenges are serious but not insurmountable. An informed and proactive public is vital to the grassroots effort that any social revolution demands. A new scientific paradigm, one that openly supports research in legitimate carbonless technologies, will help meet the energy requirements of future generations.

On March 5, 2001, the Intergovernmental Panel on Climate Change (IPCC) adopted the third and final volume of its Third Assessment Report on climate change. The 800-page summary, *Climate Change 2001: Mitigation*, examines mitigation strategies for decreasing global greenhouse gas emissions. Three years in the making and extensively reviewed by more than 400 governmental and nongovernmental experts, the authoritative assessment strongly emphasizes a more intelligent use of the planet's dwindling energy resources, as well as

> Two new studies provide the strongest evidence yet that greenhouse gases are causing the Earth's oceans to warm, further strengthening the case that global warming is real and is being caused at least in part by air pollution.[A]
>
> *Washington Post*

better stewardship of our precious natural capital, such as forests, oceans and aquifers, which represent multigenerational equity.[1] Inefficient economic production and manufacturing systems must be revised to eliminate gross waste and planned obsolescence, thereby increasing our wealth while simultaneously reducing resource use. Reducing toxic air pollution, cutting anthropogenic (human caused) carbon dioxide ($CO_2$) production, and mitigating our influence on climate change will require a long-term, intergenerational effort involving scientific, technical, environmental, and economic solutions, as well as social progress and commitment. The global scope and economic impact associated with the quickly approaching peak in inexpensive oil production and distribution, as well as widespread concern regarding rapid climate change, involves complex interactions and large-scale processes that can seem overwhelming. The successful transition to a more sustainable society will require a conscientious effort by all of us, not just government and industry.

Intelligent, comprehensive national energy policies (both short- and long-term), dedicated energy conservation, and mandated energy efficiency measures can help reduce the United States' gluttonous demand on regional and global resources. Congress and federal and state agencies should establish, update, and enforce energy efficiency standards for appliances and equipment used in American homes, businesses, and factories. In the energy market in general, and in electricity particularly, a familiar principle is that a dollar invested in conservation is worth as much as the same investment in new generation. Every gallon of gasoline or kilowatt-hour of electricity made available to other consumers and businesses because of efficiency improvements is worth exactly as much as the same amount of energy produced by a new oil well or power plant. Best of all, energy efficiency gains come without contributing to ecological disruption or emitting greenhouse gases. In

*Air pollution from fossil fuel use has immediate, local, and global impacts on public health.*

general, it's cheaper and better for the environment to save fuel than to buy it, no matter what kind it is. In response to the price spikes of oil and gasoline during the 1970s, most energy-intensive devices manufactured in the US between 1975 and 1985 doubled in efficiency — especially cars, buildings, refrigerators, and lighting systems.[2] When petroleum prices crashed in 1986 and then remained dirt cheap throughout much of the 1990s, US oil consumption increased but energy efficiency priorities stalled. In consequence of poor efficiency standards, the United States currently emits about twice as much $CO_2$ per person as Germany does. When compared to the Japanese, Americans waste nearly $200 billion a year in energy costs because we do not employ the same efficiency practices in businesses and homes.[3]

There is no doubt that higher efficiency standards combined with renewable energy systems will play a major role in the US as the nation tries to wean itself from costly and unreliable imported petroleum. But that transition will take years, and efficiency alone will not alleviate the world's growing energy demands. The importation of petroleum costs the US $60 billion a year, plus another $40 to $50 billion to support and protect this far-flung energy resource. The economics of fossil-fuel energy are huge; the market is estimated at $4.5 trillion per year and historically, the fluctuating price and availability of oil has been the driving force behind America's economic expansion or contraction. The global oil market has become highly integrated and a major disruption of supply anywhere in the world will raise prices and tighten supply in the United States. No amount of investment in domestic fossil fuel resources will significantly change this fundamental reality, including the controversial plan to drill in Alaska's Arctic National Wildlife Refuge (ANWR).[4]

It took only 10 years for Alaska's oil-producing North Slope to reach its midpoint of depletion in 1988, and production has since fallen by nearly 50%. A similar fate awaits the oil companies waiting to drill ANWR. US President George W. Bush's Energy Secretary, Spencer Abraham, readily admits that the roughly 10 billion barrels of oil expected to be found there would be the equivalent of less than a year's worth of US consumption if used all at once, but he argues

> The United States accounts for about 25% of global oil consumption but has only 3% of proven global oil reserves.[B]
> — World Resources Institute

> Let us set as our national goal, in the spirit of Apollo, with the determination of the Manhattan Project, that by the end of this decade we will have developed the potential to meet our own energy needs without depending on any foreign energy source.
> — Richard M. Nixon —
> September 7, 1973.

that production would actually occur over many years, all the while helping to supply the nation's growing demand for oil.⁵ Energy experts rightly point out that it takes many oil fields in various regions to produce the huge volumes of petroleum that America guzzles every day. The Alaskan crude would certainly contribute significantly to US oil production once pumping began, but the windfall will be relatively short-lived, and domestic demand for oil will continue to grow unless new energy policies are implemented.

Drilling in the Arctic National Wildlife Refuge is a key part of President Bush's new supply-side national energy policy, which is designed to increase domestic energy supplies and reduce US dependence on foreign oil imports. But exposing the Arctic to the environmental risks of oil production will not alleviate the long-term problems stemming from increasing reliance on imported oil, nor can it succeed as an energy security strategy. The United States should decrease its heavy reliance on foreign oil by reining in its level of consumption, not through more drilling or mining in environmentally sensitive ecosystems. The US consumes about 20 million barrels of oil a day, more than two-thirds of which is used in transportation. In 2001, for the second time in three years, the average fuel economy of new passenger cars and light trucks sold in the United States dropped to its lowest level since 1980. The fuel performance for 2001 model-year passenger cars improved slightly, but once again the fuel-efficiency of light trucks declined. The US government requires that passenger cars meet an average fuel economy standard of 27.5 miles per gallon, and light trucks 20.7 mpg. Light trucks include pickups, minivans, and the popular sport utility vehicles (SUVs). Although technology exists to improve gasoline mileage without any sacrifice in the way people drive, federal miles-per-gallon standards have not changed in more than a decade.

Voters should support appropriate zoning to restrict suburban sprawl, which can reduce the frequency and distances traveled by car; encourage more investment in mass-transit systems; and demand more aggressive and proactive policy adjustments by the federal government, such as higher minimum efficiency standards on automobiles, especially light

> Conservation may be a sign of personal virtue, but by itself it is not a sufficient basis for a sound, comprehensive energy policy.ᶜ
> — Vice President Dick Cheney

> Improving the gasoline mileage of the nation's new vehicles by just three miles per gallon would displace more petroleum than the Arctic National Wildlife Refuge is expected to produce.ᴰ
> — Los Angeles Times

trucks and SUVs. In response to the oil crisis in the mid-1970s, US legislators mandated a doubling of the average fuel efficiency of new automobile fleets to the still current 27.5 miles per gallon. By 1985, the United States was 32% more oil efficient than it had been in 1973. If the US had remained at the old 1973 levels of efficiency, Americans would have used the equivalent of 13 million barrels of oil more than they actually did in 1985.[6] According to calculations by the Natural Resources Defense Council, raising the fuel efficiency of new passenger vehicles to an average of 39 miles per gallon over the next decade would save more than *15 times* as much oil as is economically recoverable from beneath the Arctic National Wildlife Refuge.[7]

Nationally, electricity is generated by a variety of processes, but coal, uranium, and, increasingly, natural gas, are the principal fuels. All of these methods of electricity generation are nonrenewable and contribute significantly to environmental degradation.[8] Howard Geller, former executive director of the American Council for an Energy-Efficient Economy, proposes that the:

> "US government should provide financial incentives to support innovative energy efficiency measures; stimulate expanded energy efficiency programs on the part of states and utilities; increase energy efficiency investments by federal agencies; and expand energy efficiency research and development, information and promotion programs."[9]

In the United States, where consumer and industry demand for petroleum products, natural gas, and quality electricity is increasing each year, only a small fraction of households or businesses have upgraded their energy efficiency to the maximum degree that is cost-effective. The IPCC report asserts that fully half of the potential greenhouse gas reductions achievable by 2020 can be met by using technologies available today to improve energy efficiency in buildings, transportation, and industry.

One fundamental problem in the US is that because the cost of energy is relatively cheap, American's undervalue the energy they consume every day; but when utility prices soar, or supply reliability falters, everyone is affected. California

> Motor vehicles are now arguably nature's principal antagonist in the Northwest, and — despite popular perceptions to the contrary — the real alternative to cars is not better transit; it's compact neighborhoods.[E]
> — *Northwest Environment Watch*

> Coal burning plants actually emit more radiation than nuclear plants in normal operation.[F]
> — *Physics: Extended with Modern Physics*

residents were hoping that the free market and deregulation would reduce the cost of electricity and bring lots of gain without much pain, but dramatic price spikes and rolling blackouts spoiled the party. A National Federation of Independent Business poll questioning 523 small business owners across California reported that nearly one in five employers with up to ten workers are seriously considering uprooting their business and moving out of the state. Sixteen percent said that they would cut employer-paid benefits to make up for their increased energy expenses.[10] Rolling blackouts cost the California economy $2.3 billion during a two-week period in January 2001, and, according to Alan Blinder, a former vice chairman of the Federal Reserve Board, California's energy bill for year 2001 is likely to be in the range of $45 to $50 billion.[11] In March 2001, Energy Secretary Abraham told the US Chamber of Commerce that in addition to expected electricity shortages in California and New York City, states in the Midwest, Southeast and northern Plains face possible power disruptions. Southern Nevada endured its first electricity blackout on July 2, 2001, in what officials called a "wake-up call." It has become obvious that the energy crisis is spreading and is no longer just a "California problem."[12]

*California is implementing an $800 million energy conservation program.*

Energy conservation and efficiency improvements are not only economically sound, they also have a huge potential to reduce the United States' long-term energy needs.[G]
— *Los Angeles Times*

Energy conservation is an effective and relatively easy partial solution that shows personal concern for the environment and also your wallet. And it makes a difference. According to the US Department of Energy (DOE), California is one of the most energy-efficient states in America, ranking 47$^{th}$ in the nation in per capita energy use and 49$^{th}$ in the nation in per capita electricity consumption, but there is still room for improvement. In February 2001, California businesses and consumers slashed electricity use by 8% (2,578 megawatts) — enough juice to power more than 2.5 million homes. In an effort to decrease electricity demand by a total of 10%, California's Governor, Gray Davis, recently outlined an $800-million energy conservation program, the largest such effort ever launched by any state.[13] Historically, energy conservation in the US has been price-driven and often temporary, but, if Americans are truly concerned about breathing clean air, the long-term risk associated with high-level radioactive waste, and leaving a well-cared-for planet to future generations, they will embrace conservation as a first promising step toward a more sustainable lifestyle.

Most of the world's energy comes from oil, coal, and natural gas — fossil fuels that, when burned, emit some 22 billion tons of carbon dioxide into Earth's atmosphere each year. Many scientists fear that rising levels of $CO_2$ and other greenhouse gases (GHGs) in the atmosphere are contributing to rapid global warming with its inherent risk of catastrophic climate change.[14] In 1997, more than 160 countries sent representatives to Kyoto, Japan, to attend the United Nations Framework Convention on Climate Change (UNFCCC)[15] The legally binding UNFCCC, first adopted at the 1992 Earth Summit in Rio de Janeiro, aims to stabilize atmospheric concentrations of greenhouse gases at safe levels. The Kyoto Protocol was negotiated by an international community serious about addressing the issues of air pollution, anthropogenic $CO_2$ production, and human-induced climate change. A planet-scale perspective on atmospheric and oceanic health and the serious consequences associated with rapid climate change have made limited, national approaches obsolete. Communities all around the world are tired of paying the health and environmental costs associated with carbon-based fuels and are collectively concerned about increasing evidence of emission-induced global warming. Kyoto marked the first time that the world's industrial nations committed to binding limits on the heat-trapping gases that scientists say threaten the planet's climate.

Under the 1997 Kyoto Protocol, nearly forty industrialized nations are committed to cutting their greenhouse gas emissions to an average of 5.2% below 1990 levels by 2008 to 2012. The US agreed to cut its $CO_2$ emissions to 7% below 1990 levels. Under this global policy, economically developed nations would face a strict timetable of mandatory reductions in greenhouse gas emissions, and less developed

> Establishing values for natural capital is a first step toward incorporating the value of ecosystem services into planning, policy, and public behavior.[H]
> — *Natural Capitalism*

> Except for nuclear war or a collision with an asteroid, no force has more potential to damage our planet's web of life than global warming.[I]
> — *Time*

countries would be encouraged to reduce their emissions but would not actually be required to do so.[16] (The six targeted greenhouse gases are carbon dioxide, methane, nitrous oxide, and three kinds of fluorinated gases.) The Kyoto Protocol in its present form mandates that developing countries do not have to adopt the same emissions targets or deadlines as the industrialized countries, a critical point considered unacceptable by the US government. The protocol's entry into force requires that 55 countries ratify — not just sign — it and that the emissions of these countries represent at least 55% of the total for all Annex I countries. Since the United States emits roughly one-quarter of all GHGs released into the atmosphere, ratification by the US is considered essential if the protocol is to succeed.[17]

Indicative of substantial disagreement associated with the current emission reduction contract, more than 100 countries have signed the Kyoto agreement, but no developed state has yet ratified it. In early 2001, President George W. Bush pulled the US out of the already tentative international treaty because of concern that it would hurt the US economy and American workers. Although the science behind global warming still contains substantial uncertainty, that uncertainty cuts both ways, in that the negative effects of climate change could be even worse than computer models predict. Reducing air pollution by seriously cutting global GHG emissions will affect everyone to some degree, but fears of economic ruin are unfounded. The IPCC "Summary for Policy Makers" released March 3, 2001, warns: "Climate change mitigation will both be affected by, and have impacts on, broader socio-economic policies and trends, such as those relating to development, sustainability and equity." But the panel also suggests multiple benefits associated with the transition from fossil fuels to renewable energy systems:

> "Climate mitigation policies may promote sustainable development when they are consistent with broader societal objectives. Some mitigation actions may yield extensive benefits in areas outside of climate change: for example, they may reduce health problems; increase employment; reduce negative environmental impacts like air and water pollution."

> The only way scientists can combat these doubts is through an open and well-reviewed process whereby the public can have trust in what the scientists are saying, especially when the stakes are very high, as in the global warming debate.[j]
> — *Los Angeles Times*

One study found that installing carbon-dioxide scrubbers on an older coal-burning power plant in Ohio would cost several hundred million dollars, would sap one-third of the plant's power output, and would double or triple the cost of electricity, but there is little evidence to support the fear that better energy efficiency and pollution control will hurt the economy or that Americans will lose jobs.[18] Mainstream studies actually indicate that market-oriented policies to protect the climate by saving energy can raise American living standards and benefit the economy.[19] Despite vociferous protest by environmentalists and many Western European leaders, Bush administration officials explained that environmental ministers had dominated the negotiations that led to the 1997 protocol, choosing targets that were symbolically tough but hopelessly unrealistic. The Bush administration had a point. At the time of the Kyoto negotiations, US emissions already were up nearly 10% from 1990 levels, and they have risen about 1.2% a year since then. Because electric power plants last for 30 to 50 years (subsystems may be replaced on a shorter time-scale), most of the facilities that will be burning fossil fuels from 2008 to 2012 are already in place today or will be going on-line soon. Both Congress and the White House realize that it is very unlikely that the US can deliver on its promise of reducing emissions to below 1990 levels in the next decade.[20]

In the US, where the protocol is considered a bad contract and an economic risk, the House and Senate must ratify the controversial agreement by a two-thirds vote. But that is unlikely when Congressional leaders warn that the quantified emissions targets and timetables embody flaws so severe that they cannot be fixed by incremental adjustments. Critics charge that certain tenets in the protocol may actually hinder its goal of the long-term reduction of GHG emissions. If the US refuses to get on board, or other countries fail to reach the targeted emission reductions by 2012 — an increasingly likely scenario given the present trends — growing cynicism and broken deadlines may undermine the spirit of international cooperation that is critically needed if a new and revised agreement is to succeed. The protocol's stipulated goals could become ceilings and regarded as "good

> The Kyoto protocol is far from perfect, and if we had to do it again, the treaty would likely look quite different. But the criticism of Kyoto has been disingenuous.[k]
> —*Washington Post*

enough" although environmental scientists warn that the reductions mandated in it are insufficient to prevent, or even appreciably slow, global warming. Perhaps most importantly, the "quantitative emissions goals for $CO_2$ and other GHGs involve great uncertainty, considerable interpretation, and deeply contentious ethical disputes," which could mire down negotiations for years, pre-empting any chance for a viable global treaty.[21]

Despite the United States' unilateral withdrawal from the Kyoto Protocol, the troubles plaguing the controversial contract do not preclude the need for a strong, catalyzing action among the world's worst contributors to GHG emissions and air pollution. International government institutions succeed best when they can facilitate and encourage cooperation among nations, as opposed to wielding coercive power.[22] The US government must think globally and stay involved, setting priorities and articulating a coherent policy framework that focuses on promoting renewable technologies and greater international energy cooperation. The Bush administration is conducting a Cabinet-level review of US climate change policy and is considering what policies should be pursued domestically and internationally to address increasing concentrations of greenhouse gases in the atmosphere. The sooner America, as the world's richest and most polluting industrialized nation, takes the lead in devising effective new energy policies, the better off everyone will be.[23]

Unfortunately, while politicians and lobbyists dicker and protect the fossil-fuel industry from emission reduction mandates, they miss one very important point. The same emissions from industrial and transportation activities that threaten humanity with rapid climate change are already killing us today. An estimated 10,000 people in the US die each year from effects of pollution from coal-burning power plants.[24] New scientific research suggests that smog not only hurts the lungs, but it can also cause sudden death in some people with existing heart conditions. Evidence shows that tiny airborne particulates found in smog may decrease a person's ability to regulate heartbeat and suggests that death can occur as rapidly as 24 hours after exposure.[25] Emitted particulates pose a major hazard for human health, with seri-

> Prudhoe Bay oil fields generate twice as much air pollution as Washington, D.C., and the area suffers more than 400 spills a year or oil or oil-related pollution.[L]
> — Sierra Club

> Oil spills, labor exploitation, and political destabilization [are] endemic to the global oil trade, not to mention smog, global warming, and other consequences of the West's reliance on hydrocarbon fuels.[M]
> — Harper's

ous adverse effects traced mainly to small particles with an aerodynamic diameter of 10 microns or less which can reach deep into the lungs. Smaller particles go deeper into the lungs or can even enter the bloodstream. Polluted air can cause immediate or acute effects ranging from asthma attacks to death in those whose lungs are already weakened. Every year, the US spends as much as $100 billion on health costs related to the effects of polluted air.[26]

Fossil-fuel-generated emissions and particulates are a global problem and, therefore, require a global solution in the international spirit of Kyoto. These tiny particles can remain aloft for days or weeks, and travel thousands of miles. Researchers have tracked carbon particulates from specific smokestacks in Beijing, China, to Mauna Loa, Hawaii. The first study ever to look at the global short-term health impact of fossil fuel consumption indicates that the benefits of reducing $CO_2$ go substantially beyond averting potential disruptions of the Earth's climate. Even relatively small reductions in emissions worldwide could prevent 700,000 premature deaths a year by 2020. Implementing climate policies now would yield immediate benefits locally and globally by reducing particulate air pollution, by slowing the build-up of greenhouse gases, and by protecting public health. Over the next two decades, at least 8 million deaths could be avoided.[27]

The Kyoto Protocol represents only the first attempt at negotiating a global contract concerning greenhouse gas emissions, but make no mistake — the world community is dead serious about reducing GHGs, and, eventually, the wealthy industrialized countries will have to broker and sign some sort of deal. Energy conservation and greater energy efficiency in industry and society will help us meet future carbon reductions, but will they suffice as we adapt to a world without cheap oil? Chemists, engineers, physicists, and inventors are working feverishly to rapidly develop better energy systems

Annual US health costs due to exposure to conventional gasoline vehicle emissions are estimated to be $20 - $50 billion.[N]
— *Environmental Science & Technology*

*This coal-fired power plant on the Yangtze River in China dramatically illustrates the potential health risks facing developing countries in the absence of climate policies.*

> Scientists at five of the country's national laboratories calculate that with energy-efficient technologies, the country could reduce new electricity demand by between 20% and 47%.[o]
> — *The New York Times*

> The White House energy report is expected to call for the construction of 1,300 to 1,900 new power plants in the next 20 years.[p]
> — *The Wall Street Journal*

so that we don't pollute the biosphere or ravage all of the planet's non-renewable natural capital in the transition to renewable, and eventually, future energy sources. The stewardship of the planet's limited natural resources is an important responsibility that world governments and most industries have not yet taken very seriously.

There are major institutional obstacles to developing and implementing new technologies that compete with the well-established fossil-fuel industry. For instance, older energy systems that over decades have been subsidized with billions of dollars of government funding serve to perpetuate staid government bureaucracies and sustain government-funded national laboratories as well as academic projects. Political, corporate, and financial impediments stymie change at nearly every step in the process. The American media plays a huge part in disseminating information, generating awareness, and sculpting public opinion about complex issues like energy policy, environmental impacts, and rapid climate change. There is a lot of conflicting information being bantered about, and it's often hard to tell where the factual science ends and the hyperbole begins. Heated public debate in the media and in the US Congress over the veracity of climate change science and an energy crisis in the western US that threatens to spread, have slowed federal efforts to clean up the nation's air and reduce GHG emissions.[28] In fact, the US has fallen significantly behind other countries in its ability to simulate and predict long-term shifts in climate. According to a wide range of scientists and recent federal studies, advanced climate research in the United States is fragmented among a medley of agencies, strained by inadequate computing power, and starved for the basic measurements of real-world conditions that are needed to improve simulations.[29]

On March 13, 2001, President George W. Bush announced his decision to reverse a previous pledge to legislate limits on carbon dioxide emissions from US power plants. President Bush's flip-flop on carbon dioxide emissions elicited vigorous protest by politicians and environmentalists in the US and abroad.[30] It is critically important that the US take a leading role in the reduction of toxic and dangerous greenhouse gases. The US government needs to remove the economic, legal, behavioral, and institutional barriers that can discourage consumers and companies from exploiting clean-energy technologies. Emissions from US power plants alone exceed the combined emissions from 146 countries, roughly three-quarters of the countries in the world.[31]

Bush's policy shifts have been victories for automakers and appliance manufacturers that feared new $CO_2$ emission restrictions would increase the cost of electricity, but the reversal was especially sweet for the coal industry. Oil, coal, and natural gas power generators all release carbon dioxide, but coal-fired plants emit the most $CO_2$ per megawatt produced. Coal-fueled plants, which currently supply 55% of US electricity, would find it significantly more difficult than oil and natural gas to reduce carbon dioxide emissions.[32] Despite increasing public and political support to replace coal-fired power plants with cleaner-burning natural gas generators, President Bush envisions using more coal, not less. The President fears that relying too heavily on natural gas for power generation will drive up its cost. Coal is an abundant natural resource in the US, and, if President Bush gets his way, it will play a very important and strategic role in America's energy future.[33] The new strategy proposes funding the DOE's existing Clean Coal Technology Program with $2 billion over 10 years.[34] The goal is to use new technology and efficiency improvements to reduce significantly the environmental impact of burning coal to make electricity, but it is yet to be seen whether the modifications will be implemented and really work as advertised. Current generation "clean coal" technologies have had undesirable side effects including added water demands, large amounts of sludge and other solid wastes, and decreased energy efficiency, which increase emissions of other compounds — including carbon dioxide.[35]

In order to give clean, renewable, and emerging carbonless technologies a bigger piece of the power supply pie, subsidies to the fossil fuel industry must end, and the playing

*It is critical that the US take a leading role in the global reduction of greenhouse gases.*

field must be equalized.[36] For example, consider the remarkable history of the semiconductor/computer industry. The US government and military paid premium prices for early microchips, which gave semiconductor companies the time and money to improve the design and production technology of the circuit — eventually reducing costs for everybody. The US government can and should commit to purchasing a fixed and growing percentage of its power from renewable sources so that the financial might of the federal government can help create a vigorous market for new clean technologies. The boost could be quite significant; after all, through its departments and agencies, the federal government spends about $8 billion a year on energy — probably more than any other single consumer in the world.[37]

The Department of Energy draws the roadmap to America's energy future. Not since the 1977 "National Energy Plan" has the DOE addressed national energy policy in an organized report, but in April 1998, the DOE released the Comprehensive National Energy Strategy (CNES). This document contains the DOE's game plan for the years ahead and is a blueprint of the nation's future energy mix. An independent review of the CNES indicates that the long-term outlook does not appear very promising for US efforts toward reducing GHG emissions over the next decade and more. In fact, the CNES fossil fuel consumption forecasts indicate increasing carbon emissions through 2010 to 44% higher than 1990 levels, which is 51% higher than the original Kyoto Protocol mandate. The CNES fails to indicate the advantages of reducing carbon emissions and the cost-effective benefits of doing so, even after a 1997 five-laboratory DOE report indicated that the paybacks will meet or exceed the investment in three out of four sectors. A vigorous national commitment to develop and deploy cost-effective, energy-efficient and low-carbon technologies could reduce carbon emissions at least to 1990 levels. The DOE study estimated that the "cost" in dollars per ton of carbon emissions is negative, meaning that they more than pay for themselves.[38]

When it comes to promoting pioneering research at the frontier of emerging carbonless technologies, the scientific and academic communities are handcuffed by established

> Electricity restructuring, growing international markets, public support, and continuing technical advances will lead to increased use of renewable electric technologies, which will replace coal-based, nuclear power plants, and perhaps, natural gas power plants.[Q]
>
> — *The Washington Quarterly*

protocols that reward conservatism and stonewall the careers of those daring to think "outside of the box." In these self-policed and "publish-or-perish" professional environments, those professors, graduate students, and post-doctorate researchers who try to pursue highly innovative lines of research are rooted out and shut down. Unless individuals successfully bid and obtain funding — no easy task — they are essentially ostracized and soon relegated to the sidelines of their profession, with serious obstacles to promotion, retention, and tenure. Therefore, it is often the retired scientist who has less to lose in terms of reputation or funding, that is more open to new ideas. Unfortunately, history indicates that most of the world-changing ideas have come from the younger scientist or theoretician. In addition to the devastating character-assassination associated with those that boldly pursue frontier science, friends and colleagues will often avoid contact for fear of their own loss of reputation and guilt by association with a scientist deemed an outcast. Therefore, the proponent of the anomalous claim is isolated from further debate and dialogue with the rest of the scientific community. Any scientist proposing a paradigm-challenging discovery will face a host of serious obstacles, such as difficulties with obtaining funding or passing peer-review in respected scientific journals.

The pressure and scope of professional conformity in the sciences has long been recognized and studied over the years. In his 1962 book, *The Structure of Scientific Revolutions*, Thomas S. Kuhn wrote:

> "We have already seen, however, that one of the things a scientific community acquires with a paradigm is a criterion for choosing problems that, while the paradigm is taken for granted, can be assumed to have solutions. To a great extent these are the only problems that the community will admit as scientific or encourage its members to undertake. Other problems, including many that had previously been standard, are rejected as metaphysical, as the concern of another discipline, or sometimes as just too problematic to be worth the time."[39]

Science, in clearing away the fog of myth and mysticism that shrouded the world in the Dark Ages, has exposed not only sharply delineated islands of knowledge, but also boundless seas of ignorance.[R]

— *The Fermi Solution*

The technical sector of the US government research community is drawn from the same institutionalized universities and private industries that instill scorn on scientific adventure beyond the accepted parameters. Professional papers published and fundraising are the major requirements in this realm, both of which are controlled within well-established and largely accepted scientific constraints. Managerial government scientists must compete for funding and annual budgets from top level officials who represent the aims and perceptions of conservative scientific leaders. Like graduate students and university professors, individual scientists must also publish or perish, and they too are constrained by the influential reviewers and editors of the scientific journals. Because of this entrenched institutional bias, the scientific community is itself often the greatest enemy of breakthrough innovation.

For the most part, the organized scientific community varies from highly resistant to openly hostile toward novel scientific research. At some levels, this is quite understandable. There are some very misguided and unorthodox energy system inventors and scientists out there, who in the guise of furthering emerging technology, actually contribute to the problem rather than the solution. Some are more interested in selling "dealerships" and "stocks," and few have actually submitted their alleged over-unity device or experiment to rigorous testing by an independent, government-certified test laboratory. Good science requires accurate measurements and effective criticism. By accenting the activities of "the crazies," the conventional scientific community often convinces non-technical government managers and personnel that the unorthodox alternative scientific community is comprised only of lunatics, charlatans, stock-scam artists, and misguided crank inventors.[40]

The history of science is replete with rejections of novel discoveries that challenged the dominant paradigm. Indeed, each new major advance in science starts with an anomaly that is unacceptable at first. The anomalies are important because they inspire important new questions and innovative thinking that can lead to dramatic scientific breakthroughs. Few subscribed to Copernicus' ideas for almost a

> It just takes a lot of time, work, ingenuity, and guts to eventually figure a way to cheat the laws of physics.[5]
> — *The Car That Could*

century after his death, and Newton's work was not generally accepted for more than 50 years after he published his *Principia*. Contemporary scientists laughed when Benjamin Franklin proposed that lightning was a form of electricity. Before Darwin, the paradigm that preceded evolutionary theory was natural theology, in which each creature was considered to be perfectly adapted to its environment. Darwin observed that some species were less well adapted to their environment and explained these anomalies on the basis of natural selection, an ongoing evolutionary process. His new concept drew criticism from the usual suspects, but Darwin was optimistic. Near the end of his revolutionary book, *On the Origin of Species*, describing the evidence for evolution, Charles Darwin wrote:

> "Although I am fully convinced of the truth of the views given in this volume… I by no means expect to convince experienced naturalists whose minds are stocked with a multitude of facts all viewed, during a long course of years, from a point of view directly opposite to mine…but I look with confidence to the future, to young and rising naturalists, who will be able to view both sides of the question with impartiality."[41]

In a 1993 paper published in the *British Homoeopathic Journal*, biophysicist Beverly Rubik wrote:

> "Conventional scientists attempt to explain the anomalies within the framework of the dominant paradigm, while a smaller, usually younger group of scientists attempts to develop an alternative paradigm. The crisis is resolved by a dramatic change of perspective, a paradigm shift. A struggle typically ensues that may result in the overthrow of the old paradigm, but usually takes years until the presently established scientists retire. What was an anomaly earlier now becomes the expected result. Textbooks are written in such a way that they even disguise the very existence of the revolution that generated them."[42]

All US politicians are reliant on their constituency for reelection, and the Chief Executive is guaranteed only four years in office, which makes it difficult for any one administration, no matter how popular or effective, to influence long-

> Quantum mechanics and chemistry have well defined limitations, while the processes underlying *new* energies clearly go beyond these limitations.[T]
> —*Infinite Energy*

term national energy policy. Governments limited to next-election thinking and corporations focused on the next-quarter perspective cannot operate on the planetary time-scale that truly reflects the multiple decades and even multiple centuries that it may take to ameliorate the negative effects of greenhouse gas buildup or to address other issues like biodiversity and population growth. One of the key roles in deciding how Americans will slice their energy pie resides with the United States Department of Energy (DOE). The DOE's 1998 CNES document states:

> "The advances spawned through American innovation will range from improvements seen directly in our everyday lives — much more efficient light bulbs, cars, appliances — to new approaches for large baseload energy sources. We must engage the talent in our universities and national laboratories to advance basic science and engineering research and to partner with the private sector to develop and deploy new technologies. This is a central component of a modern, forward-looking energy strategy."[43]

> Economists favor heavy taxes on fuel, but that's a political no-no.[U]
> — *Business Week*

One of the Department of Energy's missions is to discover, develop, and commercialize novel sources of energy and propulsion. However, there has been an inherent difficulty in incorporating dramatically new approaches to science and technology programs due to various factors, not the least of which has been continual budget pressures and the focus on near-term science and technology requirements. Critics have complained that, historically, the DOE and its agencies have slammed the door shut on novel ideas in new technologies, but there is at last a quiet revolution underway. Indeed, it might be the long-awaited paradigm shift.

For years NASA's Advanced Concepts Office (ACO), which is part of the Office of Space Access and Technology created in August 1994 at the approval and direction of the NASA Administrator, has solicited and reviewed novel propulsion schemes. NASA's Breakthrough Propulsion Physics (BPP) Research Project is part of a comprehensive strategy for advancing space propulsion through the year 2025. This strategy, called the Advanced Space Transportation Plan, spans the near-term technology improvements for launchers all the way through seeking the breakthroughs that could revolutionize space travel and enable interstellar voyages. NASA has actively funded scientific efforts pursuing innovative, breakthrough-oriented research in emerging propulsion technologies, but now, if adequate funding is appropriated, the DOE may finally be making a move towards exploring future energy conversion/storage and transportation systems.

To the DOE's credit, a new office has been established within the organization to serve in a strategic research management function, similar to that which the ACO serves within NASA Headquarters. The Office of Advanced Energy Concepts (OAEC) is managing the new *Breakthrough Energy Physics Research* (BEPR) Program. The BEPR Program is a joint exploratory research effort between the Office of Energy Efficiency and Renewable Energy, the Office of Science, and the Office of Nuclear Energy, Science and Technology.[44] The goal of the OAEC is to identify, research, develop, and push new, far-reaching concepts that have the potential to advance significantly a cross-section of DOE energy and transportation programs. Among the principal goals, objectives, and strategies, the BEPR Program will research, develop, and validate emerging theories, new ideas, and anomalous effects that can lead to breakthrough advances in energy storage and conversion, as well as novel transportation systems with the potential to greatly reduce or eliminate the need for energy or thermal sources. To ensure the maximum return on investment and to prevent duplication of efforts, the BEPR Program Plan will coordinate closely with the NASA BPP Research Project to ensure that the same specific ideas are not explored, thereby allowing the two federal agencies to test effectively a broader range of potential breakthrough areas.

If successful, the BEPR Program will go a long way toward securing US leadership in the breakthrough carbonless energy and propulsion technologies that will define the 21$^{st}$ century. Among the goals of the United States' 1992 Energy Policy Act: are 1) To maintain the technological competitiveness of the United States and stimulate economic growth through the development of advanced technologies; 2) To consider the obstacles inherent in private industry's develop-

> A series of technological breakthroughs — and more importantly, a critical mass of scientific ideas — has begun to coalesce around a new model for an energy system that would better the needs of the near future, while enabling power producers as well as consumers to lessen their impact on the environment.[v]
> — *Wired*

> Worldwide, it would take 3,000 nuclear plants — a tenfold increase — to replace all coal plants, yet that increase would reduce carbon emissions by only 20%, while enormously expanding risks that materials from nuclear power plants would be applied to making weapons.[x]
>
> — Nuclear Control Institute

ment of new energy technologies and the steps necessary for establishing or maintaining technological leadership. The BEPR Program was initiated to fill a void in R&D activities that the US government wants to accomplish. If funded and implemented, the BEPR Program will accept proposals for small-scale, exploratory, high-risk research in the fields of science and engineering related to DOE's energy and transportation missions. This research approach is unique in that it begins with novel ideas but seeks credible supporting theories and/or hitherto unexplained effects in the peer-reviewed scientific literature as a basis for initial research exploration. This organization is empowered to work in an independent fashion and to conduct research as free as possible from the administrative burdens of a larger bureaucracy, and it is committed to engaging the brightest minds available to assist in this endeavor.

The United States in the next 25 years faces major challenges in meeting ever-increasing demands for energy and transportation goods and services. Along with mounting concerns over the continuity of nonrenewable fossil fuel supply for transportation and electricity generation, coupled with increasing concerns of global environmental impacts, the solution to this difficult and long-recognized problem will require the development of some revolutionary concepts. To help avert a potential energy crisis in the coming decades, scientists and researchers around the world are pursuing development of innovative and, sometimes, exotic energy systems to somehow tap subtle forces of nature that they barely understand. In years ahead, newly emerging energy processes and technologies could lead to pollution-free, decentralized electricity generation and carbonless propulsion systems.

Many potential energy and transportation breakthroughs have been reviewed earlier in this book, such as Zero-Point Field (ZPF) Energy Extraction, Mach's Principle and Impulse Engines, Anomalous Gravitational/Inertial Mass Effects, and Low Energy Nuclear Reactions (LENRs). Other emerging theories and anomalous effects that are current in the scientific literature include Nuclear Isomer Decay (NID) Phenomena; Advanced Electromagnetic Theory; Gravito-electrodynamics Effects; and Hyper-fast Travel within General

Relativity. The exotic nomenclature may sound as if it was lifted from some dime-store sci-fi magazine, but the theories and science behind these technologies represent the seeds of future mainstream science.

In order to minimize the potential risk from published theories or reported effects that may lack rigor as to their scientific basis, the Office of Advanced Energy Concepts will implement an initial ranking methodology based on published merit. All initial rankings generated from the database will be reviewed by the Laboratories Scientific Oversight Committee and adjustments made as needed, realizing that some peer-reviewed journals are more rigorous than others and some researchers follow the scientific method more closely than others. The OAEC will also develop a Breakthrough Energy Effects Database, which will list all emerging theories and anomalous effects from the scientific literature that relates to DOE mission areas of interest. Development of the database will include links to known citation indices and cross-references on the theories and effects. The database will be maintained for as long as the BEPR Program is authorized and funded. If this much-needed program receives the approval and funding it requires, it is hoped that a potential energy breakthrough can be developed on a faster time-scale than incremental research would have traditionally supported, and some solutions to society's ever-compounding energy and transportation problems could be acquired much sooner.[45]

Once the appropriate potential technologies have been identified by the BEPR Program, the US must consider approving and funding a coordinated effort similar to the $20 billion Manhattan Project, a 1940s program to build the first atomic bomb. In 1942, scientists in both England and the United States were convinced that a nuclear weapon based on the principle of fission could be constructed within a period of three to four years. Yet the complex process of sepa-

*Ruggero Santilli (left) developed new mathematics from which new physical theories may be constructed. Shang-Xian Jin (right) developed the mathematical model that describes the HDCC phenomenon.*

Albert Einstein's theory of gravitation has passed a series of observational tests, but nobody has figured out how to make it compatible with the laws of quantum physics that govern the subatomic world.[W]
— Washington Post

rating fissionable U$^{235}$ or producing man-made plutonium, also fissionable, was expensive and unproven. Initially, the bomb design and experimental testing were conducted in scattered university research laboratories in the US and England. In 1943, a team of scientists was formed at Los Alamos under the direction of physicist J. Robert Oppenheimer, who argued that the most effective way to develop the new atomic technology was to allow the civilian scientists free discussion and an open exchange of ideas. Oppenheimer got his way, and his legendary community of brilliant "crackpots" proved to the world that he was right. Dialogue and debate in an open forum is the best approach to any scientific endeavor — vital ingredients that are sorely lacking in today's new energy field.[46]

> The effort to construct the world's most power laser is at least $1 billion over budget and four years behind schedule, raising disquieting questions about the way large science projects in the United States are managed.[Y]
> —*Los Angeles Times*

It is hoped that the new technology systems that are developed in the 21$^{st}$ century will be small-sized units of simple design that can be mass-produced for the growing global energy market, especially in the developing world. But unbridled faith in technological solutions to all of the environmental and energy issues present and future generations are facing is a dangerous and potentially myopic vision. There will always be popular and persuasive technological optimists who believe that the human mind has an unlimited capacity to solve all our problems, but technology is no substitute for civilization's essential natural resources such as food, forest, land, water, energy, and biodiversity. Political leaders, economists, clergy, and ministers — all of us — must be realistic as to the limits of technology to help humans feed themselves and provide other essential resources.[47] Technology has provided in the past and will continue in the future to provide improvements in food production, transportation, medicine and health, and in other important sectors of the human experience, but it must be remembered that technology relies on natural resources; it does not create them. The finite size of resources, ecosystems, the environment, and the Earth have led some resource scientists to recognize that the frequently used term "sustainable growth" is an oxymoron.[48]

Much of the planet's natural capital is made up of finite resources, nonrenewable bank accounts that are being quickly depleted as an increasingly industrialized and growing world population demands more and more. Energy is the key that unlocks all other resources, and it is crucial to human prosperity. In a comprehensive study of renewable energy sources, a well-supported analysis suggests that:

> "Figures commonly quoted on costs of generating energy from renewable sources can give the impression that it will be possible to switch to renewables as the foundation for the continuation of industrial societies with high material living standards. Although renewable energy must be the sole source in a sustainable society [unless new alternative energy systems are discovered and developed], major difficulties become evident when conversion, storage and supply for high latitudes are considered. It is concluded that renewable energy sources will not be able to sustain present rich world levels of energy use and that a sustainable world order must be based on acceptance of much lower per capita levels of energy use, much lower living standards, and a zero growth economy."[49]

Current projections by the DOE predict that world energy consumption will increase 36% by the year 2010. Extrapolation would project an almost doubling of global energy consumption over that of today by 2025. Within the next half-century, the world requirements for energy and transportation as well as those of the United States will greatly increase as the global population continues to grow dramatically. The current world population of more than 6 billion is projected to grow to 8.0 billion by 2025, and there will be more than a 50% increase to the present population by 2050 (currently projected at 9.3 billion).[50] This steady growth in population will increase the demand for energy and transportation resources worldwide and will have a significant negative impact on traditional, nonrenewable energy resources. Current energy resources and inefficient energy conversion methods will, at some future time, fail to fulfill the combined energy needs of the world and the US in both economic and environmentally acceptable ways.

In March 1998, the International Energy Agency, for the first time, forecast a possible date of the peak of world oil production stating: "…a peaking of conventional oil pro-

> Americans once considered wilderness something to conquer: We're only now, imperfectly and gradually, coming to regard wilderness as something to protect.[z]
> — *Bucks County Courier Times*

duction could occur between years 2010 and 2020."⁵¹ Twenty years gives precious little time for those countries that rely primarily on the revenues of exported oil to support their health, social, and employment systems, and ultimately government stability, to adapt to decreasing petroleum production even as their population grows. Two decades is not much time for a nation like the United States, which relies heavily on cheaply priced imported oil, to maintain its own economic vitality in a world of diminishing oil supplies. Nevertheless, the production peak is coming, and the planet's human population continues to increase at an alarming rate. The combination of mechanization, petrochemicals, and genetic engineering produced the "green revolution" that so greatly increased agricultural productivity during the 20th century and continues to support a huge world population today. It's important to realize that petrochemicals and mechanization are derived from oil and natural gas, two nonrenewable and diminishing resources. Oil and gas fuel agricultural machinery, and they are the raw material base for fertilizers that increase crop yields, as well as for the pesticides that protect vital crops from insects, diseases, and weeds.

It's been said, "Modern agriculture is the use of land to convert petroleum into food."⁵² Thanks to oil, just 2% of the working US population provides most of the food for this nation, which is also the world's largest grain exporter. In fact, the world's principal grain exporters are the US, Canada, Europe, Australia, and Argentina — all highly dependent on petroleum-based industrial agriculture.⁵³ Geologist Walter Youngquist points out two salient facts:

> "There are now two trends clearly on a collision course: First, population is growing at the astounding rate of nearly a quarter of a million a day, and is highly and increasingly dependent on oil and natural gas for food production. Second, the end of the petroleum supplies are clearly in sight." In the postpetroleum paradigm, "All possible economic energy sources will have to be used, but replacing oil in its great energy use versatility probably will not be completely possible. Replacing the role of both oil and gas in agricul-

> The problem of population size is politically sensitive and therefore largely avoided in discussions, but the energy problem cannot be sustainably solved if the demand target is a continually growing population.ᴬᴬ
> — Walter Youngquist

tural production will be the most critical problem, and may not be entirely solvable."[54]

Overpopulation is the root cause of many kinds of environmental deterioration. Climate change, rain forest destruction, desertification, and most types of pollution are linked to an expanding world population and increased pressure on limited resources. The world population is currently growing by 83 million people per year, with predictions that the total could climb to as high as 10 billion by the year 2050. In the next 50 years, the US population will jump to 400 million people, when it will be the only developed country among the world's 20 most populous nations. In 1950, at least half of the top ten most populated countries were industrial nations. In 2000, there were still three, including Japan and Russia.[55] Global population growth and/or growth in the rates of consumption of resources cannot be sustained. World energy production per capita peaked in 1979 and has been declining ever since.[56]

The wealthy industrialized countries along with the United Nations must initiate major comprehensive educational, technical, and outreach programs in the areas of social responsibility, contraception, and family planning. A world campaign for the reduction of birth rates is more important than any program of foreign aid and investments.[57] Some countries, both developed and developing, have adopted policies aimed at slowing population growth by encouraging family planning. India's National Population Policy 2000 report recognized that stabilizing the population is an essential requirement for promoting sustainable development with more equitable distribution or resources. Every effort helps. India, with a population about 2.5 times the size of the European Union population, is growing much faster than some earlier predictions suggested. In 2000, the EU had a natural increase of 343,000 people. India added that many people in the first week of 2001. If you add migration to Europe, the net growth would be about 1.2 million. India grew that much in the first three weeks of the year.[58]

To be in the midst of change yet oblivious to it is characteristic of the human condition. The growth of world population may have peaked in 1962-1963, at an estimated 2.2%,

> If political leaders are wise enough to capitalize on the opportunity provided by the coming together of this extraordinary constellation of circumstances, the population problem could be solved within two generations.[BB]
> — *Critical Masses*

but the goal of zero population growth and a more equitable and sustainable level of energy use eludes the citizens of planet Earth. The education, employment, and the changing roles of women, as well as access to birth control, have played an important part in reducing fertility. Studies have shown that economic development and family planning programs contribute significantly to the decline in fertility rates. Population growth decreases naturally when women are able to make informed, voluntary decisions about the size and spacing of their families. Worldwide, about a third of all pregnancies each year — 80 million — are unintended. In the United States, nearly half of all pregnancies are unintended — one of the highest rates among all industrialized nations.[59]

*There is an urgent need to address economic activity, population growth, and environmental protection as interrelated issues.*

Surveys show that unabated sprawl equals longer commutes, snarled traffic, dwindling open space, and further fraying of inner-ring communities and the city.[CC]

— *The Philadelphia Inquirer*

National and world leaders must look to the future, to a time when the population and the sustained carrying capacity of the planet reach parity.[60] Planners, politicians, and economists must make the difficult social and moral decision as to what is acceptable as a reasonable standard of living. If humans don't try to find the balance, nature will respond as it has before, indiscriminately and painfully.

Alternative Energy Institute, Inc., is dedicated to educating the public that an energy and environmental crisis is at hand. The crisis is real, and, if society heeds the warning, disaster may be forestalled, and the crisis may be abated. In the transition from a world based on polluting nonrenewable hydrocarbons to a sustainable one, there must be deliberate constraints on growth. The implications are staggering. The world's population has doubled since 1940, but human water consumption has quadrupled. Almost two-thirds of the world's agricultural land has been degraded in just the past 50 years. Seventy percent of the world's major marine fisheries are depleted.[61]

A relatively smooth transition to a society in equilibrium is possible, provided there are major reductions in population growth and material consumption and output, but a disturbing flaw has been noted by human ecologists:

> "The continued disregard for the problem of overpopulation is generated by the current imperialistic attitude of the human species toward the rest of nature that sees humans as distinct from the natural realm. However, for their own sake, humans have to learn as soon as possible how to include in their cultural identity, values and moral obligations that are based on respect for the rest of the biosphere to which they belong. In fact, this is the only way to guarantee a fair opportunity to future generations for a quality of life."[62]

Population growth in the next two to three decades is possibly the world's most serious problem, reducing our chances for a successful transition to sustainability while maintaining quality of life. Exploring the idea of optimum population at the 1993 forum for Negative Population Growth, Paul Werbos stated:

> *Facts do not cease to exist because they are ignored.*
> — Aldous Huxley

> "The present mix of fuels and energy technologies is not sustainable in the long-term, even if population were dramatically reduced. In the long-term, fossil fuels will run out. Even now, our present ways of using energy have led to unhealthy levels of ozone in almost every American city, to toxic pollution leaking into ground water and into natural bodies of water. In the next decade or two, oil imports may return to being a crisis-level problem, as demand increases and domestic supply decreases, almost inevitably. Sooner or later, we must also deal with the problem of greenhouse warming, which is associated with all realistic uses of fossil fuels."[63]

Corralling runaway population growth, deadly pollution, and nonrenewable resource depletion will not be quick, painless, or easy. Not acting in the face of crisis has been likened to rearranging the deck chairs on the Titanic. But, in one very important way, the metaphor does not represent the realities of the citizens on planet Earth. For us, there is no seasoned captain who has the best interests of his crew and passengers in mind. It is also important to remember the intentional class structure of the Titanic; survivors came primarily from the upper-deck classes. But where are the lifeboats on planet Earth? And where will the survivors go?[64]

Conservation, resource and energy efficiency, technologically advanced power and transportation systems, and more sustainable lifestyles represent real hope for the future. But supplying ever-increasing amounts of energy for the 83 million people who are added to global population every year is like trying to hit a target accelerating into the distance. Because the United States consumes 25% of the world's energy, Americans especially must learn to live more responsibly and with awareness that today's positive actions will offer future generations their only chance at a reasonable and fulfilling lifestyle. It is unwise and unreasonable to rely on technology to offer all the solutions to a wasteful, selfish, and overpopulated world community. Even so, it will take innovative scientific advances in clean energy systems, coupled with a demographic trend toward zero population growth, to help balance a quality human experience with a healthy biosphere.

If scientists and engineers are going to succeed in replacing our dirty and unsustainable fossil-fuel-based economy with clean, decentralized power systems, inventors, academics, and government researchers must learn to initiate and maintain an open and constructive dialogue. Healthy skepticism is important, as is an emphasis on keeping science an open system of inquiry. All too often, the work of pioneering frontier scientists represents isolated, individual efforts. By contrast, most quality science involves collaborative efforts. Researchers must produce well-designed experiments that are clearly communicated in the scientific literature and that show replication of phenomena. Nothing less will be tolerated by the skeptics of voodoo science, nor should it be.

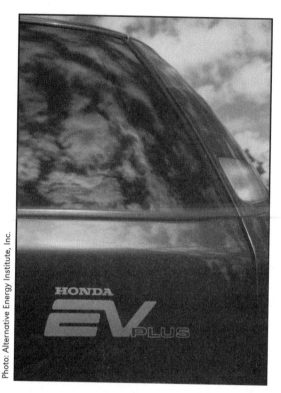

*The transportation revolution is here today.*

Photo: Alternative Energy Institute, Inc.

In her analysis of the challenge of anomalies at the frontiers of science, Beverly Rubik clearly explains the obstacles that new energy researchers face:

> "Major changes in science have never been brought about by isolated experimental findings, but by collective evidence. Thus, it is crucial for scientists who dare to venture into tributaries of the mainstream or into uncharted terrain to come together to enter into dialogue and share their data, to find that what may seem as isolated anomalies fit together to form the rudiments of an emerging paradigm. It is important to look at the problems of our science and the gaps in our knowledge. We must continually ask new questions and never be satisfied with the old ones, nor with the answers that have come to pass. Scientists must continually be motivated by the 'mother' of all questions: what facets of nature remain undiscovered because what we consider to be theoretical certainties prevent the posing of new challenging questions?"[65]

Although Americans born in 1950 saw per capita oil production quickly double in a few short decades, those born in 2000 are likely to see it cut in half, dropping below 1950 levels.[DD]

— *Beyond Malthus*

The world is at a critical crossroads. How we act in the next decade will mean the difference between passing on an impoverished planet, polluted and down to a barebones supply of energy-related natural resources, or a world on the mend, heading brightly into a future with endless possibilities. Who wants to explain to their grandchildren that our generation consumed oil so prodigiously that it polluted the biosphere, changed the weather, and negatively affected global climate patterns? Who will explain to them that gone forever is much of the planet's inexpensive oil supply, a vital commodity that has driven America's economy for more than 100 years, and provides medicine, plastics, and fertilizer to heal and feed the world's booming population? Hydrocarbon resources are a one-time gift from the planet and are now being consumed as if there were no

*Young people are relying on us to make the right decisions for their future.*

tomorrow, with little consideration for future generations. Now is the time to make the decisions that will facilitate the transition from the pollution-ridden fossil fuel age to a future of clean and virtually limitless energy technologies. If not us, who? If not now, when?

> The story of the energy revolution does not end here, but this book must. Please visit AEI's website at:
> <www.altenergy.org/book>
> for updates, what you can do to help, and the latest information concerning the topics covered in *Turning the Corner: Energy Solutions for the 21st Century*.

## AEI's Solutions Summary

- Increase government-mandated vehicle fuel efficiency requirements that reflect an understanding of dwindling global oil supplies, and which will help reduce air pollution and its related health costs of $100 billion a year.

- Implement appropriate zoning to restrict suburban sprawl to reduce the frequency and distances traveled by car. To revitalize our cities, reduce automobile pollution and energy-wasting traffic congestion, we need to realign a tax structure where federal tax deductions for home mortgage interest subsidize homeowners (suburbs) over renters (city). Taxing land more than buildings will reduce taxes for homeowners.

- More investment in mass-transit systems which can be partially supported by taxing carbon in fuels and motor vehicle emissions. The real alternative to cars is not better transit, it's compact neighborhoods.

- Governments should provide financial incentives to support innovative energy efficiency measures. Energy efficiency programs on the part of states and utilities should be expanded, and a national energy conservation program implemented. Support energy efficiency research and development, information and promotion programs.

- In order to give clean renewable technologies like wind, biomass gasification, and solar power a bigger piece of the energy supply pie, governments must end subsidies for fossil fuels and instead reward non-polluting technologies. The US currently provides more than $20 billion in federal subsidies to the oil, coal, and natural gas industries every year. Remove subsidies to industries that extract raw material from the Earth and damage the biosphere.

- Increase resource productivity to obtain the same amount of utility or work from a product or process while using less material and energy. Shift taxes away from people to the use of resources.

- Among the goals of the 1992 US Energy Policy Act: To consider the obstacles inherent in private industry's development of new energy technologies and the steps necessary for establishing or maintaining technological leadership. The BEPR Program has been initiated to fill a void in research and development, to encourage professional scientific dialogue, and to create a comprehensive database for emerging energy and propulsion technologies. This innovative research program deserves federal funding and interagency support.

- Global population growth and/or growth in the rates of consumption cannot be sustained. The wealthy industrialized countries along with the United Nations should initiate major comprehensive educational, technical, and outreach programs in the areas of social responsibility, contraception, and family planning.

- The US must rejoin the ongoing international negotiations that address global air pollution and greenhouse gas emissions. As the world's largest polluter, the US should become the world leader in reducing greenhouse gas emissions/biosphere pollution.

- Go to <www.altenergy.org> to see what you can do to save energy in your home, car, and community.

# FOOTNOTES

1. <www.unep.ch/ipcc/> Summary for Policy Makers — In February 2001 the Intergovernmental Panel on Climate Change finalized two comprehensive assessments: one on observed and projected changes in climate, the other on climate change impacts, vulnerability, and adaptation. That work set the stage for the third and final IPCC assessment, which reviews the many technologies and policies that are available for reducing or limiting greenhouse gas emissions in order to minimize future climate change.

2. Paul Hawken, Amory Lovins and L. Hunter Lovins, *Natural Capitalism: Creating the Next Industrial Revolution* (New York: Little, Brown and Co., 1999), p. 253. This book chronicles remarkable opportunities for saving both money and resources through the ingenious application of novel technologies and business practices. Amory and L. Hunter Lovins are the founders and co-CEO's of Rocky Mountain Institute, a nonprofit resource policy center, startup incubator, and advisor to dozens of leading firms. Their efficiency innovations have won many of the world's top awards. The Rocky Mountain Institute website is: <www.rmi.org>

3. *Ibid.*, p. 57.

4. Martha Caldwell Harris, "US Energy Woes: Think Globally," *The Sacramento Bee*, March 6, 2001, p. B-7. Harris is a senior fellow at the Atlantic Council of the United States. This opinion appeared in the *Los Angeles Times*.

5. David Westphal, "Abraham: Crisis Could Sap Economy, Lifestyle," *The Sacramento Bee*, March 20, 2001, p. A-12.

6. Daniel Yergin, *The Prize: The Epic Quest for Oil, Money* (New York: Simon & Schuster, 1991, 1992), p. 718. Daniel Yergin is an authority on world affairs and the oil business. Winner of the 1992 Eccles Prize and reproduced as a PBS series, *The Prize* recounts the panoramic history of oil and the struggle for wealth and power that has always surrounded this vital liquid carbon.

7. Greg Easterbrook, "Fuel economy not in W.'s 'balanced' energy policy," *The Sacramento Bee*, p. L-1. Gregg Easterbrook is a senior editor for the *New Republic* and BeliefNet.com and is a visiting fellow in economics at the Brookings Institution. He wrote this article for the *Los Angeles Times*.

8. Paul L. Leventhal, "Too Much Risk Bedevils Nuclear Power Plants," *The Sacramento Bee*, May 19, 2001, p. B-7. Paul L. Leventhal, President of the Nuclear Control Institute, was co-director of the Senate investigation of the Three Mile Island nuclear accident. Article reprinted from the *New York Times*. "The nuclear industry's safety and security claims are often misleading. Influential lobbyists for the US nuclear industry are forcing the Nuclear Regulatory Commission to begin the process of granting life extensions to American's

aging supply of 104 power reactors, despite a rash of forced shutdowns due to equipment failures caused by aging." There have been at least eight such shutdowns over the past 16 months, according to an analysis of NRC data by the Union of Concerned Scientists. Their website is <www.ucsusa.org>. Leventhal writes: "Worldwide, it would take 3,000 nuclear plants — a tenfold increase — to replace all coal plants; yet that increase would reduce carbon emissions by only 20%, while enormously expanding risks that materials from nuclear power plants would be applied to making weapons. And since reserves of uranium ore are limited, millions of kilograms of plutonium, equivalent to hundreds of thousands of bombs, would have to be separated from wastes each year to help fuel so many reactors in the future."

[9] Editorial by Howard Geller, "Here's the Drill: Learn to Conserve." *The Sacramento Bee*, March 8, 2001, p. B-7.

[10] Gilbert Chan, "Energy Crunch May Push Firms Out of California," *The Sacramento Bee*, March 6, 2001, p. D-2. Poll conducted by Mason-Dixon Polling & Research of Maryland.

[11] Gray Davis, "The Need for Electricity Price Caps," *The Sacramento Bee*, June 1, 2001, p. B-7. Gray Davis, Governor of California in 2001, notes that "California's problems stem from a fundamentally flawed 1996 state electricity deregulation law. The utilities were required to sell off half their fossil-fuel-fired power plants while they were barred from entering into contracts for a long-term supply of cheap electricity. The result is an unregulated sellers market for electricity, susceptible to naked manipulation."

[12] Westphal, *op. cit.*, p. A-12.

[13] <www.governor.ca.gov> "California Energy Use Down Eight Percent," March 3, 2001. California Governor Gray Davis cited demand reduction and energy efficiency efforts as key factors in the decline.

[14] Brain Fagan, *The Little Ice Age* (New York: Basic Books, 2000), pp. 213–214. Brian Fagan is a Professor of Archaeology at the University of California at Santa Barbara. This book views history through the lens of climate change by distilling a huge range of source material, from the dates of long-ago wine harvests and the business records of 14th century monasteries to the latest chemical analysis of ice cores.

[15] <www.altenergy.org/2/kyoto/kyoto.html> Alternative Energy Institute background profile on the 1997 Kyoto Protocol.

[16] <www.oecd.org> Organization of Economically Developed Countries

[17] Eugene B. Skolnikoff, "The Role of Science in Policy," *Environment* (June 1999), Vol. 41, No. 5, pp. 17–45. Eugene B. Skolnikoff is an emeritus professor of political science at the Massachusetts Institute of Technology (MIT).

18 *The Sacramento Bee*, June 17, 2001, p. A-5. Reprinted from the *New York Times*.

19 Hawkens, Lovin & Lovin, *op. cit.*, p. 242.

20 David G. Victor, "A Post-Kyoto World," *The Sacramento Bee*, March 30, 2001, p. B-7. Article reprinted from the *New York Times*.

21 Frank N. Laird, "Just Say No to Greenhouse Gas Emissions Targets,"*Issues in Science and Technology* (Winter 2000-01), pp. 45 – 52 . Laird is an associate professor of technology and public policy at the Graduate School of International Studies, University of Denver.

22 Concerns regarding climate change have resulted in substantial national and international institutions that are facilitating progress on this global problem. The World Meteorological Organization established the Intergovernmental Panel on Climate Change in 1988, a decade before the United Nations FCCC, which now functions as a secretariat for both the FCCC and the Kyoto Protocol. By preparing extensive databased reports on climate change, the IPCC helps to build scientific consensus on the technical aspects of this issue. Climate change in IPCC usage refers to any change in climate over time, whether due to natural variability or as a result of human activity. This usage differs from that in the Framework Convention on Climate Change, where climate change refers to a change of climate that is attributed directly or indirectly to human activity that alters the composition of the global atmosphere and that is in addition to natural climate variability observed over comparable time periods.

23 Janet Larsen, "US and Japan Sink Emissions Reductions Plans," *World Watch* (November/December 2000), pp. 12–13. Two of the world's top four greenhouse gas emitters, the United States and Japan, have indicated that they plan on using a loophole in the protocol that allows countries to count carbon sinks such as farmlands and forest against their carbon dioxide releases. The United States, which is by far the world's largest emitter of $CO_2$ from fossil fuels, wants credit for vegetation sinks as a way of minimizing the need to change how Americans use energy, but forests and soil retain carbon only temporarily as they are prone to unanticipated release by fire or insect infestation. Because the European Union has fewer forests and less agricultural land, they oppose the inclusion of carbon sinks, but other countries have jumped ship to join the Americans and Japanese, including Canada, Russia, Australia, and New Zealand.

24 Richard Wolfson and Jay M. Pasachoff, *Physics: Extended with Modern Physics*, (Glenview, Illinois & London, England: Scott, Foresman/ Little, Brown Higher Education, 1989), p. 1199. Burning coal generates nitrogen and sulfur oxides, carbon monoxide, particulate matter, hydrocarbons, arsenic, cadmium, chromium, lead, zinc, dioxins, polycyclic aromatic hydrocarbons, polychlorinates biphenyls (pcbs), and benzene.

25 *Weatherwise*, November/December 2000, p. 9. Author's Note: More than 600 coal-fired plants around the country don't meet modern air-quality standards because they were grandfathered under the original Clean Air Act in 1970. But instead of being shut down, as lawmakers expected, many of the old plants built more than 30 years ago continue to operate.

26 Hawkens, Lovin, and Lovin, *op. cit.*, p. 58.

27 <www.wri.org/health/note-ghg.html> "The Hidden Benefits of Climate Policy: Reducing Fossil Fuel Use Saves Lives Now." December 1997.

28 On March 13, 2001, President Bush announced his decision to reverse a previous pledge to legislate limits on carbon dioxide emissions from US power plants. Although Bush maintained his position of requiring all power plants to meet clean air standards by reducing emissions of sulfur dioxide, nitrogen oxide and mercury, he cited a Department of Energy report that found that new limits on $CO_2$ emissions would lead to an even more dramatic shift from coal to natural gas for electric power generation and significantly higher electricity prices compared to scenarios in which only sulfur dioxide and nitrogen oxides were reduced. The EPA does not currently regulate $CO_2$, a gas that many scientists say is a key contributor to global warming.

29 Andrew C. Revkin, "US Lags in Climate Sciences," *Santa Cruz Sentinel*, June 11, 2001, p. A-6. Article reprinted from the *New York Times*: "Many climate experts say that the problems are deep-rooted, and that a clearer picture of the local and global impact of coming climate shifts will emerge only if there is a substantial shuffling of the scientific bureaucracy and permanent support for basic monitoring of climate-influencing factors like the ebb and flow of greenhouse gases." Revkin also notes: "While Britain and Japan have poured tens of millions of dollars into computing centers focused on long-term climate research, budgets for similar efforts in the United States have been flat at best, and the work is done at dispersed research centers run by a variety of federal agencies."

30 Douglas Jehl and Keith Bradsher, "Bush Says 'energy crisis' caused emission flop-flop," *The Sacramento Bee*, March 15, 2001, p. A-8. Article reprinted from the *New York Times*. Bush's policy shift is a classic example of a short-term political decision in the US cutting support for a long-term, international and collaborative effort to reduce greenhouse gas emissions. It is a troublesome feature in US political circles where campaign fund raising and lucrative lobbying go hand-in-hand.

31 <www.wri.org/wri/climate/bush_response.html> Total national emissions from large developing countries like Korea, Mexico, South Africa, Brazil, Indonesia, and Argentina added together barely meet US utility emissions levels. The power sector contributes roughly one-third of US greenhouse gas emissions, twice more than all the emissions from India. US emissions in total are still more than double those from China.

32. Elizabeth Shogren, "Bush No Longer Backs Carbon Dioxide Curbs," March 14, 2001, *The Sacramento Bee*, p. A-6. Article reprinted from *Los Angeles Times*.

33. <www.energy.gov> President Bush's new National Energy Policy developed by the National Energy Policy Development Group. Refer also to the AEI website regarding the Institute's critical evaluation of the Bush administration's strategy.

34. <www.lanl.gov/projects/cctc/> Clean Coal Technology Compendium.

35. Seth Dunn, "King Coal's Weakening Grip on Power," *World Watch* (September/October 1999), p. 15.

36. *Rachel's Environmental & Health News* #714, December 21, 2000, p. 5. A report by Friends of the Earth (FoE) points out that taxpayers currently provide billions of dollars worth of unnecessary support to polluting industries each year. Article "Paying for Pollution." ISBN 0-913-890-82-0 <www.foe.org/eco/>

37. Ronald Brownstein, "Here Comes the Sun, Feds Willing," *The Sacramento Bee*, p. L-1. Brownstein is a national political correspondent for the *Los Angeles Times*, from which this article is reprinted.

38. <www.integrity-research.org> You can purchase a copy of *Energy Crisis: The Failure of the Comprehensive National Energy Strategy* by Thomas Valone. This 176-page report analyzes the DOE 1998 CNES.

39. Thomas S. Kuhn, *The Structure of Scientific Revolutions*, Second Edition (Chicago, University of Chicago Press, 1962), Vol. II, No. 2, p. 37. Kuhn argues that once a paradigm has been accepted in a given area, its practitioners are unlikely to permit themselves to be without some such common reference point again. Kuhn's concept of paradigm has been heavily criticized, but for many the concept of paradigm remains vitally important. The *Dictionary of the History of Science* states that the Kuhnian paradigm is "a significant contribution to that view of human action stressing its social nature and dependence on tradition and precedent, in contrast to the view stressing individuality and abstract principles of rationality."

40. Thomas E. Bearden, *"The Unnecessary Energy Crisis: How to Solve it Quickly,"* Position Paper by Bearden, June 24, 2000, p. 23. "Such laboratories are private and professional testing companies, where the US government has certified their expertise and qualifications, their testing to US government standards, their use of calibrated instruments, and the experience and ability of their professional test engineers and scientists. Aerospace firms routinely and widely use such labs. A Test Certificate from such a lab is acceptable by the courts, the US Patent and Trademark Office, the US government — which requires it on many contracts — and by the US scientific community."

[41] Charles Darwin, "*On the Origin of Species*," authorized edition from 6th English ed.; New York, 1889, pp. 295-296.

[42] Beverly Rubik, "The Perennial Challenge of Anomalies at the Frontiers of Science," *British Homoeopathic Journal*, Volume 83, July 1994. Paper for Proceedings of the 1993 GIRI Symposium, Montpellier, France. <www.stockton-press.co.uk/bbj> According to science sociologist Marcello Truzzi, an anomaly is something that actually occurs (that is, something both perceived and validated); is not explained by some accepted scientific theory; is perceived to be something which is in need of explanation; contradicts what we might expect from applying our accepted scientific models. Studies on the psychology of science suggest that scientists have a resistance to acknowledging data that contradict their own hypothesis.

[43] Department of Energy "Draft Comprehensive National Energy Strategy," January 30, 1998, p. 10.

[44] Phil Carpenter, David Hamilton, Dave Goodwin, et. al., Breakthrough Energy Physics Research (BEPR) Program Plan Office of Energy Efficiency & Renewable Energy, Office of Science, and the Office of Nuclear Energy, Science and Technology — Joint Exploratory research (October 2000 — *Draft for Agency Comment Only*), p. 6. Executive Summary: The OAEC's program direction is provided through congressional legislation dating from the 1970s, and most recently the Energy Policy Act of 1992 (Public Law 102-486), which authorizes enhanced research programs and new demonstration programs. To meet the congressionally legislated goals expeditiously, the Breakthrough Energy Physics Research (BEPR) Program Plan describes the research that OAEC plans to undertake to progress the scientific base in support of potential breakthroughs that could significantly expand the nation's preeminence in science and technology. Achieving the goals expressed in the BEPR Program is furthered through the use of laws and regulations relating to intellectual property. Patent and copyright protection of intellectual property associated with the new science and/or technologies that may be developed under the BEPR Program Plan encourages exploitation of such knowledge by enhancing the competitive position of the partners involved in the researches. The degree of intellectual property ownership provided to the research partners arising under the plan will be determined on a case-by-case basis and is commensurate with cost-sharing amounts.

[45] *Ibid.*, BEPR — *Draft for Agency Comment Only*.

[46] Edited by Robert C. Williams and Philip L. Cantelon, *The American Atom: A Documentary History of Nuclear Policies from the Discovery of Fission to the Present 1939-1984* (Philadelphia, PA: University of Pennsylvania Press, 1984), pp. 24–25.

47 David Pimentel and Mario Giampietro, "Implications of the Limited Potential of Technology to Increase the Carrying Capacity of Our Planet," *Human Ecology Review* (Summer/Autumn, 1994), Vol. 1, pp. 248-251.

48 Albert A. Bartlett, "Reflections on Sustainability, Population Growth, and the Environment," *Population and Environment: A Journal of Interdisciplinary Studies*, Volume 16, No. 1, September 1994 copyright *Human Sciences Press, Inc.*, p. 7. Bartlett notes: "…political leaders use the term 'sustainable' to describe their goals as they worked hard to create more jobs, to increase population, and to increase rates of consumption of energy and resources."

49 F. Trainer, "Can Renewable Energy Sources Sustain Affluent Society?" *Energy Policy*, 1995 Vol. 23, no. 12, pp. 1009-1026.

50 <www.altenergy.org/2/population/population.html> Reference to AEI's data on overpopulation.

51 Walter Youngquist, "The Post-Petroleum Paradigm — and Population," Reprinted from *Population and Environment: A Journal of Interdisciplinary Studies*, Vol. 20, No. 4, 1999, pp. 302–303. Youngquist correctly observes: "It is not when the last drop of oil is pumped, but rather the peak of production (maximum daily amount) after which there is an irreversible decline in oil production, which is important. Then all social and economic programs based on oil income will have to be curtailed."

52 Albert Bartlett, "Forgotten Fundamentals of the Energy Crisis," *American Journal of Physics*, 1978, Vol. 46, No. 9, p. 880.

53 Brian Fleay, *The Decline of the Age of Oil*, Pluto Press Australia Limited, 1995. Fleahy, an engineer and Associate of the Institute of Sustainability and Technology Policy at Murdoch University in Western Australia, spent his professional life with the Water Authority of Western Australia, mainly in the operation and maintenance of Perth's surface and ground water sources. Fleay's most recent article, "USA's Triple Energy Whammy in Electric Power, Natural Gas & Oil" dated January 10, 2001, can be read at: <http://groups.yahoo.com/group/dieoff/message/5> Overview: "The USA is in a major electric power crisis. Peak power demand is exceeding reliable generation and transmission capacity, especially in California and on the west coast. The industry is trying to overcome the crisis by installing gas turbines just when the supply of natural gas in North America has reached a peak, creating in turn a natural gas supply crisis. The only short term solution is to reduce consumption of both electricity and natural gas, bringing energy-efficiency back on the agenda."

54 Youngquist, *op. cit.*, p. 306.

[55] Barbara Crossette, "U.N. Study Predicts US Population will Jump," *The Sacramento Bee*, February 28, 2001, p. A-1. Reprinted from the *New York Times*.

[56] Richard Duncan, "The Peak of World Oil Production and the Road to the Olduvai Gorge," *Pardee Keynote Symposia*, Geological Society of America, Summit 2000, Reno, Nevada, November 13, 2000, p. 1.

[57] Kenneth Boulding, Foreword to Thomas R. Malthus, *Population, The First Essay*, Boulder, Colorado. Colorado Associated University Press, 1971. Collected papers, Vol. II, pp. 137–142.

[58] Crossette, *op. cit.*, p. A-1.

[59] <www.zpg.org> Zero Population Growth Fact Sheet. "Women's Empowerment in all spheres of life is key to reducing population growth and improving the lives of all people. Future population growth depends on the ability of today's young women and men to make informed, responsible decisions about their health, their families, and their lives. Today there are more young people than ever before in history. Almost half of the six billion people on Earth are under age 25. More than one billion young women and men aged 15-24 are entering their childbearing years."

[60] <www.prb.org/press/transform.html> Population Reference Bureau: "Of the 83 million people added to global population each year by the difference between births and deaths, only one million are in the industrialized countries. In 1950 there were twice as many people in the less developed countries. By 2050 that difference could be almost six to one. The developing world's population is projected to increase by 2.9 billion by 2050, compared with only 49 million in the more developed countries."

[61] *World Resources 2000-2001: People and Ecosystems: The Fraying Web of Life*," produced by the Millennium Ecosystem Assessment, an international effort to measure the planet's health.

[62] Pimentel and Giampietro, *op. cit.*, p. 250.

[63] <www.dieoff.org/page63.htm> Paul J. Werbos, "Energy and Population: Transitional Issues and Eventual Limits." Werbos is a Program Director at the National Science Foundation. He was with the Energy Information Administration (EIA) of the US Department of Energy (DOE) for ten years. He served on the interagency Task Force on Models and Data convened to respond to the problems identified in the Global 2000 Report to the President.

[64] Robert W. Kates, "Labnotes from the Jeremiah Experiment: Hope for a Sustainable Transition," *Annals of the Association of American Geographers*, 1995, p. 637.

[65] Rubik, *op. cit.*, p. 40.

## SIDEBAR FOOTNOTES

A "Studies Bolster Global Warming," *Washington Post* article reprinted in *The Sacramento Bee*, April 13, 2001, p. A-9.

B James J. MacKenzie, February 15, 2001, Press Release: "WRI study reveals oil from Arctic National Wildlife Refuge Coastal Plain will not alleviate increasing US dependence on foreign sources."

C Joseph Kahn, "Cheney: No Time for Conservation," *New York Times* article reprinted in *The Sacramento Bee*, May 1, 2001, pp. A-1, A-14.

D Gregg Easterbrook, "Fuel economy not in W.'s 'balanced' energy policy." Article reprinted in *The Sacramento Bee*, April 1, 2001, pp. L-1, L-2.

E *This Place on Earth 2001: Guide to a Sustainable Northwest*, Northwest Environmental Watch (Seattle, 2001), p. 12.

F Richard Wolfson and Jay M. Pasachoff, *Physics: Extended with Modern Physics* (Glenview, Illinois: Scott, Foresman, Little Brown Higher Education, 1989), p. 1199.

G Burton Richter, "Reducing demand is way to go," *Los Angeles Times* article reprinted in *The Sacramento Bee*, May 27, 2001, p. L-1.

H Paul Hawken, Amory Lovins, and L. Hunter Lovins, *Natural Capitalism* (Boston, New York, London: Little, Brown and Company, 1999), p. 155.

I Michael D. Lemonick, "Life In The Greenhouse," *Time* (April 9, 2001), p. 23.

J James Wells, "The Science is Solid: Time to Act," *Los Angeles Times* article reprinted in *The Sacramento Bee*, May 13, 2001, pp. L-1, L-6.

K Roger S. Ballentine, "Politics Blinding Us to Kyoto's Significance," *Washington Post* article reprinted in *The Sacramento Bee*, July 9, 2001, p. B–5.

L Sara Callaghan Chapell, "Don't Drill in Alaska's Wildlife Refuge," *The Sacramento Bee*, March 28, 2001, p. B-7.

M Ken Silverstein, *Harper's* (July 2001), p. 69.

N Heather L. Maclean and Lester B. Lave, "Environmental Implications of Alternative-Fueled Automobiles: Air Quality and Greenhouse Gas Tradeoffs," *Environmental Science & Technology* (No. 2, 2000), Vol. 34, p. 225.

O Joseph Kahn, *New York Times* article reprinted in *The Sacramento Bee*, "Report Disputes Bush's Policy," May 6, 2001, p. A-6.

P Jeanne Cummings, "Cheney Pitches Energy Policy to Labor Leaders as Jobs Plan," *The Wall Street Journal*, May 15, 2001, p. A-28.

<sup>Q</sup> Stanley R. Bull and Lynn L. Billman, "Renewable Energy: Ready to Meet Its Promise?" *The Washington Quarterly* (Winter 2000) Vol. 23, p. 229.

<sup>R</sup> Hans Christian von Baeyer, *The Fermi Solution: Essays on Science* (Mineola, New York: Dover Publications, Inc., 1993), p. 20.

<sup>S</sup> Michael Shnayerson, *The Car That Could: The Inside Story of GM's Revolutionary Electric Vehicle* (New York: Random House, 1996), p. 110

<sup>T</sup> Ruggero Maria Santilli, "AquaFuel, An Example of the Emerging New Energies and the New Methods for their Scientific Study," *Infinite Energy* (Issue 19, 1998), p. 75.

<sup>U</sup> Peter Coy, "The US Needs an Oil Policy — One Flexible Enough for the Future," *Business Week* (October 9, 2000), p. 44.

<sup>V</sup> Steve Silberman, "The Energy Web," *Wired* (July 2001), p. 116.

<sup>W</sup> *Washington Post* article reprinted in *The Sacramento Bee*, May 2, 2001.

<sup>X</sup> Paul L. Leventhal, "Too Much Risk Bedevils Nuclear Power Plants," *The Sacramento Bee*, May 19, 2001, p. B-7.

<sup>Y</sup> Robert Lee Hotz, "Huge Livermore Laser Project Dimmed by Overruns, Doubts," *Los Angeles Times* article reprinted in *The Sacramento Bee*, November 24, 2000, p. B — 5.

<sup>Z</sup> Ellen Goodman, "Worshipping Oil Barrels: America's false god," *Bucks County Courier Times*, March 11, 2001, p. 19-A.

<sup>AA</sup> Walter Youngquist, *Alternative Energy Sources: Water and Energy, the Basis of Human Society: Are They Globally Sustainable Through the 21$^{st}$ Century?* Kansas Geological Survey Open-File Report 2000-51.

<sup>BB</sup> George D. Moffett, *Critical Masses: The Global Population Challenge* (New York: Penguin Books, 1994), p. 292.

<sup>CC</sup> Editorial, "Walk the Walk," *The Philadelphia Inquirer*, May 13, 2001, p. E-4.

<sup>DD</sup> Lester R. Brown, Gary Gardner, and Brian Halweil, *Beyond Malthus: Nineteen Dimensions of the Population Challenge* (New York: W. W. Norton & Company, 1999), p. 48.

# Bibliography

## Fossil Fuels, Nuclear Power, & Global Warming

BOOKS

Colin J. Campbell, *The Coming Oil Crisis*. Essex, England: Multi-Science Publishing Company and Petroconsultants S.A., 1997. This 210-page paperback discusses how much conventional oil remains to be produced and its depletion pattern. It explains how to properly interpret published petroleum reserve numbers, many of which Campbell considers spurious or distorted by vested interests.

Walter Youngquist, *GeoDestinies: The Inevitable Control of Earth Resources over Nations and Individuals*. Portland, Oregon: National Book Company, 1997. Excellent overview of the historical relationship between civilizations and minerals. Also covers strategic minerals, free trade versus geologic provinces, earth resources, the future and the potential for achieving a sustainable society.

Nebojsa Nakicenovic, Arnulf Grubler, and Alan McDonald, *Global Energy Perspectives*. Cambridge: Cambridge University Press, 1998. The International Institute for Applied Systems Analysis and the World Energy Council present six alternative long-term futures covering a wide range of scenarios involving coal, oil, natural gas, nuclear, renewable energies, etc.

Robert C. Williams and Philip L. Cantelon, *The American Atom: A Documentary History of Nuclear Policies from the Discovery of Fission to the Present 1939-1984*. Philadelphia: University of Pennsylvania Press, 1990. An excellent collection of primary documents relating to the history of nuclear power and weapons programs. Compelling insight into the Manhattan Project, the hydrogen bomb, proliferation, etc.

Robert L. Peters and Thomas E. Lovejoy, *Global Warming and Biological Diversity*. New Haven, Connecticut: Yale University Press, 1992. Discusses in readable detail and with scientific credibility the consequences of global warming for ecosystems and considers the synergies between climate change and human activities.

Ross Gelbspan, *The Heat Is On*, Cambridge, Massachusetts: Perseus Publishing, 1998. Well-documented resource on the subject of rapid climate change and global warming.

PERIODICALS

*Hubbert Center Newsletter*, published by M. King Hubbert Center for Petroleum Supply Studies. The Hubbert Center has been established as a nonprofit organization for the purpose of assembling and studying data concerning global petroleum supplies and disseminating such information to the public. The M. King Hubbert Center is located in the Petroleum Engineering Department at the Colorado School of Mines, Golden, Colorado, 80401-1887.

*Issues in Science and Technology*, a quarterly journal published by the National Academy of Sciences, National Academy of Engineering, and the University of Texas at Dallas. The magazine is published to inform public opinion and to raise the quality of private and public decisionmaking by providing a forum for discussion and debate. It is open to all responsible points of view, and the material reflects only the views of the authors, not the policy of any institution.

WEBSITES

<www.wri.org> — *World Resources Institute* provides information, ideas, and solutions to global environmental problems.

<www.dieoff.org> — This provocative website hosted by Jay Hanson offers an extensive selection of professional position papers and articles by a variety of experts that address the major concerns relating to nonrenewable fuels, renewable resources, pollution, population, and energy. Comprehensive and compelling argument indicating that the end of cheap oil signals the end of civilization as we know it.

<www.eia.doe.gov> — *Energy Information Administration*. Official energy statistics from the United States government.

## Renewable Energy & Energy Efficiency

BOOKS

Paul Hawken, Amory Lovins, and L. Hunter Lovins, *Natural Capitalism*. New York, London: Basic Books, 1999. The authors advocate resource productivity by redesigning industry on biological models that result in zero waste. Amory and L. Hunter Lovins are founders and co-CEOs of Rocky Mountain Institute <www.rmi.org>, a nonprofit resource policy center.

Douglas R. Pratt, *The Real Goods Solar Living Sourcebook: The Complete Guide to Renewable Energy Technologies and Sustainable Living*, Whiter River Junction, Vermont: Chelsea Green Publishing Company, 1999. Extensive catalog of renewable and sustainable products, along with information, introductions to concepts, and design guidelines.

Paul Gipe, *Wind Power for Home and Business: Renewable Energy for the 1990s and Beyond*, Whiter River Junction, Vermont: Chelsea Green Publishing Company, 1993. Considered the bible for small to midsize wind power applications, Gipe provides detailed and comprehensive engineering and educational information on all aspects of wind energy.

Michael Shnayerson, *The Car That Could: The Inside Story of GM's Revolutionary Electric Vehicle*, New York and Toronto: Random House, Inc., 1996. A highly readable insider's account of corporate giant General Motors' top-secret electric car project. Shnayerson's explanations of the technical terms are clear and concise, and his understanding of the GM politics make for a good read.

Matthew Stein, *When Technology Fails: A Manual for Self Reliance & Planetary Survival*, Santa Fe, New Mexico: Clear Light Publishers, 2000. *When Technology Fails* is the first book to offer basic instructions and recommended resources for the wide range of skills and technologies required for self-reliant living on or off the grid.

## PERIODICALS

*World Watch: Working for a Sustainable Future*. A bimonthly, not-for-profit magazine that tracks key indicators of the Earth's well being. Articles are written by the staff of the Worldwatch Institute <www.worldwatch.org> Worldwatch monitors and evaluates changes in climate, forest cover, population, food production, water resources, biological diversity, and other key trends.

*Solar Today*. Award-winning magazine published by the American Solar Energy Society (ASES). Covers all solar technologies, from photovoltaics to climate-responsive buildings to wind power.

## WEBSITES

<www.internationalfuelcells.com> — *International Fuel Cells* (IFC) is the world leader in fuel cell production and development for commercial, transportation, residential, and space applications. IFC is the only company in the world producing commercial stationary fuel cell systems. Your questions regarding fuel-cell technology will be answered here.

<www.worldwatch.org> — *Worldwatch Institute* is a nonprofit public policy research organization dedicated to informing policy makers and the public about emerging global problems and trends and the complex links between the world economy and its environmental support systems.

<www.ucsusa.org> — *Union of Concerned Scientists* is an independent nonprofit alliance of 50,000 concerned citizens and scientists across the United States. They publish a quarterly journal and unite scientists, engineers, and citizens as a positive political force for change.

## Energy per Capita & Population Challenge

BOOKS

Lester R. Brown, Gary Gardner, and Brian Halweil, *Beyond Malthus: Nineteen Dimensions of the Population Challenge*, New York: W.W. Norton & Company, 1999. The burden of enormous populations is making itself felt as governments struggle with the need to educate children, create jobs, and deal with the environmental effects of population growth. The authors call for immediate expansion of international family planning assistance and new investment in educating young people — especially women — in the Third World, helping to promote a shift to smaller families.

George D. Moffett, *Critical Masses: The Global Population Challenge*, New York: Penguin Books, 1994. Every day the human family grows by a quarter of a million people, many of them born into the very countries that can least afford to provide them with jobs, housing, and nutrition. *Critical Masses* serves to give warning but also provides a positive prescription for change.

Donella H. Meadows, Dennis L. Meadows, and Jorgen Randers, *Beyond the Limits: Confronting Global Collapse, Envisioning a Sustainable Future*, Whiter River Junction, Vermont: Chelsea Green Publishing, 1993. *Beyond the Limits* shows us where we are headed and explores alternatives to change the course from collapse to sustainability.

WEBSITES

<www.hubbertpeak.com/duncan/index.html> — A collection of professional research papers and letters by Richard Duncan, Director of the Institute on Energy and Man, that address the trends of world energy production, population growth, and their impact on human society.

<www.zpg.org> — Zero Population Growth (ZPG) is a national nonprofit organization working to slow population growth and achieve a sustainable balance between the Earth's people and its resources. ZPG seeks to protect the environment and ensure a high quality of life for present and future generations.

## New Physics & the Emerging Energy Paradigm

BOOKS

Ervin Laszlo, *The Whispering Pond: A Personal Guide to the Emerging Vision of Science.* Boston, Massachusetts: Element, 1996. Laszlo clarifies the shifting paradigm in science and offers a rich overview of the evolution of consciousness and the challenges facing science. The reader is introduced to the search for a unified theory and the emerging insight of new physics.

Hans Christian von Baeyer, *The Fermi Solution: Essays on Science.* Mineola, New York: Dover Publications, Inc., 1993. Von Baeyer takes readers on a witty and intelligent journey through the workings of the Universe and into the minds of the creators of today's scientific worldview. The author offers a unique perspective on a wide range of subjects as he pursues his central concern: What do the breakthroughs and revolutionary insights at the cutting edge of science mean for ordinary people?

Eugene F. Mallove, *Fire from Ice: Searching for the Truth Behind the Cold Fusion Furor.* USA: Infinite Energy Press, 1999. Mallove's book is the most accessible overview of the cold fusion debacle. Recommended reading for anyone who is interested in the nature of scientific controversy and scientific change. The book may be purchased at the *Infinite Energy* website at <www.infinite-energy.com>.

Todahiko Mizuno, *Nuclear Transmutation: The Reality of Cold Fusion*, Concord, New Hampshire: Infinite Energy Press, 1998. A frank and open account about Mizuno's cold fusion experiments in Japan, where pioneering work on loaded, solid-state proton conductors was conducted as well as key new studies of transmutation products. Mizuno is an Assistant Professor of Nuclear Engineering, College of Engineering, Hokkaido University.

PERIODICAL

*Journal of New Energy: An International Journal of New Energy Systems.* A quarterly journal published by the Fusion Information Center, Salt Lake City, Utah. The *JNE* is devoted to publishing professional papers with experimental results that may not conform to the currently accepted scientific models. Papers with experimental data are preferred over theoretical papers.

WEBSITES

<www.altenergy.org> — *Alternative Energy Institute, Inc.*, (AEI) website is a great educational resource for students, policymakers, and the interested public to learn about what's new in all aspects of the energy field from fossil fuels and nuclear power, to renewable resources like wind and solar energy, and onward to breakthrough energy and propulsion technologies for the 21st century. The site offers unique inventor and conference profiles and an overview of where energy systems are heading, and it's loaded with live Internet links for quick and efficient research. Sign up for the free bi-weekly electronic newsletter.

<www.padrak.com/ine/> — *The Institute for New Energy* (INE) is a nonprofit technical organization whose primary purpose is to promote research and educate society on the importance of alternative energy. The INE believes that one key to solving the environmental holocaust is with New Energy Technologies and that this technology should be researched, developed, and released to the public and industry to save the planet. INE offers a detailed and comprehensive review of New Energy topics, such as "over unity" and "free energy" machines, magnetic motors, cold fusion, zero-point energy, homopolar generators, and many other intriguing concepts.

<www.integrity-research.org> — The *Integrity Research Institute* (IRI) is a nonprofit organization dedicated to research and development of emerging energy sources that are carbonless future energy technologies, not contributing to global warming, yet boosting economic growth worldwide. The IRI website offers for sale an extensive selection of books, conference proceedings, journals, professional papers, and videos on advanced energy/propulsion theories and technologies.

<www.infinite-energy.com> — *Infinite Energy* is the foremost international magazine for cold fusion and new energy technologies. It is a technical magazine with outreach to the general public, providing material of interest to all people. To maintain the highest editorial standards, *Infinite Energy* is written and edited by scientists, engineers, and expert journalists. The website offers for sale the latest books and videos specializing in the cold fusion field.

# Index

## A

Abraham, Spencer
    US energy secretary, 327
Acid rain
    Norway, 27
    US and Canada, 26
Acidic precipitation
    utility plants, 26
Adirondack Council
    environmental group, 28
Advanced Electromagnetic Theory, 344
Air pollution
    annual health costs in US, 335
    California motor vehicle exhaust, 126
    EPA cost estimates, 21
    global impacts on health, 335
    health costs from smog and particulates, 334
    long-term reduction, 326
    respiratory infections, 20
    US annual deaths from burning coal, 334
    US emissions, 337
    volatile compounds in vehicle exhaust, 317
Alpha Foundation's Institute for Advanced Study
    Evans, Myron and Sachs, Mendel, 221
Alternative Energy Institute, Inc.
    CNES analysis, 286
    solutions summary, 355, 356

Alvarez, Louis
    muon-catalyzed cold fusion, 187
American Council for an Energy Efficient Economy
    "green guide", 25
American Wind Energy Association
    wind power advocacy, 116
Anomalous Gravitational/Inertial Mass Effects, 6
Arctic National Wildlife Refuge
    fuel efficiency versus drilling, 329
    oil drilling, 327
Aristotle
    erroneous concept, 241
Arrhenius, Svente
    Swedish chemist, 30
Atomic Energy Commission
    basic unit, 82
    established, 82

## B

Babbitt, Bruce
    US dam construction, 138
Ballard Power Systems
    fuel cells, 66
Ballinger, Ronald
    outright rejection, 191

Bearden, Thomas
　electrical loads powered by vacuum energy, 219
　Lorentz electrodynamic theory, 220
Bibliography, 367
Bicycles
　replacing autos, 22
Big Bang
　cosmic microwave background radiation, 216
　theory of specific heat, 239
Biomass, 155
　sources, 156
　waste to energy facilities, 156
Biomass energy
　environmental costs, 159
　US production, 156
Biomass fuel feedstocks
　reduced emissions, 156
Biomass gasification
　economic and risk evaluation, 157
Biomass Resource Assessment Task
　biomass assessment program, 160
Biomass-to-gas
　David Wallman's technology, 282
Biorefinery
　biomass to ethanol, 158
Black Holes, 246
　gravitational force, 247
Blacklight Power, Inc.
　reducing hydrogen below ground-state, 280
Brahe, Tycho
　Danish astronomer, 242
Breakthrough Energy Effects Database, 345
Breakthrough Energy Physics Research Program, 343, 344
Breakthrough Propulsion Physics Program, 342
　advancing space propulsion, 342
Brigham Young University
　confirming evidence energetic tritons, 192
　low-energy fusion experiments, 188
British Petroleum
　investment in solar technologies, 111
Brown, Paul
　betavoltaic battery, 276
　photo-deactivation technology, 277
　radioactive waste remediation, 196

Brown, Thomas Townsend
　"antigravity" device, 261
　electrogravitic force, 254
Bush, George W.
　2002 National Energy Plan, 292
　2002 US federal budget slashes renewable funding, 296
　Bush says Kyoto will hurt US economy, 332
　energy policy, 89
　policy shifts, 337
　promoting Arctic drilling, 327
　reverses pledge to reduce emissions, 336
　supply-side economics, 328
　US pulls out of Kyoto Protocol, 299

# C

CAFE
　Corporate Average Fuel Economy, 23
　standards, 23
California
　air pollution, 22
　crop losses, 22
California Air Resources Board
　emission-free vehicles mandate rollback, 318
　more stringent than Federal standards, 317
　zero-emission vehicle mandate, 126
Caltech
　excess heat data errors, 190
Campbell, Colin
　geologist, 33
Carbon dioxide
　global warming, 16
　plant growth, 32
　sequestration research, 51
Carbon emissions
　global supply, 10
Carbon Era
　US consumption disproportionate, 308
Carbon Man
　era, 6
Carbon sinks, 52
Casimir force
　force of attraction, 210

Casimir, Hendrick
    vacuum fluctuations, 213
Center for Renewable Energy and Sustainable Technology
    July 2000 report on 50 years of subsidies, 313
Centralized power generation
    waste and inefficiency, 293
Chandra X-ray Observatory, 249
Chemically-assisted nuclear reactions
    various techniques, 199
Chernobyl
    deaths, 80
    fatalities, 13
    reactor meltdown, 10
Chevron
    methane hydrate drilling, 69
China
    environmental costs, 50
Clarke, Arthur C., 193
Clean Air Act, 20
    1990 Amendments, 12
    non-attainment, 21
    regulations, 27
Clean Air Trust
    vehicle emissions, 24
Clean Coal Technologies
    $2 billion program, 337
    R&D timeline, 294
Close, Frank
    cold fusion critic, 193
Co-generation power generation, 293
Coal
    ancient biomass, 47
    atmospheric disruption, 49
    carbon sequestration, 52
    Clean Coal Technologies, 49, 337
    consumption and gross imports, 55
    exports and emissions, 56
    global supply, 10
    hydrocarbon, 9
    Industry subsidies, 53
    reserves, 48
    reserves and production, 54
    smogs, 48
    social cost, 48
    statistical data resources, 57

Cold Fusion
    chemically assisted nuclear reactions, 183
Cold fusion
    by electrolysis, 189
    history, 185
    laboratory replication errors, 191
    leading scientists Will and O'M Bockris, 200
    Los Alamos National Laboratory, 192
    multiwatt excess output, 192
    problem with cold fusion process, 283
    research papers, 192
    tritium production, 192
Cold fusion experiments
    anomalous heat, 192
    US Naval Air Warfare Center, 192
    Yale and MIT, 190
Comprehensive National Energy Act
    tax credits for biomass, 156
Conference on Future Energy (COFE)
    photo-remediation, 196
Conservation of energy, 178
Copernican theory, 242
Corporate Average Fuel Economy, 23
Correa, Paulo
    new energy research, 226
Crude oil
    Carbon dioxide emissions, 39
    Consumption/gross imports, 37
    Exports, 38
    fact sheet, 36
Cygnus X-1
    binary star system, 248

# D

DaimlerChrysler
    New Electric Car, 128
Dalton, John
    British chemist, 185
Dammann, Wil
    biomass gasification process, 157
Dark matter, 250
Darwin, Charles
    new scientific paradigm, 341
Darwin, Sir Charles G.
    "the fifth revolution", 175

Department of Energy
  1,300 new power plants, 295
  Breakthrough Energy Effects Database, 345
  Comprehensive National Energy Strategy (CNES), 285, 338
  energy strategy, 342
  transmutation using particle accelerator, 197
  US solar testing, 107
  world energy consumption projections, 347
Dirac, Paul
  energy states of hydrogen, 213
Duncan, Richard
  OPEC production, 33

# E

Einstein, Albert
  cosmological constant, 216
  equivalence principle, 245, 261
  gravitational fields, 246
  mass and energy equation, 212
  scientific theories, 244
  Special Relativity theory, 185
  theory of relativity, 211
  zero-point-energy, 212
Einstein-Stern
  theory of specific heat, 212
Electric car technologies
  current status, 128
Electric Power Research Institute
  cold fusion funding, 194
Electric Validum
  charge-cluster technology, 283
Electricity
  California blackouts in 2001, 310
  economic costs of blackouts, 330
  key to new energy economy, 297
  not a primary energy source, 319
  on-site generation systems, 310
  regional power disruptions, 330
  reliability in the information age, 309
  time-of-use metering, 315
  US demand projection, 314

Electricity generation
  zero-pollution, low-cost technology, 275, 276
Electrogravitics, 254
Electromagnetic Theory, 6
Electromagnetism
  history, 252
Electron clusters
  Ken Shoulder's HDCC research, 282
Electrostatic motor, 281
Energy
  global market value, 327
  US energy waste and inefficiency, 315
Energy conservation
  effective partial solution, 330
  Energy Star appliances and electronic devices, 316
  leaking electronic devices consume 10%, 316
  low-emission vehicles, 317
  most realistic short-term solution, 315
  new energy efficiency measures, 326
  new fuel-efficient, low-emission vehicles, 316
Energy consumption
  United States Geological Survey, 103
Energy crisis
  energy/carbon conflict, 291
Energy efficiency
  1970s gasoline shortage in US, 327
  gains come without pain, 326
  per capita, 330
  in relation to other countries, 327
  US improvement between 1973 and 1985, 329
Energy inefficiency
  suburban sprawl, 328
  US vehicle fuel economy standards, 328
Energy Information Administration
  energy facts, 12
Energy Policy Act, 103
Energy Revolution
  Industrial Age, 9
Environmental Defense Fund, 24
Environmental Protection Agency
  Acid Rain Program, 27
  tougher emission standards, 24

Ethanol
    federal subsidies, 158
    uneconomical, 158
Ethanol production
    inefficient use of resources, 158
European Union
    investing in renewable energy, 312
EVs
    "electric validum", 223

## F
Federal Energy Regulatory Committee
    dam relicensing policy, 142
Feynman, Richard, 179
    ZPE calculation, 210
Fission
    energy source, 82
Flavin, Chris
    technical challenges for biomass, 160
Fleischmann, Martin
    electrochemist, 177
    press conference, 183
Ford Motor Company
    hydrogen fueled engines, 130
Fossil fuel industry
    obstacles to change, 336
    subsidies, 301
Fossil fuels
    chemicals, 9
    cost taxpayers billions of dollars, 312
    environmental and health costs ignored, 310
    environmental impacts, 16
    finite reserves, 15
Frank, F. C.
    muon-catalyzed cold fusion, 186
Franklin, Benjamin
    electrostatic motor, 281
Fuel cell technology
    efficiency, 126
    methanol, 65
    process, 125

Fuel cells
    costs per watt, 126, 130
    difficulties with hydrogen fuel, 319
    electricity production, 125
    NASA, 128
    obstacles, 318
    technology, 129, 319

## G
Galileo
    first experimental evidence, 241
Gasoline
    smog-causing emissions, 24
Gelbspan, Ross
    global warming, 31
Geothermal energy, 147
    closed-loop system, 150
    history, 148
    limitations, 151
    potential, 149
    US geothermal generation, 151
Glacier National Park
    glacial melting, 29
Global energy crisis
    midpoint of depletion, 175
Global warming, 28
    Antarctic, 30
    climate data, 29
    politics, 336
    veracity of climate change science, 336
Goodstein, David
    cold fusion defense, 191
Graneau, Peter
    liberating energy from ordinary water, 277
Gravitation
    expansion of Universe, 240
    fundamental force, 240
    history, 240
    mysterious force, 240
Gravitational Constant, 240
Gravito-electrodynamics Effects, 344
Gravity
    Graviton Model, 251
    history, 241
    new theory, 252

Gravity modification
    Breakthrough Propulsion Physics Program, 258
    NASA research, 255, 259
Gravity shielding
    Podkletnov, Eugene, 256
    superconductive plates, 255
Greenhouse gas emissions
    global warming, 331
    projected rise in US, 338
    six targeted gases, 332

# H

Harkin, Tom
    government cost overruns, 284
Harkins, William D.
    American chemist, 185
Hawking, Stephen
    black hole theory, 247
    theory unifies black hole physics, 248
HDCC energy production, 283
Heisenberg, Werner
    uncertainty principle, 207, 217
    zero-point-energy, 212
High-density charge cluster (HDCC)
    EV borehole production, 225
    evidence of excess energy, 221
    evidence of low-energy nuclear reaction, 225
    International research efforts, 224
    potent micron-sized plasmoids, 222
    US Patents, 224
Hot fusion
    financial investment, 187
    technological process, 186
Hybrid-electric vehicles, 318
Hybrid-gas-electric vehicles, 25
Hybrid vehicles, 127
Hydrocarbon Man, 13
    oil consumption, 16
Hydroelectric dams
    environmental costs, 140
Hydroelectric energy
    benefits, 138
    global use, 140
    uncertain future, 142

Hydroelectric Licensing Process Improvement Act, 143
Hydroelectric power
    efficiency, 139
Hydrogen
    benefits, 133
    as a fuel, 123
    future fuel, 101
    limitations, 131
    production and applications, 124
    technological challenges, 132
Hydrogen fuel cells
    history, 124
    potential, 124
Hydropower
    carbon emission displacement, 143
    history, 137
    leading source of renewable energy, 139
Hydropower
    endangerment of fish species, 141
Hyper-fast Travel within General Relativity, 344

# I

Impulse Engines, 6
India
    solar power program, 109
Inertial confinement fusion, 187
Institute for Advanced Studies
    the search to tap ZPE, 217
Institute for Energy and Environmental Research
    land and water contamination, 86
Intergovernmental Panel on Climate Change
    2001 climate change mitigation report, 325
    climate change mitigation, 332
    man-made methane emissions, 67
    predictions, 29
International Energy Agency
    forecasting peak of production, 347
International Thermonuclear Experimental Reactor, 187
    US pulled out in 1998, 284
Iraq
    oil reserves, 35

Iran
    natural gas reserves, 70

# J
Japan
    investing in renewable energy, 312
    solar program, 109
Jefimenko, Oleg
    electrostatic motor, 281
Jet Propulsion Laboratory
    boost in satellite velocity, 240
Joint European Torus
    commercial hot fusion decades away, 188
Jones, Stephen
    cold fusion research, 189

# K
Kepler, Johannes
    planetary movements, 242
King Coal
    global energy provider, 48
King, Moray, B.
    primary hypothesis for ZPE, 226
    references to ZPE dynamics, 227
Koczor, Ron
    technical paper on gravity modification, 259
Kuhn, Thomas
    The Structure of Scientific Revolutions, 339
Kurachi, Takayuki
    Japanese cold fusion experiment, 188
Kyoto Protocol
    agreement needs revisions, 333, 334
    carbon dioxide emission reductions, 331
    greenhouse gas emissions, 90
    international climate treaty, 331
    projected emissions beyond treaty, 338
    US complaints, 333
    US pullout of treaty, 332
    US pulls out, 299

# L
Lamb, Willis
    Lamb shift, 213

Lamoreaux, Steven
    confirmed existence of Casimir force, 214
Lewis, Nate
    Caltech scientist, 190
Li, Ning
    gravity modification pioneer, 258
Light
    wavelengths, 248
LIMPET
    500 kW wave-power plant, 309
Low-energy nuclear reactions, 184

# M
Mach, Ernest
    Mach's Principle, 260
Mach's Principle, 6
    combined gravitational attraction, 261
    origin of inertia, 260
Mach's Principle & Impulse Engines, 260
Mallove, Eugene
    "overwhelmingly compelling evidence", 193
Manhattan Project, 82
    history, 345, 346
Marshall Space Flight Center
    gravitational physics research, 250
    NASA gravity modification experiments, 258
Massachusetts Institute of Technology
    Plasma Fusion Center, 190
    receives US Patent relating to cold fusion, 194
    researchers attack cold fusion, 193
Maxwell, James Clerk
    proved light an electromagnetic wave, 253
Mead, Frank
    US Patent for zero-point energy antenna, 208
Methane hydrates, 67
    sea floor sediments, 68
Mexico
    solar power program, 109
Michell, John
    gravity and light waves, 245
Michelson, A.A.
    aether and speed of light experiment, 211

Michelson-Morley experiment, 211
Mills, Randall
   energy from hydrogen, 280
Mizuno, Tadahiko
   Japanese cold fusion experiment, 188
Moray, Henry
   new energy research, 226
Morley, E.W.
   aether and speed of light experiment, 211
Motor vehicles
   air pollution, 21
   emissions, 22
   gasoline consumption, 23
MTBE
   gasoline additive, 158

# N

NASA research
   Advanced Concepts Office, 342
   Breakthrough Energy Physics Research Program, 343
   search for breakthrough science, 343
National Energy Technology Lab
   coal mining limitations, 52
National Ignition Facility
   over budget, 284
National Oceanic and Atmospheric Administration
   predictions, 30
National Renewable Energy Laboratory
   biomass gasification research, 157
National Climatic Data Center
   weather extremes, 30
Natural capital
   multigenerational equity, 326
   "sustainable growth?", 346
Natural Energy Laboratory of Hawaii
   experimental OTEC facility, 170
Natural gas
   BTU production, 63
   consumption and gross imports, 73
   depletion, 69
   deposits, 62
   environmental costs, 71
   exports and emissions, 74

Natural gas *(continued)*
   global warming, 66
   global demand, 61
   global reserves/production, 72
   green revolution, 348
   increasing demand, 307
   industrial importance, 63
   methane hydrate drilling, 11
   North American production trend, 307
   Persian Gulf reserves, 70
   statistical data resources, 75
   US consumption/production, 64
   US production, 70
   wet gas and dry gas, 62
Natural gas vehicles
   efficiency, 65
Neighborhood Electric Vehicle
   energy efficient/non-polluting, 127
New energy
   Breakthrough Energy Effects Database, 345
   changing the scientific paradigm, 341
   funding obstacles, 279
   institutional bias, 340
   misguided inventors, 340
   new Manhattan Project, 345
   potential energy and transportation technologies, 344
   professional ostracism, 339
   scientific conformity, 339
New energy breakthrough technology
   obstacles, 301
New energy field
   lack of national data base, 286
   overunity claims, 285
New energy R&D
   obstacles, 300
Newton, Sir Isaac
   gravitational force, 243
   lack of acceptance, 341
   law of inertia, 243
   theory of gravitation, 242
Nitrogen oxide emissions
   US domestic production, 26
Non-OPEC
   oil production, 17

Non-renewable resources, 9
Nuclear battery, 276
Nuclear energy
    better fission technology, 314
    federal subsidies, 313
    German taxpayers revolt, 314
    global capacity, 80
    hazards and environmental costs, 83
Nuclear Energy
    hazards and limitations, 79
Nuclear energy
    industry costs, 313
    newly-designed power plants, 294
    public and political resistance, 313
    public fear, 80
    technological difficulties, 80
    transmutation of radioactive waste, 314
    US number one user, 314
Nuclear energy industry
    power plants, 80
Nuclear Energy Institute
    nuclear power advocates, 90
Nuclear energy phase-out
    trend in Europe, 89
Nuclear fuel cycle, 83
Nuclear Isomer Decay Phenomena, 344
Nuclear power
    hazards and limitations, 12
Nuclear power plants, 80
    Asian production, 90
Nuclear reactor processes, 87
Nuclear reactor technologies
    breeder reactor, 88
Nuclear Regulatory Commission
    waste responsibility, 86
Nuclear waste burial
    deep geologic storage, 91
Nuclear Waste Policy Act
    costs, 92
Nuclear waste vitrification
    waste storage process, 91

# O

Ocean Thermal Energy Conversion
    environmental costs, 171
    process, 169

Ocean warming
    infectious diseases, 31
Office of Advanced Energy Concepts, 343
    ranking methodology, 345
Office of Energy Efficiency and Renewable Energy, 343
Office of Nuclear Energy, Science and Technology, 343
Office of Power Technologies
    renewable energy strategy, 133
Oil
    being rapidly depleted, 305
    Big Oil diversifying energy sources, 311
    contamination, 19
    cutting gasoline consumption, 317
    diminishing returns on exploration, 306
    economic support, 30
    geopolitics, 32
    global consumption, 18
    global petroleum exports, 306
    global supply, 2
    green revolution, 348
    intrinsic qualities, 308
    Non-OPEC production near peak, 307
    peak production, 292
    petroleum products, 306
    politics, 19
    provides 97% world's transportation needs, 317
    US consumption, 292
    US daily consumption, 328
    US domestic production, 17, 19
    US historic production, 306
    US imports cost, 327
    US production, 11
Oil Era, 11
OPEC
    exports by country, 33
    oil production, 17
Oppenheimer, J. Robert
    supported open exchange of ideas, 346
Oren Lyons
    Native American spiritual elder, 7
OTEC designs
    obstacles, 170

OTEC processes
   technology, 169

## P

Papp, J.
   ZPE research, 226
Patterson, James
   Clean Energy Technologies, 195
Patterson Power Cell
   Electrochemical energy device, 195
   low-level nuclear transmutation, 195
   US Patent, 195
Persian Gulf
   proved natural gas reserves, 11
   world energy prices, 20
Persian Gulf oil
   cost to US, 12
Petroleum
   percent world energy, 11
Photovoltaic cells
   PV cells, 106
Photovoltaic panels
   manufacture and economics, 111
Photovoltaic systems
   operating costs, 108
Planck, Maxwell
   second theory of blackbody radiation, 212
Podkletnov, Eugene
   experimental results, 257
   superconductors, 256
Polder, Dik
   Casimir-Polder effect, 214
Pons & Fleischmann
   forced to resign, 194
   press release, 189
Pons, Stanley
   electrochemist, 177
   press conference, 183
Population
   environmental damage, 298
   oil consumption, 34
Population growth
   energy demand, 347
   environmental destruction, 349, 350
   family planning, 349
   increasing energy demand, 299

Population growth *(continued)*
   key per capita production trends, 298
   most serious problem, 351
   projection in worldwide oil demand, 299
   United States, 349
   zero population growth, 350
President Clinton
   alternative energy research, 18
President Roosevelt
   Manhattan Project, 185
Proton Exchange Membrane
   PEM fuel cell, 129
Puthoff, Hal
   the motion of charged particles generate ZPE, 216
   ZPE electromagnetic waves, 253

## Q

Quantum gravity, 263
Quantum mechanics
   explains physical world, 263
   vacuum-energy frequencies, 210
Quasars, 247

## R

Radiation
   gamma, beta, alpha, 81
Radioactive decay
   dangerous by-products, 84
Radioactive waste
   disposal, 80
Radioactivity, 81
Rapid climate change
   global warming, 2
Redding, Jim
   CETI executive, 196
Renewable energy
   limitations, 100
   role in energy production, 99
   sufficient to sustain current levels of energy use, 347
   US commitment, 338
Renewable Energy Policy Project, 103
   July 2000 report on 50 years of subsidies, 313

Research and Development investments
   US R&D spending trend, 177
Richardson, William
   laser project well over budget, 284
   US Energy Secretary, 86
Roentgen, Wilhelm
   discovered X-rays, 248
Rubik, Beverly
   mechanics of paradigm shift, 341
   obstacles to new energy research, 353
Rutherford, Ernest
   alpha particle bombardment experiments, 185

## S

Sakharov, Andrei
   gravitation a secondary effect, 209
   muon-catalyzed cold fusion, 186
Saudi Arabia
   petroleum reserves, 35
Schulz, Jurgen
   Omicron Vector Field Model, 252
Searl, John Roy Robert
   "Searl Effect" technology, 261
Shell Oil
   investment in solar technologies, 111
Shoulders, Ken
   pioneer of HDCC technology, 223
   problem with cold fusion process, 283
   US Patent for energy from ZPE, 224
Shoulders, Steve
   HDCC research pioneer, 223
Silvertooth, Ernest
   laser interference experiment, 212
Solar energy
   benefits, 109
   California consumers, 112
   California solar-energy projects, 310
   economic argument, 112
   funding obstacles, 312
   International programs, 109
   solar radiation, 105
   subsidies and taxes, 111
Solar/hydrogen production, 131
Solar power
   costs per watt, 108

Solar technologies
   history, 106
   NASA, 107
   solar cells, 106
Solid Oxide fuel cell, 129
Soviet Union
   natural gas reserves, 70
   proved natural gas reserves, 11
Space-time, 244
Space travel
   experimental mass reduction, 260
   VASIMR, 278
Speed of light, 245
Spitzer Jr., Lyman
   magnetically confined plasmas, 187
Sport Utility Vehicles
   emissions, 23
SRI International
   cold fusion research, 194
Stars
   evolving phases, 246
Storms, Edmund, 178
   published examples of excess heat production, 198
   replicates Takahashi's results, 193
Strategic minerals, 34
Sulfur dioxide
   emissions, 27
Swartz, Mitchell
   finds faults in original MIT calorimetric testing, 194

## T

Takahashi, Akito
   Osaka University, 193
Tandberg, John
   chemist-physicist, 185
Technology
   as solution, 346
The Geysers
   California geothermal facility, 150
Thomas, Jr., John A.
   Inverse G-Vehicle, 261
Three Gorges Dam
   problems, 141

Three Mile Island, 13
Tidal and ocean thermal energy, 165
Tidal energy
  global facilities, 166
  limitations, 166
  world's first commercial power plant, 309
Tidal energy potential, 167
Tokamak
  hot fusion reactor, 186
Tokamak Fusion Test Reactor, 188
Torr, Douglas
  gravity modification pioneer, 259
Torsion fields
  Russian research, 222
Tracy, James
  Omicron Vector Field Model, 252
Twain, Mark, 32, 35

# U

UN Framework Convention on Climate Change
  Kyoto, Japan, 331
  Kyoto Protocol, 299
Unified Field Theory, 239
United Nations
  Iraqi oil-embargo, 35
United States
  Energy Policy Act, 343
  energy policy politics, 341
  federal subsidies to non-renewables, 312
  greatest contributor of greenhouse gas emissions, 296
  least efficient major economy, 295
  reliance on oil, 348
  responsibility to future generations, 353
  responsibility to reduce consumption and emissions, 352
United States Geological Survey
  methane hydrate reserves, 67
Uranium
  global reserves and production, 95
Uranium-235, 82
  radioactive element, 83
Uranium isotopes
  reactor processes, 87

Uranium mining
  cancer, 84
Uranium ores
  extraction, 83
Uranium reserves
  estimates, 85
Uranium tailings
  radioactive ore, 84
US Energy policy
  2002 National Energy Plan, 292
US Fish and Wildlife Service
  toxic waste, 86
US health costs
  associated with air pollution, 4
  associated with engine exhaust, 65
US Magnetic Fusion Energy Engineering Act
  Congress funds hot fusion, 188
US oil
  imports, 18

# V

Valone, Thomas
  distortion and virtual particles, 215
Van der Waals force
  attraction between neutral atoms, 213
Van Flandern, Thomas
  gravity propagates faster than lightspeed, 251
Voodoo science, 285

# W

Wallman, David
  Carbo-hydrogen gas, 157
  carbon-arc gasification process, 281
Watkins, James
  Secretary of Energy Admiral, 190
Wave energy
  demonstration facilities, 168
Wave energy potential, 168
Weinberg, Steven
  minimal energy from space, 210
Wesson, Paul
  virtual photons, 218
  ZPE research justified, 219
Western Fuels Association
  pro-energy video, 31

Wheeler, John, 179
    black holes, 246
    ZPE calculation, 210
Wind energy
    global generation capacity, 116
    history, 115
Wind energy industry
    global expansion, 117
Wind industry
    US industry growth, 118
Wind power
    double-digit growth, 308
    efficiency, 118
    environmental costs, 119
    increasing global generating capacity, 309
    projections, 117
Woodward, James
    acceleration without exhaust, 260
    tapping inertia for space travel, 259
World energy production
    per capita, 175
World Resources Institute, 51
Worldwatch Institute
    nuclear energy cost analysis, 93

## X
X-rays, 247

## Y
Yangtze River
    Chinese dam project, 140
Youngquist, Walter
    population and energy demand, 348
Yucca Mountain
    High-Level Nuclear Waste Repository, 91
    seismic stability, 92

## Z
Zero-Point Energy
    electromagnetic zero-point field, 227
    energy from the vacuum, 207
    history, 211
    kinetic energy, 209
    Milonni, Peter and Shih, M.L., 227
    NASA funding for ZPE, 209
    source of energy, 178
    ZPE-powered nano-motors, 227
Zero-Point-Field Energy Extraction, 6